# FISH PHYSIOLOGY

*VOLUME X*
Gills

*Part A*
Anatomy, Gas Transfer, and Acid–Base Regulation

# CONTRIBUTORS

ROBERT G. BOUTILIER

CHARLES DAXBOECK

NORBERT HEISLER

THOMAS A. HEMING

G. M. HUGHES

GEORGE K. IWAMA

PIERRE LAURENT

STEFAN NILSSON

JOHANNES PIIPER

DAVID RANDALL

PETER SCHEID

# FISH PHYSIOLOGY

*Edited by*

## W.  S.  HOAR
DEPARTMENT OF ZOOLOGY
UNIVERSITY OF BRITISH COLUMBIA
VANCOUVER, CANADA

## D.  J.  RANDALL
DEPARTMENT OF ZOOLOGY
UNIVERSITY OF BRITISH COLUMBIA
VANCOUVER, CANADA

## *VOLUME X*
## Gills

*Part A*
Anatomy, Gas Transfer, and Acid–Base Regulation

1984

ACADEMIC PRESS, INC.

(Harcourt Brace Jovanovich, Publishers)
Orlando   San Diego   San Francisco   New York   London
Toronto   Montreal   Sydney   Tokyo   São Paulo

ACADEMIC PRESS, INC.
Orlando, Florida 32887


*United Kingdom Edition published by*
ACADEMIC PRESS, INC. (LONDON) LTD.
24/28 Oval Road, London NW1 7DX

Library of Congress Cataloging in Publication Data

Hoar, William Stewart, Date
    Fish physiology.

    Vols. 8-    edited by W. S. Hoar, D. J. Randall,
and J. R. Brett.
    Includes bibliographies and indexes.
    CONTENTS: v. 1 Excretion, ionic regulation, and
metabolism.--v. 2. The endocrine system.--[etc.]--
v. 10. Gills. pt. A. Anatomy, gas transfer, and acid-
base regulation. pt. B. Ion and water transfer (2 v.)
    1. Fishes--Physiology--Collected works.  I. Randall,
D. J., joint author.  II. Conte, Frank P.
III. Brett, J. R.  IV. Title.
QL639.1.H6              597'.01              76-84233
ISBN 0-12-350430-9 (v. 10A)
ISBN 0-12-350432-5 (v. 10B)

PRINTED IN THE UNITED STATES OF AMERICA

84 85 86 87    9 8 7 6 5 4 3 2 1

# CONTENTS

## 4. Model Analysis of Gas Transfer in Fish Gills
### Johannes Piiper and Peter Scheid

## 5. Oxygen and Carbon Dioxide Transfer across Fish Gills
### David Randall and Charles Daxboeck

## 6. Acid–Base Regulation in Fishes
### Norbert Heisler

## Appendix: Physiochemical Parameters for Use in Fish Respiratory Physiology
### Robert G. Boutilier, Thomas A. Heming, and George K. Iwama

# CONTRIBUTORS

Numbers in parentheses indicate the pages on which the authors' contributions begin.

ROBERT G. BOUTILIER (401), *Department of Zoology, The University of British Columbia, Vancouver, British Columbia V6T 2A9, Canada*

CHARLES DAXBOECK (263), *Pacific Gamefish Foundation, Kailua-Kona, Hawaii 96745*

NORBERT HEISLER (315), *Abteilung Physiologie, Max-Planck-Institut für Experimentelle Medizin, D-3400 Göttingen, Federal Republic of Germany*

THOMAS A. HEMING* (401), *Department of Zoology, The University of British Columbia, Vancouver, British Columbia V6T 2A9, Canada*

G. M. HUGHES (1), *Research Unit for Comparative Animal Respiration, University of Bristol, Bristol B881UG, United Kingdom*

GEORGE K. IWAMA (401), *Department of Zoology, The University of British Columbia, Vancouver, British Columbia V6T 2A9, Canada*

PIERRE LAURENT (73), *Laboratoire de Morphologie Fonctionnelle et Ultrastructurale des Adaptations, Centre National de la Recherche Scientifique, 67037 Strasbourg, France*

STEFAN NILSSON (185), *Department of Zoophysiology, University of Göteborg, S-400 31 Göteborg, Sweden*

JOHANNES PIIPER (229), *Abteilung Physiologie, Max-Planck-Institut für Experimentelle Medizin, and Institut für Physiologie, Ruhr-Universität Bochum, D-4630 Bochum, Federal Republic of Germany*

DAVID RANDALL (263), *Department of Zoology, The University of British Columbia, Vancouver, British Columbia V6T 2A9, Canada*

*Present address: Department of Medicine, UCLA School of Medicine, Los Angeles, California 90024.

PETER SCHEID (229), *Abteilung Physiologie, Max-Planck-Institut für Experimentelle Medizin, and Institut für Physiologie, Ruhr-Universität Bochum, D-4630 Bochum, Federal Republic of Germany*

# PREFACE

The fish gill is an intriguing tissue because of its multifunctional nature. Gills are involved in ion and water transfer as well as oxygen, carbon dioxide, acid, and ammonia exchange, and there are many interactions between these processes. These interactions lead to exchange of information among groups of scientists who otherwise might not meet, because they are brought together by their mutual interest in gills. Many aspects of gill structure and function have been studied, and our understanding of these different systems is reasonably well advanced. This is not to say that there are not vast gaps in our knowledge. Only a relatively few species have been studied in detail, goldfish, eels, and trout being most prominent because of availability and ease of handling. Freshwater fish are much more studied than marine species for the same reasons. The range of freshwater is very broad, and we are only at the beginning of understanding the differences in gill structure and function that, for example, allow some fish to flourish in soft water but restrict others to hard waters.

This volume attempts to review the structure and function of fish gills and also makes some attempt, particularly in the final chapter, to review some of the methodology used in studying gills. The terminology concerned with the gills has grown with the studies and is often confusing. We have attempted, with the help of Drs. Hughes, Laurent, Nilsson, and many others, to arrive at some terms and abbreviations to be used with fish gills. It is always difficult to change habits but some uniformity is always an aid to understanding, particularly for those just entering the field. We would urge the adoption of these terms and abbreviations, even though we cannot persuade all contributors to do the same.

Many people advised and helped us in editing this text; in particular, the chapters were reviewed by many people, and we are most grateful for all help given. We hope the end result is a work on gills that will be of use to those interested in this fascinating organ for many years to come.

W. S. HOAR
D. J. RANDALL

# TERMS AND ABBREVIATIONS

There is a profusion of terms associated with fish gills; we have attempted herein to standardize the terminology and abbreviations, in the hope that these will be generally accepted and used by all in the field.

| Term | Abbreviation | Synonym/Notes |
|---|---|---|
| Branchial (Gill) arch | B | Gill bar refers to arch |
| Filament | F | Primary lamella; gill rod refers to filament |
| Lamella (s) | L | Secondary lamella |
| Lamellae (pl) | | |
| Proximal lamellae | | Lamellae proximal to arch |
| Distal lamellae | | Lamellae distal to arch |
| Interlamellar space/water | | Space/water between lamellae |
| Inhalent water | I | Do not refer to afferent and |
| Exhalent | E | efferent water. |
| Prelamellar water | | Water that has not passed over lamellae |
| Postlamellar water | | Water that has passed over lamellae |
| Pillar cell | | |
| Epithelial cell | | |
| Interstitial cell | | Avoid terms like stem cell, pillar cell II etc. unless better evidence indicates function of these cells. |
| Erythrocyte | RBC | |
| Interstitial space | IS | Not lymphatic space |
| Ventral aorta | VA | A = artery |
| Dorsal aorta | DA | |
| Suprabranchial artery | SBA | |
| Carotid artery | CA | |
| Coeliacomesenteric artery | CMA | |
| Afferent branchial artery | af.BA | af. = afferent |
| Afferent filament artery | af.FA | Afferent primary (lamella) artery |
| Afferent lamellar arteriole | af.La | Afferent secondary (lamella) arteriole<br>a = arteriole |
| Efferent lamellar arteriole | ef.La | Efferent secondary (lamellar) arteriole<br>ef. = efferent |

*(continued)*

*Continued*

| Term | Abbreviation | Synonym/Notes |
|------|-------------|---------------|
| Efferent filament artery | ef.FA | Efferent primary (lamellar) artery |
| Efferent branchial artery | ef.BA | |
| Branchial vein | BV | V = vein |
| Dorsal branchial vein | DBV | |
| Ventral branchial vein | VBV | |
| Inferior jugular vein | IJV | |
| Anterior cardinal vein | ACV | |
| Posterior cardinal vein | PCV | |
| Central venous sinus | CVS | Not filament sinus or *veno-lymphatic* sinus, not lymphatic. |
| Anterior-Venous anastomosis | AVas | Avoid shunt because this implies function, as = anastomosis |
| Efferent side | ef.AVas | |
| Afferent side | af.AVas | |
| Lamellar *basal* blood channel | | Blood channel at base of lamella |
| Lamellar *marginal* blood channel | | Blood channel in free edge of lamella |
| Filament epithelium | | Primary (lamellar) epithelium |
| Lamellar epithelium | | Secondary (lamellar) epithelium = respiratory epithelium |
| Leading edge[a] (refers to water) | | Filament efferent side (refers to blood) |
| Trailing edge[a] (refers to water) | | Filament afferent side (refers to blood) |

[a]Leading and trailing edge refer to the direction of water flow over the gill arch, filament, or lamella.

# CONTENTS OF OTHER VOLUMES

# GENERAL ANATOMY OF THE GILLS

*G. M. HUGHES*

Research Unit for Comparative Animal Respiration
University of Bristol
Bristol, United Kingdom

## I. INTRODUCTION

The gills form a highly characteristic feature of fishes, and their presence has a marked effect on the anatomy and functioning of the rest of the animal. Although their origin among early chordates may have been with particular reference to feeding, nevertheless, it is to exchanges with the environment, particularly $O_2$ and $CO_2$, that they have become most adapted. Their evolu-

1

tion may be considered in relation to this function, and their adaptation to particular environments provides a fascinating study in comparative functional anatomy and physiology. During this evolution it is apparent that the existence of a large surface exposed to the external medium with a thin barrier separating the internal and external media inevitably leads to exchanges in addition to $O_2$ and $CO_2$, and consequently, these have had a modifying influence on the evolution of these organs and in certain circumstances other functions have become more important than that of respiration.

It is only in the early stages of most living fish that the gills are clearly visible: external gills form important respiratory and indeed nutritive organs in some elasmobranchs (Needham, 1941). In adults the red gills are normally covered up and enclosed within elaborations of the gill slits. These slits originate in the pharyngeal region as perforations between the alimentary canal and the lateral body surface, the enclosed mesoderm developing into the gill arches. The large number of gill slits among agnathan fishes becomes reduced to a more or less constant number among true fishes, in which typically there are five slits on each side of the animal. Anterior to these branchial organs there are the hyoid, mandibular, and premandibular gill arches of the primitive head, which can be recognized during embryonic development (Goodrich, 1930; Horstadius, 1950). The slit between the premandibular and mandibular arches becomes incorporated in the mouth, and only in certain orders is a slit present between the mandibular and hyoid arches. In most elasmobranchs and some primitive actinopterygians this is confined to the dorsal region where it forms the spiracle. The remaining gill slits are of the same general type and consist of increased surface foldings of the epithelium, which is perforated between the alimentary canal and outside. Spacing between these slits in primitive forms is greater than in more advanced groups, and from an anatomic point of view involves a reduction in the interbranchial septum and is one of the most important changes that has occurred in the external anatomy of gills. As a consequence there has been a condensation along the longitudinal axis in the extent of the gilled portion of the pharynx. These changes are associated with further modifications of the head skeleton leading to the evolution of gill covers of various kinds, which reach their maximum development in the evolution of the opercular bones that cover the gills of the teleost fishes.

Another important selection pressure has been in relation to the swimming habits of these animals and the necessity for the maintenance of streamlining at the head end of the fish. Economy of space and accommodation within the head are clearly important aspects of external gill morphology. The most advanced developments of this kind are found in oceanic forms such as tunas, in which streamlining of the head has been maintained in spite of the evolution of a gill system with a very large surface exposed to the water. Ventilation of these gills is associated with the swimming movements, the water entering the mouth by ram ventilation, and leaving by the opercular slits as the fish swims through the water. The proportion of water ventilating the gill system is regulated by control of the openings of the mouth and the opercular slits. It is of interest that such ramjet ventilation has also evolved among the cartilaginous fishes (Hughes, 1960a), as some sharks show this form of ventilation when they swim beyond a certain velocity. The condition in this group is completely different from that in tunas, as the operculum has not developed and externally the gill slits are clearly visible. In these forms the extent of the external openings is restricted in comparison with some primitive elasmobranchs, in which the whole gill slit is open to the outside, whereas in more advanced forms the external openings are restricted to approximately one-third at the ventral end of each gill.


## Relationship of Gills to Lungs

Typically gills are the gas exchange organs of water-breathing fishes, but in some species they are also involved in gas exchange with the air (Schlotte, 1932; Singh, 1976; Graham, 1976). Furthermore, many modifications of the gills have occurred during evolution, notably of the teleosts, in which they have become important aerial gas exchange organs. In these fishes, as in those in which a diverticulum of the pharynx forms a lung homologous with that of tetrapods, the accessory organs generally have the same basic structure of a hollow intucking of the gut lining that is ventilated tidally (Fig. 1). Such a method of ventilation inevitably leads to certain structural and functional complications, and contrasts with the more continuous flow of water through the gills (Table I). Problems of support for the much enlarged surface are different in water and air. In the "true" lungs there has been a development of a special lining layer or surfactant (Pattle, 1976, 1978; Hughes and Weibel, 1978) that is absent from the gills. Nevertheless, such a surfactant layer does not appear to be present in air-breathing organs that have developed independently of the true lung (Hughes, 1978a), and support for the gas exchange surface is achieved in a different manner. In many cases the basic lamellar structure of the gills is apparent, and perhaps the proportion of the total surface of the intucked exchange organ that is directly involved in gas exchange is less than that of the most evolved lungs, but few quantitative data are available.

In spite of the great structural variety of the parallel evolution of the air-breathing organs, a number of physiological generalizations suggested many years ago (Hughes, 1966a; Rahn, 1966) are now sufficiently well documented to justify acceptance. Thus, in most so-called bimodal breathers, the gas exchange with the water is more important with respect to $CO_2$ release,

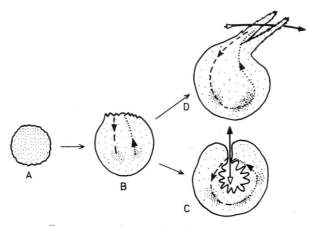

**Fig. 1.** Diagrams to illustrate general course of evolution from a small spherical organism (A) depending solely on diffusion for its $O_2$ supply to larger organisms in which localized areas of the external surface become specialized for gas exchange (B) and later differentiate into either external expansions (gills) with continuous flowthrough of water (D) or (C) air-breathing organs (lungs), which are "intuckings" of the external surface relying on tidal ventilation. In (B), (C), and (D), convection systems have evolved between the sites of gaseous exchange and the regions where the oxygen is utilized and $CO_2$ released. (After Hughes, 1984a.)

whereas the air-breathing organ is more responsible for oxygen uptake. The consequent difference in the gas exchange ratio at the two organs must lead to some differences in basic function at a molecular or fine-structural level, but these have scarcely been investigated. The relative balance in the impor-

**Table I**

Summary of Main Types of Ventilation Found among Vertebrates[a]

| Medium | Respiratory organ | Ventilatory flow | Example | References |
|--------|-------------------|------------------|---------|------------|
| Water | Gill | Continuous (Ram jet) | Tuna Shark | Brown and Muir (1970) Hughes (1960a) |
| | Gill | + or − Continuous | Trout | Hughes and Shelton (1958) |
| Water | Gill | Tidal | Dogfish Lamprey | Hughes (1960a) Hughes and Shelton (1962) |
| Air | Lung | Tidal | Lungfish Tetrapods | McMahon (1969) Gans (1976) |
| Air/water/air | Suprabranchial chamber | Tidal | Gourami | Hughes and Peters (1984) |
| Air | Lung | Continuous | Birds | Brackenbury (1972); Scheid (1979) |

[a]From Hughes (1978b).

tance of the two organs is variable and related to the particular mode of life of the fish concerned. The ventilation of some air-breathing organs has been shown to involve quite a different mechanism, in which the gas exchange surface, although in contact with air for most of the cycle, becomes flushed with water during the uptake of air at the water surface (Peters, 1978; Hughes, 1977). Such a mechanism is clearly intermediate between the typical ventilation mechanism of the gill and that of organs that are only in contact with air. Perhaps it is this feature of gas exchange organs modified from gills that has precluded the development of surfactant lining and the associated cytological structures that have been noted previously. It also supports the view that fish gas exchange organs show the greatest range in ventilatory mechanisms (Table I).

## II. DEVELOPMENT OF GILLS

Fish eggs vary in the amount of yolk they contain and consequently the timing of different developmental stages. In all cases, however, the development of the gills forms an important part of the whole process. From a respiratory point of view, it is essential that they should reach a sufficient level of morphological and functional development that they can take over gas exchange functions when the surface area: volume ratio becomes too small and gas exchange across the body surface is insufficient to meet the growing demands of the developing fish (Figs. 1 and 2). In a number of cases, development contains a "critical" period (May, 1974) during which much mortality occurs and can be a serious problem in aquaculture. One important aspect is clearly the change in nutrition of the embryo as the yolk sac is used up and other sources of energy become more important. One other feature of this problem, however, is that there are also changes in the surface area: volume ratio, particularly in the developing gill region. A morphometric study (Iwai and Hughes, 1977) has shown, for example, that the critical period in the Black Sea bream coincides with the stage at which the surface area:volume ratio of the developing gills reaches a minimum (Fig. 2). After this stage, the increasing number of lamellae ensures that the ratio increases, and hence sufficient surface is available for oxygen uptake. Coincident with development of the gills, there is also development of other parts of the respiratory and cardiovascular system and coordination of the pumping systems for water and blood flow through the gills (Holeton, 1971; Morgan, 1974).

Although such functional aspects have not been investigated a great deal until recently, there has always been much interest in the detailed morphological development of the gill system, as it constitutes one of the diag-

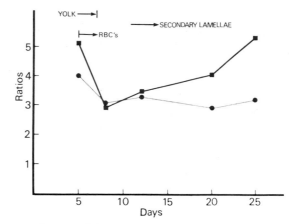

**Fig. 2.** Changes in the ratio of the intersection count to the point count for the gill arches (dotted line) and gill filaments (solid line) during developmental stages of Black Sea bream. These ratios are proportional to the surface area: volume ratio and show that for the gill filaments it is at a minimum after about 7 days, and this coincides with the sensitive period. The surface area: volume ratio for the arches declines steadily during development, because there is no further increase in surface folding. (From Iwai and Hughes, 1977.)

nostic characters of chordates. In some primitive forms (e.g., *Polypterus*) there are patches of cilia on these epithelia (Hughes, 1980a, 1982) that may represent the more continuous ciliation of ancestral chordates, which depend on ciliary currents for feeding and respiration. Early studies concerned themselves with the morphological nature of the gill epithelium and especially the development of the branchial blood vessels (Sewertzoff, 1924). During development the whole sequence is continuous; the epithelium that forms the surface of the gill clefts later becomes the surface of the filament, and still later the surface of the lamellae. Later differentiation occurs in different regions of the adult in relation to the particular microenvironmental and functional conditions. It is, however, convenient to discuss sequentially the development of the main constituents.

## A.  Branchial Arches

Differentiation of the gills proceeds from the anterior end of the embryo soon after gastrulation as gill pouches begin to evaginate from the endoderm and meet ectodermal invaginations. These two processes are followed by piercing and formation of the gill clefts, which isolate interbranchial bars between them. Much discussion has entered around the nature of the epithelium that finally develops into the gills, and some of the earlier workers (Moroff, 1904; Goette, 1878) thought that the gills developed mainly from ectoderm, whereas subsequent scholars such as Bertin (1958) and others believe that they are endodermal in origin. Some morphologists have distinguished "endobranchiate" Cyclostomata from "ectobranchiate" Gnathostomata. The precise nature of the germ layer is not now considered to be of importance, and there appears to be no consistent morphological or functional difference between ectodermal and endodermal regions of the gills.

The interbranchial bars also isolate mesodermal regions from which the coelom becomes obliterated and the mesoderm develops into musculature. At this stage, six or more primary branchial vessels arise from the ventral aorta and pass around the pharynx to become connected with the dorsal aorta. Each primary vessel becomes differentiated into afferent and efferent branchial arteries (Fig. 3A). The precise pattern of this differentiation varies between elasmobranchs and teleosts, and leads ultimately to the well-known difference between selachians—with their paired efferent branchials that join to form single epibranchial arteries that enter the dorsal aorta—and teleosts, in which there is usually a single efferent and a single afferent branchial. Modifications in this general pattern occur among different teleost groups (Muir, 1970; Adeney and Hughes, 1977), and some functional significance may be ascribed to it.

The filaments start to develop as the gill bar epithelium begins to bulge outwards and becomes invaded by a series of vascular loops, which connect between the existing afferent and efferent branchial arteries (Fig. 3A). In both teleosts and elasmobranchs, the first filaments form midway along the gill arch and are added both dorsally and ventrally over a long period extending into adult stages. Allometric growth leads to the gill arches beginning to bend backwards to an elbow shape.

Paired rows of filaments grow out in a laterocaudal direction, the anterior row appearing first. At a later stage there is differentiation of the branchial arch cartilages, so that in each septum basi-, hypo-, epi-, and pharyngobranchials arise.

The corpus cavernosum forms along each branchial arch and penetrates into the developing gill filaments (Acrivo, 1938). Details of the formation of the blood vessels vary among different groups and species, but eventually the filament loops lengthen as the afferent and efferent branchial arteries grow apart. During this early stage of filament development, the blood vessels are surrounded by mesenchyme and there is a well-defined basal lamina below the surface epithelium (Fig. 3B). The underlying collagen layer is well defined but later becomes more diffuse. Toward the tips of the developing filaments there are spaces between the mesenchymal cells and basal lamina that are packed with bundles of collagen, which are randomly oriented except where they lie within indentations of mesenchyme cells. Clusters of these mesenchyme cells differentiate into pillar cells, which

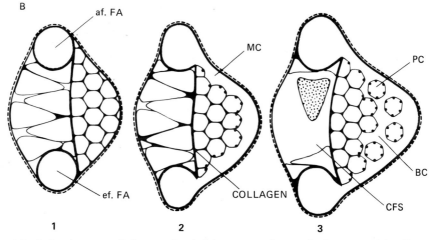

**Fig. 3.** Rainbow trout (*Salmo gairdneri*). (A) Diagrams showing the formation of the afferent and efferent branchial arteries from the primary branchial artery of a gill arch. Note the beginning of the development of the filament arteries between afferent and efferent branchial arteries. (B) Diagrams showing stages in the development of a trout lamella. Formation of the marginal channel (MC) and development of pillar cells (PC) in the main body of the lamella are illustrated. Blood channels (BC) between the pillar cells and along the margin connect the afferent (af.FA) and efferent filament arteries (ef.FA). In **1** there is an accumulation of mesenchyme cells, and in **2** the formation of the marginal channel and columns in pillar cells. In **3**, formation of the blood spaces and pillar cells is shown. Note the presence of the central venous sinus (CVS). Note marginal channel endothelial lining. (From Hughes and Morgan, 1973.)

become aligned in a single plane at right angles to the filament axis in two alternating series on the upper and lower surfaces of each filament. The differentiating pillar cells become separated from the filament mesenchyme by a thick band of collagen that joins the inner basement membrane layers of opposite sides at the base of each developing lamella (Fig. 25). Distal to this band of collagen, the mesenchyme cells continue to differentiate into pillar cells, particularly by the formation of collagen columns, which join the two basement membrane layers of the developing lamellae (Fig. 3B). Collagen columns lie within but are external to pillar cells, because the intuckings that contain the collagen are lined with pillar cell membrane.

New lamellae are added toward the tips of the filaments during the entire period of growth. Spaces within the mesenchyme around the edge of the lamellae merge to form a continuous channel joining the afferent and efferent filament arteries, and this gives rise to the marginal channels of the adult. Once blood flow through this channel begins, the cells lining its outer border develop the typical membrane-bound Weibel–Palade bodies (Morgan, 1974), which is typical of the adult marginal channel (Tovell et al., 1971). During growth the shape of the secondary lamella changes from ovoid to triangular, with the apex toward the efferent filament artery. New pillar cells are formed by division of the proximal row of mesenchyme cells, this row being completely isolated from the main body of the filament by a thick collagen sheet outside of which is the so-called pillar cell system (Hughes and Perry, 1976). The proximal cell layer is usually about one cell thick, and these mesenchyme cells sometimes have collagen columns around their outer borders. As more columns develop, the cells divide and move away to leave the smaller undifferentiated cells (Fig. 3B). In the early stages the external surface of the epithelium is folded and has short microvilli, the height and frequency of which increases to that found in the adult condition at a later stage. To begin with, the water–blood barrier is relatively thick— for example, 11 $\mu$m at 31 days in rainbow trout (Morgan, 1974), whereas after 102 days it is 7 $\mu$m. The higher proportion of chloride cells in young trout may be related to the greater osmotic problems of smaller animals having the greater surface area:volume ratio (Wagner et al., 1969).

## B. Hyoid Arch

The hyoid arch is situated immediately anterior to the first branchial arch, in front of which is the mandibular arch. In most fishes it becomes reduced and is represented by the hyomandibular and basihyal cartilages. It often plays an important role in the suspension of the jaw, which is derived from the mandibular arch. Between the hyoid arch and the first branchial arch, there is a complete gill slit; however, a prehyoidean slit, between the mandibular and hyoid arches, is only complete in a fossil group, the Aphetohyoidea (Watson, 1951). Among modern fish, it only persists as a spiracle in the dorsalmost portion and is especially well developed in bottom-living rays, where it forms the main entry for water into the orobranchial cavity and is actively closed by a valve (Hughes, 1960a). The posterior hemibranch attached to the hyoid arch persists in most elasmobranchs and some primitive bony fish as a complete hemibranch, for example in sturgeons and *Latimeria* (Table II). Among the cartilaginous fishes, it forms a distinct hemibranch that lines the anterior part of the first gill slit. Morphologically the hyoidean posterior hemibranch seems to be identical to other hemibranchs at both gross and fine-structural levels. In most fishes, development of an operculum is associated with a reduction in the spiracle, and a hyoidian hemibranch is usually absent. In such fishes as sturgeons, where the hyoid hemibranch is found attached to the inner side of the operculum, the presence of a spiracle is perhaps surprising as the pseudobranch represents the

**Table II**

Presence of Gills on Different Visceral Arches of Fishes[a]

| Genus | Mandibular arch | Spiracle | Hyoid arch | Branchial arches | | | | | | |
| --- | --- | --- | --- | --- | --- | --- | --- | --- | --- | --- |
| | | | | I | II | III | IV | V | VI | VII |
| *Scyliorhinus* | ps | + | ph | H | H | H | H | | | |
| *Hexanchus* | ps | + | ph | H | H | H | H | H | | |
| *Heptanchus* | ps | + | ph | H | H | H | H | H | H | |
| *Raia* | ps | + | ph | H | H | H | H | . | | |
| *Chimaera* | . | − | ph | H | H | H | ah | . | | |
| *Latimeria* | . | − | ph | H | H | H | H | . | | |
| *Neoceratodus* | . | − | ph | H | H | H | H | . | | |
| *Protopterus* | . | − | ph | . | . | H | H | ah | | |
| *Acipenser* | ps | + | ph | H | H | H | H | . | | |
| *Huso* | ps | − | ph | H | H | H | H | . | | |
| *Lepisosteus* | ps | − | ph | H | H | H | H | . | | |
| *Amia* | ps | − | . | H | H | H | H | . | | |
| *Polypterus* | . | + | . | H | H | H | ah | . | | |
| *Ophiocephalus* (*Channa*) | . | − | . | H | H | H | ah | . | | |
| *Anabas* | . | − | . | H | H | H | . | . | | |
| *Amphipnous* | . | − | . | . | H | . | . | . | | |
| *Opsanus* | . | − | . | H | H | H | . | . | | |
| *Dibranchus* | . | − | . | . | H | H | . | . | | |

[a]Abbreviations: ah, anterior hemibranch; H, holobranch; ph, posterior hemibranch; ps, pseudobranch (external).

gill of the slit anterior to the hyoid arch and is often associated with the spiracle, which is well developed in elasmobranchs.

## C. Pseudobranch

The pseudobranch represents the gill of the slit that lies between the mandibular and hyoid arches. In fact, it is the posterior hemibranch of the mandibular arch. It has been recognized for a long time that this gill is often supplied with blood that has already been oxygenated in the hemibranch of the hyoid or first branchial arch; therefore, it was considered to be non-respiratory and hence was named "pseudo." Many different functions have been ascribed to the pseudobranch (e.g., chemoreceptor, mechanoreceptor, secretory), but as yet there is no certain indication of any function common to all fishes. Muller (1839) distinguished two morphological types, the first being free pseudobranchs in which lamellae are distinguishable and in direct contact with the water. The second type is the covered pseudobranch where the surfaces do not come into close contact with the water. Of this latter kind, the glandular type of pseudobranch is perhaps the most developed, but even in the structure of this organ it is possible to recognize the typical pillar cell organization of the lamellae (Munshi and Hughes, 1981). Some authors have described at least four types of pseudobranchs that vary in their degree of isolation from the external environment (Leiner, 1938; Bertin, 1958). It should be emphasized that the carotid labyrinths of bony fishes (Siluriformes) and the carotid labyrinth and body of tetrapods have quite a different origin and structurally show no trace of pillar cells.

One of the interesting features of the pseudobranch is that the efferent blood from it passes to the brain and in many cases to the eyes. It has been suggested that in some cases it may result in these organs being supplied with blood of even higher than normal oxygen levels. Certainly, damage to the pseudobranch soon produces blindness in a number of fishes, but it is difficult to extirpate the pseudobranch surgically and not interfere with the ophthalmic artery.

## III. GILL ORGANIZATION

The general organization of the gills is based on a system of progressive subdivision, first of all giving rise to the individual pouches separated by the gill septa, which vary in extent from group to group, being most extensive in the more primitive forms. Along their full length and on anterior and posterior surfaces of each arch are found the gill filaments, which may be regarded as an increase in surface of the interbranchial septum (Fig. 4). The

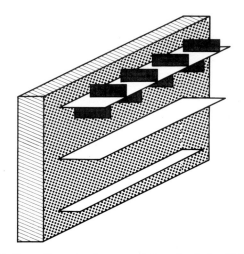

**Fig. 4.** Diagrammatic three-dimensional representation of a single gill bar of a developing fish showing the three main epithelia: (A) the gill cleft (coarsely dotted area), which becomes folded into (B) the gill filaments (white areas), which in their turn become folded into (C) the lamellae (finely dotted areas).

filaments themselves have their surface further increased by being folded into a series of lamellae. The many variations in detail of this plan are related both to the systematic position of the fish and its mode of life. In some cases the primary epithelium of both the filaments and lamellae becomes modified.

## A. Gill Septum

A gill septum separates two adjacent gill pouches, and to its surface is attached a series of filaments. In the most primitive groups (e.g., Elasmobranchii) the septum forms a complete partition between the pharynx and the outer body wall. Its extension forms a flap-valve for the next posterior slit. In more advanced groups there is a progressive reduction in the septum and the consequent freeing of the filaments at their tips (Fig. 6), so that water can pass more freely between the filaments of a given hemibranch and so enter the opercular cavity. All the filaments attached to one side of a gill septum form a hemibranch, and the two hemibranchs attached to a given branchial arch constitute a holobranch. It is clear that, where the septum is more complete, there will be some interference with the passage of water across the filament. This is less than was initially supposed, as more detailed anatomic and physiological studies have demonstrated that on the inner side of the attachment of the filaments there is a so-called septal channel (Kempton, 1969; Grigg, 1970a,b) along which the water can flow outward (Figs. 5

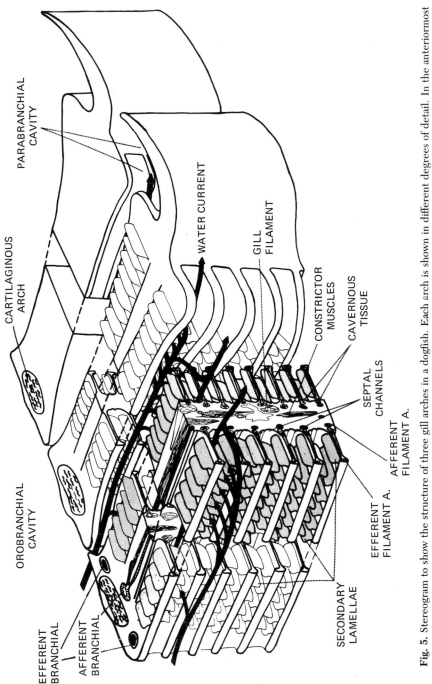

**Fig. 5.** Stereogram to show the structure of three gill arches in a dogfish. Each arch is shown in different degrees of detail. In the anteriormost arch is shown the structure of the secondary lamella and septum; in the last arch the main organization and shape of the filaments are visible, as well as the way in which extensions of the septa overlap the next posterior arch and form the valve at the exit of the parabranchial cavity. The direction of water flow between the secondary lamellae along the septal channels is illustrated in thick arrows.

PARABRANCHIAL
CAVITY

CARTILAGINOUS
ARCH

OROBRANCHIAL
CAVITY

EFFERENT
BRANCHIAL

AFFERENT
BRANCHIAL

WATER CURRENT

GILL
FILAMENT

CONSTRICTOR
MUSCLES

CAVERNOUS
TISSUE

SEPTAL
CHANNELS

AFFERENT
FILAMENT A.

EFFERENT
FILAMENT A.

SECONDARY
LAMELLAE

and 23B). Blood and water flows are countercurrent (Hills and Hughes, 1970; Piiper and Scheid, 1982), although a more crosscurrent arrangement has also been considered (Piiper and Schumann, 1967). Before leaving through the external gill slit, water passes into a series of parabranchial cavities, which are analogous to the opercular cavity of the teleost system. In the elasmobranchs, Woskoboinikoff (1932) distinguished the cavity before the gills as the orobranchial cavity, comprising the main cavity of the mouth together with parts of the gill slits from the internal slit to the point where the water flows between the filaments. In teleost fish the orobranchial cavity is more commonly referred to as the buccal cavity; furthermore, as already mentioned, the opercular cavities are partly homologous with the combined parabranchial cavities, as in teleosts the septa have become much reduced and the development of the operculum has produced a separate cavity into which the water from all of the gill slits flows. Functionally the separation between these cavities is the gill resistance, mainly constituted by the narrow gap between lamellae.

The degree of development of gill septa is best understood from a series of diagrammatic sections across individual gill arches (Fig. 6). The two extremes indicated earlier are exemplified in the elasmobranchs, with their complete septum, and in the teleosts, where the septum is very significantly reduced. In some of the intermediate groups there are varying degrees of septal development. These include, for example, the condition in *Neoceratodus* (Dipnoi) where the gill filaments of a given holobranch are held together except near their tips. A similar situation has been described for the gill of *Latimeria* (Hughes, 1980c), where the main interbranchial septum splits proximally but continues and maintains contact between the filaments of individual hemibranchs except at their tips. *Latimeria* has an operculum

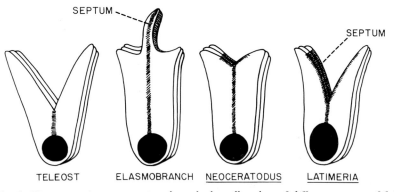

**Fig. 6.** Diagrammatic cross section through the gill arches of different groups of fishes showing varying degrees of development of the gill septum. (After Hughes, 1980c.)

that is not so well developed, and a similar situation is also found in the chimaeroid fish. Among primitive actinopterygians such as *Amia*, the interbranchial septum is well developed proximally, but the gill filaments are free for about one-third of their length. Among different groups of teleosts, there is variation in the degree of development of the septum, that has been used as a basis for gill classification (Miscalencu and Dornescu, 1970). In some forms the septum is almost complete, whereas in the more advanced perciform type the filaments are free for most of their length.

These differences are associated with other features of the anatomy shown diagrammatically in Fig. 7. Some of the associated features concern the development of the filament adductor muscles, which serve to draw together the tips of the filaments of the adjacent hemibranchs, a movement that occurs rhythmically and with particular force during coughing (Bijtel, 1949). Normally the tips of the filaments are in close contact with one another because of the elastic properties of the gill filament skeleton. However, it would seem that the different degrees of development of the septa are very much related to the systematic position of the fish.

Dornescu and Miscalencu (1968a,b) distinguished three types (Fig. 7) of gills on the basis of their studies of over 50 species from six different orders of Teleostei. The clupeiform type has a well-developed interbranchial septum, so that the filaments are only free for about one-third of their length. These workers recognized the so-called "blebs" at about one-third of the length along the efferent filament artery, which would appear to be similar to those described by Fromm (1976). The cypriniform type has a more expanded bleb, and the filament adductor muscles are mainly intrafilamental. The most advanced is the perciform type, which has a much shorter gill septum, with much of the adductor muscles running along the gill filament cartilage. It would seem that these different types with differing degrees of septal development among teleosts will influence the path of the water current. Little attention has been paid to the possible pathway of water once it has traversed the interlamellar spaces of those lamellae proximal to the end of the gill septum, it must be supposed that water flow continues along a channel analogous to the septal channel of the elasmobranch gill.

## B. Filaments

Filaments form the most distinctive respiratory structure of fish gills and are sometimes referred to as primary lamellae. However, some authors have referred to the original surface lining the interbranchial septum (Fig. 4) as the primary lamella, in which case the gill filaments formed by its folding would be secondary lamellae. The most generally used and least ambiguous

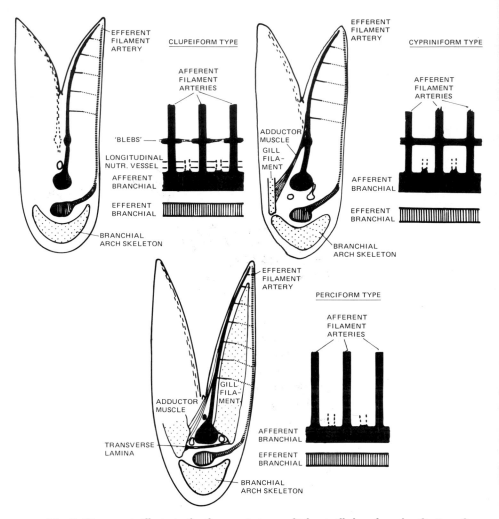

**Fig. 7.** Diagrams to illustrate the three main types of teleost gills based on classification of Dornescu and Miscalencu (1968a). A transverse section across a gill arch is shown for each type together with a medial longitudinal section to indicate the main blood vessels of the arch and gill filaments. (From Hughes, 1980c.)

term is gill filament. It also has the advantage of not involving any specific suggestion of homology or degree of branching. The shape of the gill filaments varies considerably, from being very filamental—that is, elongated— to being fairly stubby structures, but in nearly all fish the length exceeds the breadth. Although they give the appearance of being paired structures on either side of the gill septum, they usually alternate with one another, and the number of filaments in the two hemibranchs of a given holobranch are not always the same. In fact, it is more common for the number of filaments attached to the two hemibranchs that line a given gill slit to be closer in number.

In adult fish the number of filaments does not increase so markedly as during the juvenile growth period, but there is a very significant increase in the length of each of them as the fish grows. This leads to an increase in the total length ($L$) of all the filaments, which is an important morphometric dimension used in calculating gill area (Hughes, 1966b, 1984b). When measuring the length of the filament it is usual to begin from the position where the filament joins the gill arch to the tip (Fig. 8), but this is not always the full length of the filament that supports secondary lamellae. The length of the filaments along a given arch varies, and this is usually studied by making measurements of the first and last filament together with filaments at regularly spaced intervals such as every fifth, tenth, or twentieth filament, the intervals at which these measurements are made being related to the total number of filaments along a given hemibranch. The number may range from less than 50 to several hundred in large and active fish. In many cases there is a gradual increase in length of filaments from the dorsal end of the arch to about one-third of the filament number and then a gradual decrease to the ventral end of the arch. A common variation on this plan is that there is often a peak in filament length just before the main angle of the gill arch, which then increases to another maximum shortly afterwards, whereupon the fila-

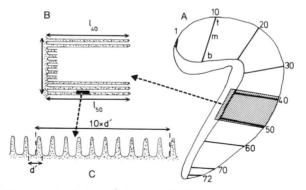

**Fig. 8.** (A) Diagram of a single hemibranch (72 filaments) of a teleost fish showing the position of filaments selected for measurement at regular intervals around the arch. Lengths of the first and last filaments are also determined. The positions for secondary lamellae selected for determination from the tip, middle, and base of each selected filament are indicated. (B) Diagram showing method for measurement of length of filaments 40 ($l_{40}$) and 50 ($l_{50}$) and the distance between filaments 40 and 50. (C) Diagram illustrating the method for measurement of secondary lamella frequency $1/d$, $= n/2$ by measuring the distance for 10 secondary lamellae. (From Hughes, 1984b.)

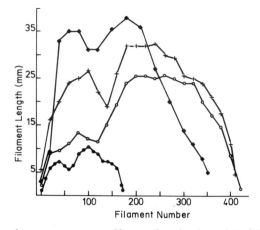

**Fig. 9.** Diagram showing variation in filament length when plotted for hemibranchs of different species of fish: mackerel, *Scomber scomber* (●), (After Hughes and Iwai, 1978.) Skipjack (♦), grouper (+), barracuda (○). (After Hughes, 1980c.)

ment length decreases to the last filament (Fig. 9). In general the length of the filaments is measured on each of the hemibranchs on one side of a fish, but in some species (e.g., flatfishes) there are differences on the two sides and it is necessary to measure the length of the filaments of all the hemibranchs on both sides of the fish. There are also differences in length of corresponding filaments on the same arch. For the anterior arches, the filaments of the posterior hemibranch are usually longer, whereas the anterior hemibranch filaments are longer in arches I and II. These differences may vary along a given arch, and this modification is very striking in the gills of *Latimeria* (Fig. 10).

Total filament length ($L$) is probably the most readily determined overall dimension of the gills and should be achieved with most accuracy. As indicated earlier, the tips of the filaments of adjacent hemibranchs are usually in close contact with one another, and this forms an important feature of the gill sieve, ensuring that little water will be shunted past the tips of the filaments. Such a shunting occurs, however, during hyperventilation, with a consequent fall in percentage utilization (Hughes, 1966b; Hughes and Umezawa, 1968). During coughing, contraction of the filament adductor muscles reduces the resistance of this pathway and enables the water current to be reversed in certain patterns of coughing movements (Hughes and Adeney, 1977). The filaments are supported by gill rays and connective tissue and vascular spaces, the detailed nature of which varies from species to species. In the elasmobranchs the vascular spaces form a special cavernous tissue on the afferent filament side only (Acrivo, 1938), which probably has a supporting hydrostatic function (Hughes, 1980b), and gill rays support the interbranchial septum. Among teleosts there is a reduction of the afferent arterial cavernous tissue, which may be represented by the "blebs," and each filament is supported by a single gill ray with its expanded edge in the plane of the filament.

The arrangement and form of the gill rays varies from species to species (e.g., between rainbow and brown trout) and has been used as a systematic character; this is also true (Iwai, 1963, 1964; Kazanski, 1964) of the gill rakers that line the gill arches mainly on the pharyngeal side. The form of the gill rays provides greatest support to the trailing edge of the filament rather than to the leading edge that faces the ventilatory water current. This position of the gill rays is related to their function as skeletal supports for the insertion

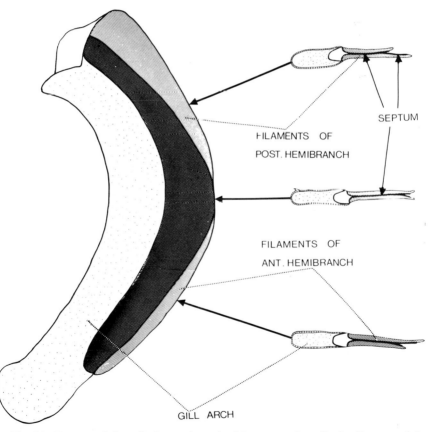

SEPTUM

FILAMENTS OF

POST. HEMIBRANCH

FILAMENTS OF

ANT. HEMIBRANCH

GILL ARCH

**Fig. 10.** Diagram of the gill of a single arch of *Latimeria*. Dorsally the filaments of the posterior hemibranch are longer than those of the anterior hemibranch, but ventrally the filaments of the anterior hemibranch are the longer. Sections across the arch at three levels show these differences in filament length. (From Hughes, 1976.)

Table III

Presence (+) or Absence (−) of Fusions between Filaments and/or
Lamellae

| Species | Filamentar fusion | Lamellar fusion |
|---|---|---|
| *Coryphaena hippurus* | − | − |
| *Scomber scombrus* | − | − |
| *Mola mola* | − | − |
| *Katsuwonus pelamis* | − | + |
| *Euthynnus affinis* | − | + |
| *Thunnus albacares* | + | + |
| *Thunnus atlanticus* | + | + |
| | (outflow surface only) | |
| *Thunnus obesus* | + | + |
| *Thunnus thynnus* | + | + |
| *Acanthocybium solandri* | + | − |
| *Terrapturus audax* | + | + |
| *Xiphias gladius* | + | − |
| *Istriophorus* sp. | + | − |
| *Amia calva* | − | + |

of the filament adductor muscles. Expansion of the gill rays in the plane of the filament is especially notable in large-gilled fish such as tunas, where their support during ramjet ventilation is especially important. Oceanic fishes also show the development of thickened supporting tissue on the leading and trailing edges of the filaments, which may coalesce to form interfilamental junctions (Table III). Thickenings without such coalescence (Fig. 16C) are found in the sunfish, *Mola mola* (Adeney and Hughes, 1977), which is not a rapid swimmer, confirming that such thickenings are related to the size of the gill. Similarly, the interfilamental junctions are only found in specimens above a certain size in some species.

The filaments are sometimes divided at their tips (Fig. 11), a condition that is more common in some species than others. It probably arises as an anomaly during development, perhaps as a result of physical damage.

## C. Lamellae

The lamellae are the most important units of the gill system from the point of view of gas exchange. The rest of the basic anatomy is directed to providing a suitable support for these structures and to enable the water and blood to come into close proximity. Essentially each lamella comprises two epithelia that are kept separated by a series of pillar cells between which the blood can flow. The direction of blood flow is opposite to that of the water and thus facilitates gas exchange (van Dam, 1938; Hughes and Shelton,

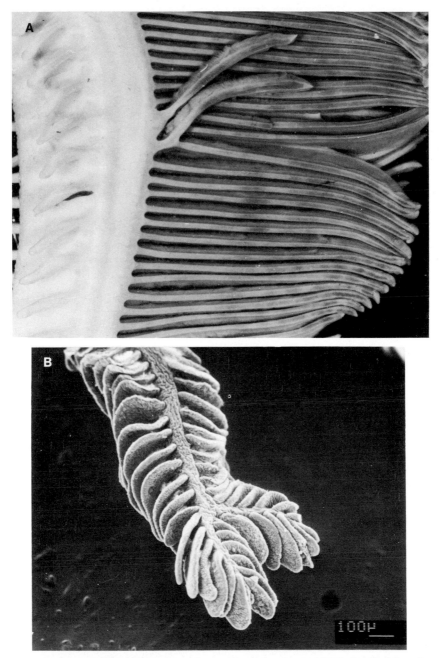

**Fig. 11.** Photographs of gill filaments that are divided at the tip. (A) Whole gill of *Huso huso* with a single divided filament; (B) scanning electron micrograph of tip of a single filament of a mudskipper.

1962). Although lamellae have apparently the same basic structure among all groups of fishes, there are nevertheless important differences in detailed anatomy. The extent to which these involve details at an electron-microscopic level is discussed in Chapter 2, especially in so far as they involve differences in relation to blood flow.

From a design point of view, lamellae are required to have a large surface area, where gas exchange can be facilitated without any excessive exchange of ions and water. A close contact between water and blood must be achieved so that oxygen uptake can occur in the limited period (contact time = about 1 sec, Hughes *et al.*, 1981) during which the water and blood are passing the lamellae. At the same time, the finer the pores in the gill sieve, the greater will be the resistance to water flow and hence the greater the expenditure of energy by the fish in moving the viscous and dense respiratory medium. In classical descriptions of the gill system, it has been thought of as being homogeneous; in fact, this superficial impression is shown to be misleading from both a structural and functional points of view. As with most respiratory systems, heterogeneity is more typical (Hughes, 1973b).

## 1. NUMBER

The number of lamellae in a fish is very large and increases steadily with body size; it may reach more than 5 million in a very active fish of 1 kg body weight. In contrast, inactive species such as the toadfish may have a very much smaller number of lamellae (Fig. 12).

## 2. SHAPE

The heterogeneity of the gill system is well illustrated by the variation in shape of lamellae from a single gill arch (Fig. 13). Heterogeneity produces problems for quantitative analyses both in relation to surface area and also of the distribution of water flow through the gill sieve. There are many variations in shape of lamellae from similar regions of a wide variety of teleost species (Fig. 14). Some elasmobranchs (e.g., *Raia*) may have horn-shaped projections at the anterior end of the lamellae, and this arrangement seems to be found in some other elasmobranchs (Cooke, 1980). In general, however, it is clear that the greatest proportion of the surface of a lamella is found toward the leading edge, that is, the edge at which the water enters the gill from the buccal cavity.

Some of the consequences of the different shapes of lamellae have been discussed by reference to gas exchange in the gill of icefish, where complications due to the oxygen–hemoglobin dissociation curve are not present (Hughes, 1972b). A general conclusion has been reached from such analyses that the distribution that maximizes lamellar area at the inlet side produces the best gas exchange situation both in counter- and cocurrent flow. For

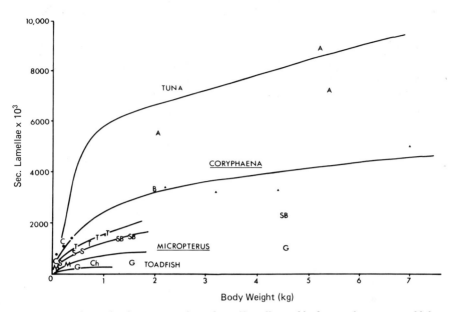

**Fig. 12.** Relationship between total number of lamellae and body mass for a variety of fishes. Lines are drawn for six of them (tuna, *Coryphaena* & mackerel, sea trout & mullet, striped bass and scup, *Micropterus*, and toadfish). In addition, points for specimens of individual fish are given for the following: A = false albacore, B = bonito, C = *Caranx*, Ch = *Chaenocephalus*, G = goosefish, M = mullet, S = scup, SB = striped bass, T = sea trout.

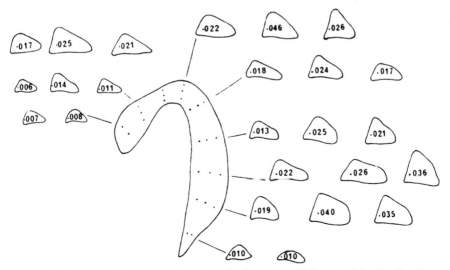

**Fig. 13.** Diagram of the anterior hemibranch of the first gill from the left side of an 8-cm trout showing regional variation in the surface area of the secondary lamellae. (After Morgan, 1971.)

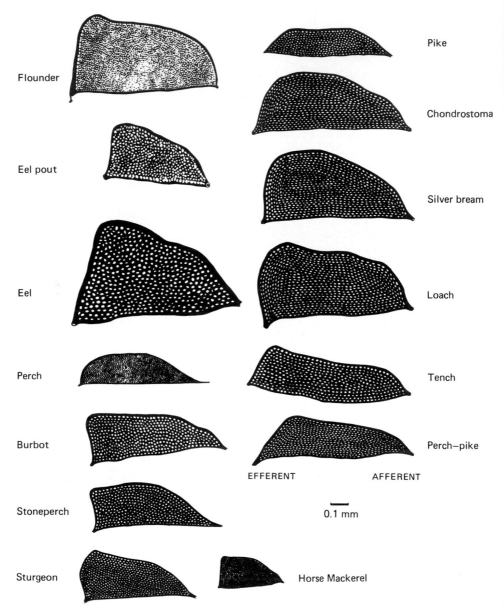

**Fig. 14.** Tracings of individual secondary lamellae of a variety of fish based on drawings by Byczkowska-Smyk (1957–1962), Nowogrodska (1943), and Oliva (1960). (Modified from Hughes, 1970a.)

some of the shapes it ensures that blood in the marginal channels at the inlet side will come into contact with water containing the maximum $P_{O_2}$. This is especially relevant, as in many species it is known that under resting conditions the blood sometimes flows differentially around only the marginal channels of the lamellae (Hughes, 1976). Furthermore, the general hydrodynamics of the situation suggest that the water flow velocity will be greatest around the marginal edges of the lamellae and be slowest in the crypt region between the bases of adjacent lamellae, where it will consequently have the greatest contact time.

Thus, even within the limits of a single lamella, it is apparent that there would be differences in gas exchange, and presumably there are adaptations in relation to the microflows of the two fluids. Such differences are extremely difficult to investigate on the water side, but on the blood side of the exchanger there is certainly evidence that the blood flow is not always by the most direct route from the afferent to the efferent filament vessel. This has been clarified for the large lamellae of some tunas (Fig. 15), where most blood flows at a definite angle to the expected water flow direction, as the adjacent pillar cells are in contact and form much more definite channels along which the red blood cells are directed (Muir and Kendall, 1968).

## 3. SUPPORT

Each lamella contains red blood cells, which can completely fill the blood channels; the combined thickness of barriers on the two sides of a lamella is 3–20 μm depending on the species. Consequently, the total thickness of

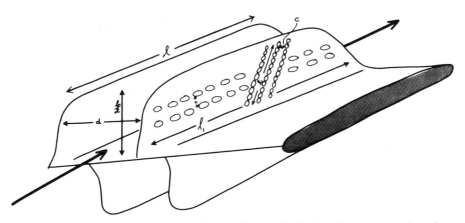

**Fig. 15.** Diagram of a single lamella of a tuna showing the blood pathways at an angle to the direction of the water flow. The relationship between water flow and dimensions for the water is $\dot{V}_G = (N \cdot \Delta P \cdot 5 d^3 b)/24l$ (Hughes, 1966) and for the blood is $\dot{Q} = (Ag \cdot \Delta p \cdot \pi c^3)/256 \mu l_1{}^2$ (Muir and Brown, 1971).

each lamella is fairly small, and it is sometimes subjected to relatively fast flows of water. The main skeletal supporting system of this structure is probably attributable to the blood contained in the channels and to the combined effects of hydrostatic pressure, viscosity, and stiffness of the red blood cells. There is also an intrinsic skeletal system of the pillar cell system, formed by the basement membrane together with the extracellular columns of collagen (Hughes and Weibel, 1972), which run in membrane-lined channels within the pillar cells. This system will also serve to resist outward distortion due to the arterial blood pressure (Bettex-Galland and Hughes, 1973) and seems to be antagonized by a contractile mechanism within the pillar cells themselves (Hughes and Grimstone, 1965; Smith and Chamley-Campbell, 1981). It would appear that the lamellae are well supported and effectively retain their orientation within the water current. In general it is believed that the orientation of the lamellae on the filaments is such that the direction of water flow is parallel to that of the lamellar surfaces, and the flow is laminar. To increase this possibility there are differences in the orientation of lamellae at different positions along the axis of the gill filaments (Wright, 1973). This would appear to ensure that there is less resistance to the water flow provided by the lamellae in their normal orientation.

## IV. MODIFICATIONS IN RELATION TO HABIT

As with other systems of the fish body, the gills show many modifications that are adaptive to the particular physicochemical conditions under which they must operate. Furthermore, as fishes from a variety of taxonomic groups frequently become adapted to similar conditions, they illustrate extremely well some of the principles of convergence and parallel evolution; perhaps the air-breathing fish provide the best examples of the latter.

### A. Fast-Swimming Oceanic Species

These fish are characterized by the presence of a large gill surface, which is the result of both an increase in total filament length and a high frequency of lamellae, and consequently very large numbers of lamellae. The lamellae themselves may, however, be relatively small and the barrier between the water and blood less than 1 μm in certain tunas (see Table IV). It is well known that these fish show ramjet ventilation, in which the main energy forcing water through the gill sieve is provided by the body musculature (Hughes, 1970b; Roberts, 1975). It is probable that this represents an economy in the overall energetics of the fish. These fish may develop into rela-

**Table IV**

The Water–Blood Barrier: Measurements Showing Range Found for Length of the Three Main Parts of the Barrier[a,b]

| Fish species | Epithelium (μm) | Basal lamina (μm) | Pillar cell flange (μm) | Total water–blood distance (μm) | Mean[c] (μm) |
|---|---|---|---|---|---|
| Elasmobranchs and benthic teleosts | | | | | |
| *Scyliorhinus canicula* | 2.38–18.48 | 0.3–0.95 | 0.37–0.71 | 5.24–19.14 | 11.27 |
| *Scyliorhinus stellaris* | 3.0–9.3 | 0.25–1.32 | 0.37–0.62 | 4.29–11.88 | 9.62 |
| *Squalus acanthias* | 3.0–22.5 | 0.3–0.6 | 0.12–0.6 | 3.43–29.55 | 10.14 |
| *Galeus vulgaris* | 1.5–22.5 | 0.11–0.66 | 0.05–0.2 | 2.31–24.0 | 9.87 |
| *Raia montagui* | 0.8–14.8 | 0.2–1.65 | 0.08–1.2 | 1.2–15.6 | 4.85 |
| *Raia clavata* | 0.5–11.5 | 0.13–0.63 | 0.03–1.13 | 3.0–11.6 | 5.99 |
| *Pleuronectes platessa* | 0.35–10.4 | 0.09–1.7 | 0.48–0.95 | 0.93–17.28 | 3.85 |
| *Solea solea* | 1.3–3.4 | 0.03–0.41 | 0.03–0.18 | 1.97–3.98 | 2.80 |
| *Solea variegata* | 1.0–12.3 | 0.13–0.33 | 0.03–0.63 | 2.06–23.1 | 5.55 |
| *Limanda limanda* | 0.13–7.0 | 0.1–0.41 | 0.063–1.0 | 0.88–8.4 | 2.53 |
| *Microstomus kitt* | 0.25–16.7 | 0.1–0.69 | 0.013–0.1 | 0.71–44.25 | 3.23 |
| Pelagic teleosts | | | | | |
| *Thunnus albacares* | 0.017–0.625 | 0.048–0.166 | 0.025–0.875 | 0.166–1.125 | 0.533 |
| *Euthynnus affinis* | 0.05–0.475 | 0.063–0.125 | 0.025–0.25 | 0.313–1.063 | 0.596 |
| *Katsuwonis pelamis* | 0.013–0.625 | 0.075–1.875 | 0.017–0.375 | 0.24–1.906 | 0.598 |
| *Scomber scombrus* | 0.165–1.875 | 0.066–1.0 | 0.033–1.75 | 0.600–3.625 | 1.215 |
| *Trachurus trachurus* | 0.38–3.4 | 0.063–0.125 | 0.30–2.33 | 0.25–3.13 | 2.221 |
| *Salmo gairdneri* | 2.075–9.25 | 0.125–1.25 | 0.20–2.4 | 3.32–9.6 | 6.37 |

[a]Measurements for elasmobranchs and benthic teleosts from Hughes and Wright (1970); those for pelagic teleosts from Hughes (1970a).
[b]The electron micrograph sections used for the measurements were not always perpendicular to the lamellar surface.
[c]The mean of about 50 measurements for the total distance.

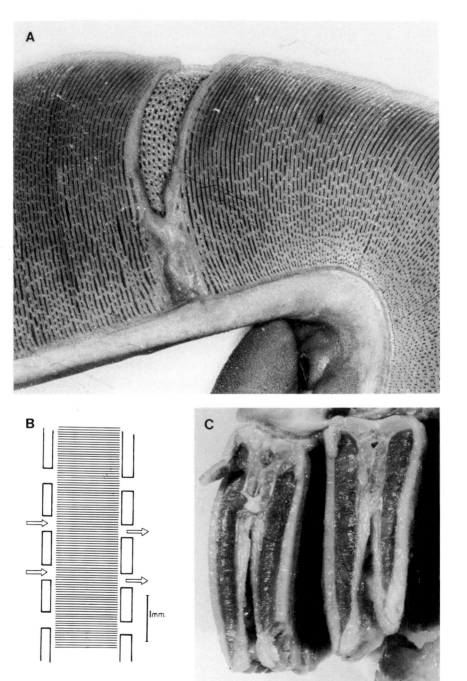

tively large specimens and swim at high speeds, and it is inevitable that sometimes foreign matter may enter the mouth and cause physical damage to the gill system (e.g., Fig. 16A). It is perhaps in response to such conditions that the various additional structures on the filaments have evolved, as they serve to consolidate the hemibranchs. In some cases (Fig. 16B) water only enters the interlamellar spaces through small pores before being collected together on the outlet side once more through another set of pores. Although a supporting and protective function has naturally been emphasized for these structures, there is also the important possibility that they effectively slow down the water velocity past the lamellae and thus extend contact time. An analysis based on the dimensions of both the water and blood parts of the gas exchanger (Fig. 15) has indicated the adaptive features in this system (Brown and Muir, 1970).

In view of the development of the interfilamental junctions in tunas, the presence of similar consolidation of the hemibranchs in the holostean *Amia* is surprising, as the habits of these fish are entirely different (Bevelander, 1934). There can be no doubt that they are independently evolved structures and apparently have a different function. A possible suggestion is that these structures serve to reduce the collapse of the gill system during air breathing of these freshwater fishes (Hughes, 1966a,b).

Another oceanic fish with special developments of the gill is the sunfish (*Mola mola*), which may reach sizes exceeding 1000 kg. The leading and trailing edges of the gill filaments are covered by a dense layer of hard tissue (Fig. 16C) in which are embedded many spines originating from a layer just beneath the outermost epithelium (Adeney and Hughes, 1977). Growth of the filaments can lead to some restriction of the space between filaments, although there is little fusion between them. One of the most interesting features of *Mola* gills is the extent of the internal gill slits through which water enters the spaces between hemibranchs. This is restricted to about one-third of the total length of the gill, and consequently, the typical teleost flow of water through the filament can only occur in this region; otherwise water entering through the slit passes either dorsally or ventrally into pouc-

Fig. 16. (A) Photograph of a single gill arch from a marlin. Several filaments have been damaged and show regeneration. The formation of interfilamental junctions leaving pores for the flow of water are clearly visible and correspond to those shown diagrammatically in (B). (B) Diagram illustrating the flow pathway of water across a filament of a tuna gill in which interfilamental junctions are developed. On the inflow side, water enters through small pores, then passes through the interfilamental space containing secondary lamellae and leaves by a corresponding series of slightly smaller outlet holes into the opercular cavity. (After Hughes, 1970b.) (C) Section of two gill arches of *Mola mola*, showing two rows of hemibranchs in each arch with a dense layer of hardened tissue covering the leading and trailing edges. (After Adeney and Hughes, 1976.)

hes from which the water can pass between the filaments of the hemibranchs. This situation closely resembles that of elasmobranchs, rather than
the typical teleost system. The opening of the internal gill slits is in fact the
only region where the branchial arches are present; the fusion together of
the hemibranches is due to connective tissue that thus prevents water from
entering the gill chamber except in the middle region. The gill ray is well
developed on the inlet side and forms a flattened structure that extends
across the full length of the lamellae. In the regions where the branchial arch
skeleton is absent (i.e., ventrally and dorsally), there is a special fusion of the
gill rays of adjacent filaments, but not with filaments of the opposite hemibranchs. However, the bases of gill rays are connected to filaments of the
neighboring hemibranch by a thin layer of connective tissue. These structures apparently support the hemibranch and replace the gill arch in these
regions, as there are no cartilaginous connections between the gill arch and
the base of the gill rays in the middle region.

## B. Fishes of Intermediate Activity

For a long time it has been recognized that it is difficult but desirable to
classify fishes according to their activity, the most controllable method being
to measure swimming speed in relation to oxygen consumption (Brett,
1972). Such a quantitative basis has not yet been completed. Nevertheless, a
broad band of species regarded as "intermediate" in activity by Gray (1954)
were as follows: *Sarda sarda, Mugil cephalus, Caranx crysos, Roccus lineatus, Archosargus probatocephalus, Chilomyceterus schoepfi, Stenotomus
chrysops, Tatutogo onitus, Prionotus strigatus, Poronotus triacanthus,
Cynoscion regalis; Palinurichthyes perciformis, Echeneis naucrates, Spher-
oides maculatus, Centropristis striatus, Peprilus alepidatus, Prionotus car-
olinus, Trichiurus lepturus, Paralichthys dentatus.* Such fishes represent
the major groups of teleosts, and their respiratory systems have the following
typical structure.

The gill arches are well developed and support filaments of average
length. The operculum covers over the whole of the gill system and communicates with the exterior via a more or less continuous slit. The lamellae are
average in size, and their frequency on each side of a filament is usually
about 18 to 25 per millimeter. Ventilation of gills usually involves rhythmic
ventilatory movements, although under certain conditions (e.g., high swimming speeds) some of them might show ram ventilation. The buccal and
opercular suction pumps are equally important in maintaining the ventilatory current and would be classified in group I of Baglioni's scheme (Hughes,
1960b).

## C. Sluggish Fishes

There are many fish, usually of benthic habits, that are relatively inactive, but this "sluggish" group would also include some fish that maintain a constant position in a condition of neutral buoyancy and make only occasional darting movements connected with feeding and mating. The latter category includes many deep-water fishes such as groupers and *Latimeria*; the icefish, lacking hemoglobin, is another example. All of these species are expected to have low oxygen consumption in relation to their body size, and their gills are relatively poorly developed. In some instances gills are restricted to three arches, as in the toadfish (Table II). The filaments are short especially in some deep-sea fishes (Marshall, 1960; Hughes and Iwai, 1978). Individual lamellae are relatively large in area, and their spacing (10–15 per millimeter) is wider than that of other species. The resistance to water flow through the gill sieve is therefore low. The opercular pumps are well developed usually because of the good development of the branchiostegal apparatus. In bottom-living forms, exit from the opercular cavity is often by a limited part of the opercular slit, which is directed dorsally. Similar restrictions of the opercular openings are found in midwater hoverers such as trigger and puffer fish. Ventilatory frequency is low, the opercular expansion phase being slow, but ejection of water from the respiratory system occupies a short portion of the whole cycle.

Flatfishes show some special adaptations related to their mode of life, especially the way in which they come to rest on one of the two sides of the body. In most cases the fish rest on the morphological right side, which is pale-colored, whereas the upper surface is darker and contains many chromatophores. Ventilation is achieved with a well-developed suction pump mainly because of the extensive branchiostegal apparatus (Schmidt, 1915; Henschel, 1941). In fact, recordings of pressure changes in the opercular cavities (Hughes, 1960b) showed slight asymmetry and the absence of an apparent reversal phase that is characteristic of many teleost fishes. It was suggested that this resulted from the active mechanism closing the opercular openings, as this would reduce the possible influx of sand or other particles from the benthic habitat. The gills themselves are well developed on both sides, although there are some differences in their detailed morphology. The upper opercular cavity is more convex and longer and accommodates somewhat better developed gills than the flatter and smaller lower opercular cavity with its moderately developed gills. Measurements of the gills on the two sides have been made by a number of authors and are indicated in Table IV.

There appear to be changes in the path of the ventilatory current especially during development (Al-Kadhomiy, 1984). The premetamorphic stages are free living, and their body configuration is similar to that of typical teleosts. In these forms the opercular cavities, gills, and bony elements concerned with gill ventilation are equally developed on both sides. Following metamorphosis there is gradual development of the asymmetry of the adult form. The life of some flatfishes seems to alternate between swimming freely in midwater and periods when they are on the sea bottom, especially during migrations (Harden-Jones, 1980). It seems possible that during the midwater phases, the ventilation is equal on both sides of the respiratory system, whereas when on the bottom they make greater use of the special channel (Yazdani and Alexander, 1967; Al-Kadhomiy, 1984), which enables water from the ventral opercular cavity to be passed into the dorsal opercular cavity and thence to pass out via the dorsal opercular opening. In such a situation, however, both sets of gills are well ventilated; thus, the observation does not necessarily lead to the expectation that they are unequally involved in gas exchange.

## D. Air Breathers

Air breathing has evolved many times among different groups of fishes and in most cases seems to have been related to a worsening of conditions for aquatic respiration. In their classical studies on the Paraguayan chaco, Carter and Beadle (1931) drew attention to groups of fishes that live near the surface and ventilate their gills with these waters of relatively high oxygen tension. Fish that habitually live in deeper waters must swim to the surface to gulp air, with a consequent increase in oxygen consumption (Singh, 1976). In more extreme conditions the aquatic environment dries up and the fish must use some form of air breathing to survive, either during migration to some other aquatic environment or as a preliminary to some resting stage. In all cases, as has been pointed out (Carter, 1957), the modifications that have evolved concern parts of the alimentary canal, where gas exchange with the air takes place. In each of these sites there is an increased vascularization associated usually with a reduction in the tissue barrier separating the air and blood.

## 1. Air-Breathing Organs

The precise anatomy ranges from suprabranchial cavities and their enclosed labyrinthine organs to forms in which special regions of the rectum have become the place where bubbles of air taken in at the mouth and passed through the alimentary canal come to rest and where gas exchange occurs (e.g., loaches, *Misgurnus*). The increase in surface for gas exchange often involves intucking of the endodermal epithelium, as in the lungs of *Polypterus* and dipnoans, which are homologous with the tetrapod lung.

Striking convergences are found, as in some catfishes (e.g., *Saccobranchus*), where a backward projection from the suprabranchial chamber forms an air sac on either side of the vertebral column (Munshi, 1962). When the "lungs" of *Saccobranchus* were investigated morphologically, it was apparent that there were similarities to the basic structure of gills, and this was substantiated by more detailed studies using electron microscopy (Hughes and Munshi, 1973a). The surface structure as revealed by scanning electron microscopy showed similarities between the microvillous surfaces of the lamellae and of the respiratory islets, whereas the "lanes" between the lamellar structures have microridged surfaces comparable to those of the gill filament epithelium (Fig. 17). Electron microscopy also showed that the labyrinthine organs of anabantoid fish, which were supposed to be homologous with gill lamellae, are not derived so directly from the gills themselves (Hughes and Munshi, 1968, 1973b). The suggested homology with gills (Munshi, 1968) rested on the recognition of the typical pillar cell structure in the labyrinthine plates, but electron microscopy revealed that these so-called pillars are intucked epithelial cells that have a similar position separating adjacent blood channels. Electron micrography also showed that the air–blood barrier was extremely thin ($<1$ μm), and consequently, such structures could have a relatively high diffusing capacity, although their surface might be fairly restricted (Table V). In other cases, however, the air-breathing organs—though well vascularized—have fairly thick barriers between the air and blood and hence could not form such important gas exchange organs.

From a physiological point of view, it has been recognized that the air-breathing organs of fish are especially important in relation to the uptake of oxygen, as this gas is in relatively short supply in the water (Singh, 1976). The release of carbon dioxide is not such an important function of the air-breathing organ, and this is much more easily facilitated in the aquatic medium because of its high solubility in water. Consequently, the gas exchange ratio for air-breathing organs and gills may be quite different, although the overall ratio is 0.8–1.0, as in many other fishes.

## 2. THE GILLS

The gills of air-breathing fishes are therefore important in gas exchange, although not so much in oxygen uptake, and in general their surface area is less than typical aquatic breathers of the same body size. In some species the gills themselves are important organs not only in water breathing but also when the fish comes out into the air. This is especially true, for example, in some mudskippers (*Periophthalmus*), which live on mudflats and mangrove swamps in many different parts of the world. When they come out of the

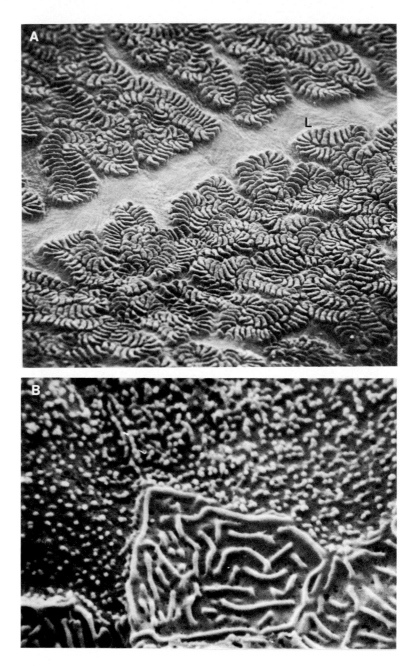

**Fig. 17.** (A) Scanning electron micrographs (SEM) of respiratory islets from the air tube of *Saccobranchus*. Notice the whorl-like appearance of the islets and the presence of a "lane" (L) separating groups of islets in which a biserial arrangement can be identified. (×120) (B) High-power SEM showing the microridged surface of the flattened cells of the lanes, which change abruptly into the microvillus surfaces of the respiratory islets. (×12,000) (After Hughes and Munshi, 1978.)

**Table V**

Diffusing Capacity of the Tissue Barrier ($D_t$) and Component Measurements for Respiratory Surfaces of *Channa* and *Anabas*[a]

| Fish species | Respiratory surface for 1 g fish (mm²) | Thickness of tissue barrier (μm) | Area g wt 100 g fish⁻¹ (mm²) | Diffusing capacity (ml min⁻¹ mm Hg⁻¹ kg⁻¹) | References |
|---|---|---|---|---|---|
| *Channa punctata* | | | | | |
| Total gills | 470.39 | 2.0333 | 71.8229 | 0.0530 | Hakim *et al.* (1978) |
| Suprabranchial chamber | 159.08 | 0.7800 | 39.1705 | 0.0753 | Hakim *et al.* (1978) |
| *Anabas testudineus* | | | | | |
| All gill arches | 278.00 | 10.0000 | 47.2000 | 0.0071 | Hughes *et al.* (1973) |
| Suprabranchial chamber | 55.40 | 0.2100 | 7.6500 | 0.0539 | Hughes *et al.* (1973) |
| Labyrinthine organ | 80.70 | 0.2100 | 32.0000 | 0.2286 | Hughes *et al.* (1973) |

[a]From Hakim *et al.* (1978).

water such fishes expand the buccopharyngeal cavity and enclose a separate volume of air with which gas exchange appears to take place. In their gill structure different degrees of stiffening of the system have been recognized (Schlotte, 1932). In many of these species cutaneous respiration is also of great importance when the fish is out of water. During the life history of some air-breathing species it has often been found that in the earlier stages the fish are almost entirely dependent on aquatic respiration, but with an increase in size and greater development of the air-breathing organ relative to the gills, there is a transition toward greater dependence on air breathing (Hughes *et al.*, 1974a).

## V.  GILL VENTILATION AND ROLE OF BRANCHIAL MUSCLES

Because of the greater density and viscosity of water with respect to air and its low content of oxygen (Hughes, 1963; Dejours, 1976), it has generally been accepted that the problem of ventilation of respiratory surfaces that faces a water-breathing species is greater than that of those that breathe air. Because of this it is usually considered that a greater portion of the standard metabolism of the fish is required for maintaining sufficient water flow, but estimates of the particular percentage are quite variable. They range from less than 1 to more than 40%, and no definite value has been established. The consensus of many studies would seem to support a figure of 5 to 10% (Hughes, 1973a; Jones and Schwarzfeld, 1974; Holeton, 1980). Similar values have been attributed to the work of the cardiac pump (Hughes, 1973a; Jones and Randall, 1978). Regardless of the precise figure, there can be little doubt that mechanisms have evolved that are extremely economical from an energetic point of view, and this leads to the great fascination of attempts to elucidate their detailed functioning. Fish would appear to have a greater range of ventilation mechanisms than any other group of vertebrates (Hughes, 1978a), the basic one being the double pump whereby a more or less continuous flow of water is maintained across the gill surfaces.

### A.  Water Pumps

Early anatomic studies of the respiratory system were based on fixed material. From a functional point of view, such studies have the serious disadvantage that the gills have a completely abnormal orientation. Consequently, standard diagrams used to illustrate fish ventilation showed the gills with large spaces between filaments through which much water could flow without having any oxygen removed from it. The largely morphological studies by Bijtel (1949) and Woskoboinikoff (1932), who emphasized the position of the gill filaments during normal ventilation, were very important, and these conclusions were confirmed by physiological measurements (Hughes and Shelton, 1958; Hughes, 1960a,b) that demonstrated that the respiratory system could be divided into two functional cavities separated by a gill resistance (Fig. 18A and B). In fact, there are three cavities in tele-osts—a single buccal cavity and an opercular cavity on each side—whereas in elasmobranchs the gills separate a single orobranchial cavity from five or

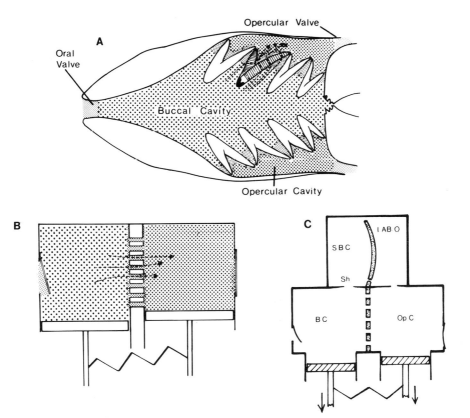

**Fig. 18.** (A) and (B) Diagrams to illustrate the double-pumping mechanism for the ventilation of fish gills. The equivalent of a simplified hydrodynamic model of the fish is shown by shading of corresponding parts. Communication between the buccal and opercular cavities is via a gill resistance, and the system is powered by changes in volume of the buccal and opercular cavities. It should be emphasized that changes in volume of these two cavities are not independent, and this mechanical coupling is indicated by a spring connecting two pumps of the hydrodynamic model. (From Hughes, 1976.) (C) Diagram of modified double-pumping model in *Osphronemus*. SBC, suprabranchial cavity; LAB O, labyrinthine organ; Sh, shutter.

more parabranchial cavities on each side of the fish. The double-pumping hypothesis was proposed because pressure changes recorded on the two sides of the gill resistance had different time courses and varied in relation to volume changes of the cavities. Apart from showing the differences in time course of pressures in the cavities, these studies using electromanometers also showed how small (about 1 cm $H_2O$) were the pressure changes during the normal ventilatory cycles. The small pressure changes contrast with the relatively large pressures recorded during feeding and some other activities involving the same basic apparatus (Alexander, 1970; Lauder, 1980, 1983). The low level of the pressure changes is indicative of the muscular activity involved and the consequent energetic economy during ventilation.

Analyses of cine films and pressure wave forms also indicated that the entrance to the typical system, the mouth, is guarded by valves (mandibu-

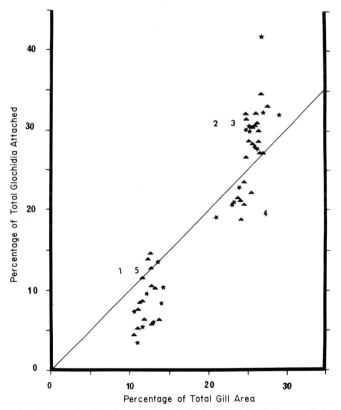

**Fig. 19.** Plot showing the distribution of glochidia larvae in the different gill slits of a tench. The percentage of the total numbers attached to a given fish are plotted in relation to the percentage of the total gill area that borders each of the gill slits (1–5). Closed triangle, experiments carried out in 1969; star; experiments carried out in 1970.

lar/maxillary), but often there are active changes in the size of the mouth opening. Similarly, in most cases movements of the opercular valves follow the pressure changes and indicate their largely passive function. At both the entrance and exit to the system, however, there are fishes in which active closing occurs, and this is important both in the normal rhythm and also during specil respiratory maneuvers such as coughing. Thus, bottom-living fishes such as flatfish have well-developed branchiostegal apparatus, which produces active closing of the opercular openings and prevents possible influx of sand particles before the expansion of the cavity occurs. Also, it has been shown that in rapidly swimming oceanic fishes such as tuna, regulation of the volume flow through the respiratory system is probably best achieved by the degree of opercular adduction. In these fish the dimensions of the elongated opercular slit also help to maintain good laminar flow of water across the body surface.

From a respiratory point of view, the most important flow concerns the water current across the gill surface, and it has been extremely difficult to obtain information concerning its detailed nature. One biological method has been to utilize the attachment mechanism of small glochidia larvae of freshwater mussels experimentally introduced into the water inspired at the mouth of the fish. Paling (1968) carried out such experiments and found fixation of the glochidia on certain hemibranchs that were related to the distribution of gill area. Similar experiments using tench showed that the distribution was not in direct proportion to the gill area but tended to be greater than expected for the second and third slits, whereas it was less than directly proportional to the surface area on the first and fifth slits (Fig. 19)

## B. Ventilation of Air-Breathing "Gills"

The gills of a number of species of fish, notably intertidal gobies and mudskippers, and eels when traversing the land, are used, and the buccal and opercular cavities are largely filled with air. The ventilation mechanism here is mainly to inflate the cavities with air and to keep them closed by active contraction of the valves at the entrance and exit of the respiratory system. Measurements of the content of the cavities have been made in eels, but few other species have been investigated (Berg and Steen, 1965). There are also cases where typical aquatic fish come to the surface, possibly under hypoxic conditions, to gulp air, much of which probably remains in the buccal cavity and presumably can serve to help increase the $P_{O_2}$ of the water as ventilation continues in the normal way. There are, however, many species of fish that gulp air and press it into contact with specific air-breathing organs, and some of these may be in the roof of the mouth, as in the electric eel (*Electrophorus*), where it forms an important gas exchange organ (Farber and Rahn, 1970). As the air will naturally stay in the roof of the mouth, it is perhaps not surprising that extensions of these cavities form the suprabranchial cavity of many air-breathing species. The surface of this cavity may be very well vascularized. Such a condition is well developed in the anabantoids, where the respiratory islets have very short tissue barriers and gas exchange is facilitated (Table V). In addition, anabantoids have expansions of the anterior branchial arches to form labyrinthine plates with increased surface area and are enclosed within the suprabranchial chamber. These organs are of great importance in these fish, and their mechanism of ventilation has developed modification of the normal buccal pressure–opercular suction pump mechanism (Fig. 18C).

In some genera (e.g., *Anabas*), air is gulped at the surface and passes into the suprabranchial chamber in the normal direction, its flow presumably controlled by differences in resistance of the gills themselves as against the resistance of the entrance to the suprabranchial chamber. This mechanism has been termed "monophasic," distinguished from the specialized mechanism called "diphasic" by Peters (1978), and is well illustrated by *Ostronemus*, the gourami. In this fish, the air within the suprabranchial chamber is replaced by water that enters from the opercular cavities and so displaces the air that is exhaled when the fish comes to the surface. Rapidly inhaled air then passes in the opposite direction through the suprabranchial chamber and so displaces the water (Fig. 18C). This unique mechanism ensures that the suprabranchial chamber gas is completely changed during each ventilatory cycle and that the respiratory surfaces in the suprabranchial cavity are bathed in water during this period. As a consequence of the high solubility of $CO_2$ in water and the thinness of barriers, it seems probable that carbon dioxide release into the water will also be aided by this novel ventilatory mechanism.

Many other mechanisms for ventilation of the air-breathing cavities of fish have been discussed by Randall *et al.* (1981), and in some cases there also seems to be the possibility of the air-breathing surface being in contact with water for at least part of the cycle. Other mechanisms, including ventilation of modified swim bladders, have suggested a whole variety of mechanisms including that of "jet streaming," which has been invoked in some interpretations of the ventilation of the anuran lung (Gans *et al.*, 1969).

## VI. GILL MORPHOMETRY

Measurements of the dimensions of respiratory organs have been made for a long time, and analogies such as that between the surface of the human lung and a tennis court vividly portray the large increase in surface that is provided for gas exchange functions. Fish gills are very much more accessible than mammalian lungs, and during recent years the investigation of these structures has proceeded very rapidly.

The term morphometry has become much more commonly used and is now frequently applied to all studies that concern the measurement of dimensions of structures within living organisms. It is also sometimes applied to gross dimensions (e.g., body weight–length relationships). Originally, however, it was used more in connection with the application of stereological methods to the investigation of gill dimensions. Another term sometimes used for nonstereological studies (Hughes, 1970a) is morphological measurements, but the shorter form, morphometry, is preferable and has now come into common usage for this whole area of study. It is impossible to give an adequate summary of all the basic techniques of stereology, and anyone wishing to extend their knowledge of this field should refer to some of the more standard texts (e.g., Underwood, 1970; Weibel, 1979). Many of the techniques used for gill and air sac morphometry are relatively simple, and a detailed knowledge of stereological principles is not always required (Hughes and Weibel, 1976). Measurement of structural entities is time-consuming, and therefore, it is important to be sure which measurements are most appropriate for the particular analysis that is envisaged. Before embarking on such a study it is often advisable to discuss the problem with a physical scientist and then choose which measurements to make; engineers often show great interest in these problems.

Many measurements are possible, but most are directed toward two main aspects. The first concerns the resistance to the flow of water and blood through the gills, and the second analyzes conditions for gas exchange, particularly the area and thickness of the tissue barriers. Although many measurements are now available, there is a great need for standardization of techniques. However, from a comparative point of view the adoption of the same technique by groups of workers for a range of different species does provide useful comparative data. For the future development of this subject more standardization is desirable, especially to obtain absolute values for comparison with other animal groups.

One of the most general problems concerns the condition of the material from which the measurements are made. Obviously it is preferable to use fresh material wherever possible, but the actual measurements are usually easier on fixed material, despite the consequent post mortem and processing changes that this introduces.

A second general problem concerns the heterogeneity of the gill system and the consequent need for representative sampling. Much more is known about the range of variation, and weighting techniques are generally used when trying to obtain overall values for the whole gill system. Fixation certainly leads to some change in dimensions, and it is important that the degree of this change should be determined for any particular material. Measurements made soon after fixation are probably the most accurate, and those that are carried out on fixed material following embedding and sectioning with an ultramicrotome do not produce the same degree of change. Fixation followed by sectioning in paraffin wax produces far greater shrinkage. Some of these problems have been discussed in more detail elsewhere (Hughes, 1984b; Hughes *et al.*, 1984).

## A. Water and Blood Flow Dimensions

During their passage through the gills both the water and blood are channeled into relatively small spaces, the dimensions of which will have an important effect on the flow characteristics, notably the relationship between pressure difference and flow. From a design point of view, the flow must not be too rapid for the exchanges required, but the resistance to that flow must not be excessive and hence the dimensions must not be too fine in relation to the power available from the ventilatory and cardiac pumps. Icefish gills illustrate this principle very well (Holeton, 1972; Hughes, 1972b).

### 1. RESISTANCES TO FLOW

The concept of gill resistance was first developed in relation to water flow through the gills (Hughes and Shelton, 1958), and as measurements in that study showed differential pressures across the resistance of about 1 cm of water, the question naturally arose whether such pressures were sufficient to ventilate the narrow spaces between the lamellae. Accordingly, measurements of these dimensions were undertaken, in particular the height ($b/2$) of the lamellae, their length ($l$) across the filament, and the distance between them ($d$), thus defining a rectangular water channel. As adjacent filaments are very close to one another, a simplification was made by doubling this space, as indicated in Fig. 20A. In fact the lamellae alternate with one another, and the possibility was envisaged that under certain conditions these spaces might be reduced in size (Fig. 20B), but no observations have been made that confirm such a possibility. The Poiseuille equation for laminar flow is usually defined in relation to a cylindrical tube, and a modification of this formula for a rectangular cross section (Hughes, 1966b) gave the following relationship:

$$q = \frac{P_1 - P_2}{\eta} \times \frac{5d^3 b}{24l}$$

where $q$ is the flow through a rectangular tube of length $l$, width $d$, and height $b$; $P_1 - P_2$ is the pressure difference, and $\eta$ the water viscosity. Ap-

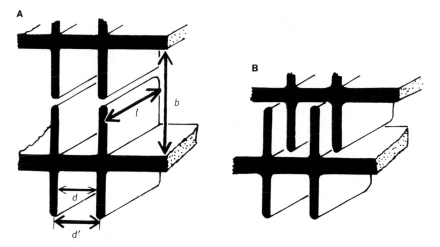

**Fig. 20.** Diagram of the lamellae attached to two neighboring filaments of the same gill arch to show dimensions used in calculations of gill resistance to water flow. In (A) the filaments are shown far apart with the lamellae opposite one another. In (B) the minimum possible size of pore is shown to result from the close interdigitation of the lamellae. (From Hughes, 1966b.)

plication of this formula to fish (e.g., tench) that had both their gill dimensions and differential pressures measured showed that the mean differential was certainly sufficient to ventilate the gills with a volume of water of the same order of magnitude as that measured (Hughes, 1966b). Measurements of these dimensions during the development of the gills in trout (Morgan, 1971) and small-mouthed bass (Price, 1931) provided valuable data not only for determining the resistance to water flow at different stages of the life cycle but also in relation to the scaling of these structures during development. Application of methods of dimensional analysis also proved fruitful (Hills and Hughes, 1970; Hughes, 1977), supporting the view that flow through the system is laminar at least up to certain sizes. The question of the nature of the water flow between the lamellae has been discussed on a number of occasions, but most authors agree that the low velocity through the channels in relation to their dimensions results in Reynolds numbers that are very small (<10) and make any nonlaminar flow extremely unlikely except in so far as it might be a result of intermittent movements of the filaments themselves. Although laminar flow is clearly beneficial from the point of view of water resistance, some authors have suggested that turbulent flow would produce greater mixing of the interlamellar water and hence be advantageous for gas exchange (Steen, 1971).

This question has become of further interest in relation to recent more detailed knowledge of the lamellar surfaces. Scanning electron micrographs have revealed the presence of many microridges and microvilli on the sur-

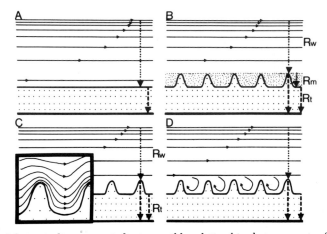

**Fig. 21.** Schematic diagrams to indicate possible relationships between water flow and the lamellar surface. (A) Laminar flow across a flat lamellar surface showing that the two main resistances to gas transfer are those in water and tissue barriers. (B) Laminar flow across a microridged surface in which thin layers of mucus fill up the spaces between the ridges. The resistance to gas transfer is here divided into either three portions between ridges ($R_w$, $R_m$, $R_t$) or two portions over the ridges ($R_w$, $R_t$). Subscripts w, m, and t refer to water, mucus, and tissue, respectively. (C) Laminar flow across a microridged surface with no mucus. The inset shows how the boundary layer would probably be maintained over all the ridged surface in view of the low Reynolds number that would operate at the dimensions of the lamella. (D) Laminar flow together with microturbulences between microridges of a surface without mucus. In this case, as in (C), the overall resistance to gas transfer consists of two components, that for the tissue ($R_t$) varying in length according to the position with respect to the ridged surface. (From Hughes, 1979b.)

faces. The patterns of these surface sculpturings are variable in different species and in different parts of individual lamellae (Hughes, 1979b). It is an interesting but unanswered question whether these sculpturings have any hydrodynamic significance or whether in fact the hydrodynamic forces present during development play a part in determining the pattern of the ridging, and so on. The possibility that the sculpturing produces microturbulence at the boundary between water and lamella has been seriously considered (Lewis and Potter, 1976), but the presence of a layer of mucus that probably fills in the spaces between the ridges (Hughes and Wright, 1970) tends to make it more likely that the actual surface across which the water flows is smooth and covered with a very fine layer of mucus that will be thicker in the spaces between the ridges (Fig. 21)

## 2. Subdivision of Water and Blood Flows

On entering the mouth the water flow is largely laminar and soon subdivides as it enters different slits on the two sides of the buccal cavity. In

teleosts one part passes anterior to the first gill arch and along the inner
margin of the operculum. This pathway is usually very narrow, but during
opercular expansion it will tend to increase in size. This increase does not
lead to an enormous increase in water flow, however, because of the total
movement of the whole branchial system, which expands during the expansion phase. This activity involves contractions of the intrinsic musculature of
the gill arches (Hughes and Ballintijn, 1965; Ballintijn, 1968; Osse, 1969) and
also to some extent those of the filaments themselves (Pasztor and Kleerekoper, 1962). Contraction of the abductor muscles tends to rotate the filaments relative to the longitudinal axis of each arch and help to maintain
continuity of the gill sieve (Fig. 22). Most water passes between the gill slits
of the first to fourth arches, but again there are differences in the last gill slit,
and as at the front of the system there are various modifications in particular
groups of fish (Table II). In teleosts typically there is only one hemibranch
for the last slit, but in elasmobranchs there are two. Such differences be-

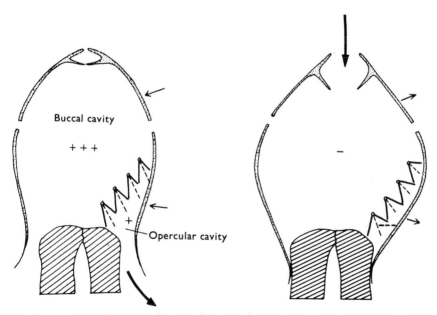

**Fig. 22.** Diagram showing, in horizontal section, the position of the gill arches and filaments
on one side during the two main phases of the ventilatory cycle as described by Pasztor and
Kleerekoper (1962). The anterior part of the buccal cavity with the oral valves is shown in
vertical section. Directions of movement of the walls of the respiratory cavities are shown by
thin arrows. Dashed lines indicate orientation of the gill arches. Pressures in the buccal and
opercular cavities are indicated relative (+ −) to that of the outside water. The pressure
difference across the gills results in flow from the buccal to the opercular cavities during both
phases of the cycle. Thick arrows indicate water flow. (From Hughes and Morgan, 1973.)

tween the individual slits produce variations in the amount of oxygen that is removed from the water as it passes along these major subdivisions of the flow pathway. The same variability extends to the flow of water between a typical pair of arches, since some of the water passes between the tips of the gill filaments, although they remain in close contact for much of the ventilatory cycle. This shunting of water has been likened to an anatomic dead space (Hughes, 1966b). During hyperventilation a larger proportion of the water flows through this pathway with a consequent lowering in the percentage utilization of oxygen from the water. The major portion of the water passes between the secondary lamellae of the gill filaments of the individual hemibranchs, and thence it is collected again into the opercular cavities.

As has been indicated previously, there are differences in the dimensions of the lamellae at different positions along the filaments, and in general we may distinguish those from the tip, middle, and base of each filament. The path of water flow for these different parts of the filament vary, and, most importantly, water passing between the basal lamellae must meet the sep-

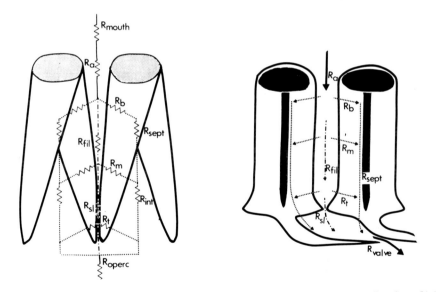

**Fig. 23.** Diagrams to illustrate the main resistances to water flow across two gill arches of (A) a teleost and (B) an elasmobranch gill. The resistances to water flow between lamellae at the base ($R_b$), middle ($R_m$), and tip ($R_t$) regions of the gill filaments are indicated. The septal resistance ($R_{sept}$) of this teleost gill is far less than in the elasmobranch gill. Resistance to water flow through the mouth of the fish ($R_{mouth}$) and between the adjacent gill arches ($R_a$) and filaments ($R_{fil}$) and outwards through the slit between tips of adjacent hemibranchs ($R_{sl}$) and finally out into the water via the opercular ($R_{operc}$) and external gill slits ($R_{valve}$) are also indicated. (–·–·), Axial flow; (····), interlamellar flow. (After Hughes, 1972c, 1973b.)

tum especially in those teleosts where this remains relatively well developed. In the perciform gills, however, this is a relatively small portion of the water. Water traversing between lamellae at the tip has a less impeded flow pathway. From such morphological features of the water pathway it is clear that there will be different resistances to water flow depending on which pathway one considers. These different resistances are indicated in Fig. 23, where differences between the elasmobranch and teleost gill are indicated. The same principles apply to all gills, but the relative flows in different portions of the filament will vary.

The dimensions of spaces between lamellae vary according to the habit of the fish. In all cases there would be a part of the water flow that is more axially placed, where the distance for diffusion of any oxygen molecules to the lamellar surfaces is greater and the water flow velocity higher compared to that of water in the immediate boundary layers. Thus, we can consider the axial flow of water as representing a sort of physiological dead space (Fig. 24), the proportion of which will depend on the dimensions of the interlamellar spaces and the velocity of water flow. The extent of the physiological dead space, as with the morphological dead space, will increase with increasing ventilation. Flow velocities in the interlamellar spaces can be estimated from measurements of the total ventilation volume and morphometry, which indicates the total number of interlamellar spaces and their dimensions. Calculations give velocities of the order of 5 cm sec$^{-1}$. Such values have not been determined accurately, although measurements of water flow velocity at least in the neighborhood of the gills indicate velocities of 10 to 20 cm sec$^{-1}$ (Hughes, 1976, 1978b; Holeton and Jones, 1975). Although the resistances of the secondary lamellar and filamental pathways are most important during ventilation, it has long been recognized that resistance of the gill arches and rakers are very important during feeding, and this has been shown in some elegant studies using rapidly responding pressure manometers (Lauder, 1983).

The situation on the blood side of the exchanger is more complex than was originally supposed, and for a detailed account of vascular pathways, reference should be made to Chapter 2. Originally it was supposed that, as with water flow, the system was homogeneous, but emphasis was given to the probability of greater flow around the marginal channels (Hughes and Grimstone, 1965) and the possibility of flow in this pathway and the proximal channels (Fig. 25) embedded in the filament being regulated relative to flow in most of the lamellar channels as a result of contractile activity within pillar cells. The efficacy of such a regulation was emphasized by Wood (1974), who calculated that a passive increase in diameter of the marginal channels of only 25% would result in the whole of the blood flow passing by this route. The absence of the contractile mechanism of the pillar cells would tend to

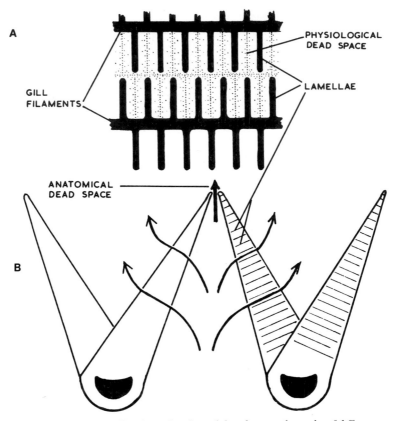

**Fig. 24.** Diagram of two gill arches of a teleost fish indicating the paths of different parts of the respiratory current. (A) Longitudinal section of two filaments. The water flows at right angles to the page. The concentration of oxygen in the interlamellar water (equivalent to alveolar air) is indicated by the density of the stippling. The water that passes through without losing any oxygen is equivalent to the physiological dead space. (B) Diagram of transverse section through two adjacent gill arches with their rows of attached filaments. The filaments normally touch one another at their tips, and most water passes between the lamellae (interlamellar water). With excessive pressure gradients across the gills, the volume of water shunted between the tips of the filaments increases. This portion of the respiratory water is equivalent to the anatomic dead space of the lung. (From Hughes, 1966b.)

produce a disproportionate increase in the marginal channel diameter with increase in blood pressure. Evidence for such a change in distribution of intralamellar blood flow has been obtained using morphometric methods (Soivio and Tuurala, 1981). Furthermore, it is also possible for blood flow to be restricted to certain lamellae depending on the conditions (Booth, 1978, 1979). Thus, on both the water and the blood sides there is recruitment of water and blood channels under different conditions. Each of these channels, however, can vary in the extent to which ventilation and/or perfusion occurs (Hughes, 1972c).

There are many variations in the proportion of blood that passes through the respiratory or intralamellar pathway, as distinct from pathways not in close contact with the water and consequently less involved in gas and ionic exchanges. Earlier suggestions of shunting of blood between the afferent and efferent blood pathways are present in the literature (Muller, 1839), but this concept was first given prominence by Steen and Kruysse (1964). Their suggestion of a direct shunt is shown diagrammatically in Fig. 26B and compares with the classical situation. It is now accepted that there is evidence for some of the other pathways such as that shown in Fig. 26C. There

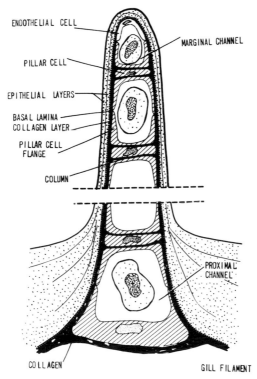

**Fig. 25.** Diagrammatic transverse section through the lamella of a teleost gill to show basic organization. The lamella is covered in two epithelial layers, within which is found the pillar cell system (PCS) consisting of the basal lamina and pillar cells, whose flanges enclose the main blood channel. The outer portion of the marginal channel is lined by endothelial cells. Note how the proximal or basal channel is embedded in the main tissue of the gill filaments. (From Hughes, 1980b.)

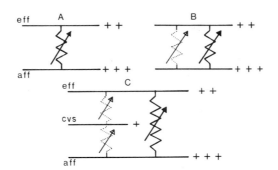

**Fig. 26.** Electrical analog diagrams to illustrate different views of the circulatory pathways between afferent and efferent filament arteries in fish. The first diagram (A) shows the classical circulation; a shunt is inserted in (B), but in (C) there is no true shunt between the afferent and efferent filament arteries as the pressure (indicated by +) in the central venous sinus (CVS) is lower than in either the efferent or afferent arteries. (After Hughes, 1979a.)

is also evidence that, in at least some species, a relatively large portion of the cardiac output may return to the heart without having any oxygen removed from it (e.g., up to 30% in the eel; Hughes *et al.*, 1982).

Thus, it is concluded that there are many variations in both the water and blood flow pathways, and although our knowledge of the morphological possibilities has increased considerably, the physiological conditions under which different pathways become emphasized are at present being actively studied. Details are given elsewhere in this volume.

## B. Gas Exchange

Transfer of oxygen across the gill surface is directly proportional to its area and inversely proportional to its thickness: $\dot{V}_{O_2} = KA/t$. The diffusion coefficient in this relationship is the Krogh permeation coefficient ($K$), which is equivalent to the product of the "true" diffusion coefficient ($d$) for the particular gas molecule and its solubility ($\alpha$) in the particular part of the pathway. Unfortunately there are few, if any, measurements of the permeation coefficient in fish tissues, and in most estimations, values obtained by Krogh (1919) for frog connective tissue have been used. Thus, much of the quantitative estimates may well need to be modified in the future if methods become available for determining the permeation coefficient of different parts of the water–blood barrier. For this reason it is important that details of different sections of the pathway should be measured wherever possible, but this involves a detailed study of transmission electron micrographs, and these are not always available. From such studies it is possible to subdivide the tissue barrier into several sections (Fig. 27). It is not known which of these is the most limiting.

## 1. SURFACE AREA

As has been indicated earlier, the surface area of fish gills is greatly increased because of the subdivision of the primary epithelium into that of the gill filaments and of the lamellae. Measurements of surface area usually involve estimation of the total surface of all the lamellae of a given specimen. The techniques involved require sampling of the filaments and lamellae with special reference to the frequency of these structures along the filaments (Fig. 8) and their average surface area. The latter measurement is the most difficult and can give rise to significant errors, especially if no account is taken of the heterogeneity of the system and a weighting technique adopted. From these basic measurements the area is given by the following relationship: gill area = $Lnbl$, where $L$ is the total filament length, $n$ is the number of lamellae per millimeter on both sides of the filaments, and $bl$ is the bilateral area of an average lamella.

It is usually this total area that is used in calculations and plots relating gill area to body mass. Some discussion has taken place regarding the extent to which the total lamellar area is equivalent to the respiratory area. It seems probable that some gas exchange can take place over this whole surface. It must be remembered, however, that the diffusion distance between the

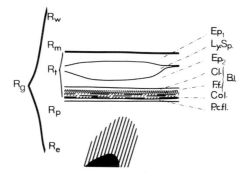

Fig. 27. Diagram to show different components of the resistance to gas transfer from water to blood in a fish gill lamella. The resistances are for water ($R_w$), the mucus layer ($R_m$), tissue ($R_t$), plasma ($R_p$), and within the erythrocyte ($R_e$). The tissue resistance ($R_t$) consists of seven different layers, namely outer and inner epithelial layers ($Ep_1$ and $Ep_2$), which in some regions are separated by lymphoid spaces (Ly.Sp). The basement membrane is composed of an outer clear layer (Cl), the middle, fine fibrous layer (F.f.), and an inner collagen layer (Col). The outer two form the basal lamina (B.l.), and the innermost layer of the tissue barrier is formed by a flange of the pillar cell (P.c.fl.). (After Hughes, 1972c.)

water and the nearest red blood cell will vary considerably according to its
position directly over a blood channel or a pillar cell. Estimates of the
proportion of the lamellar area directly above the pillar cells have given
values up to 30%, and on this basis only 70% of the total area is directly
applicable to the thinner barriers of the regions that have been referred to as
the diffusion channels.

The total area of the lamellae is much greater in more active fishes such
as tuna and is far less in sluggish specimens such as toadfish (Fig. 28).
Because the relationship between lamellar area and body mass is not a direct
one, it is important to take account of the body size when making such
comparisons. Accordingly, data are usually given for different body sizes
from a regression line and can be calculated for any given body mass using
the relationship, $X_1 = X_2 (W_1/W_2)b$, where $X_1$ is the value required at
weight $W_1$, and $X_2$ is the known value at weight $W_2$; $b$ is the slope of the
log/log regression line.

For a range of medium-size fish, the gill area of a 200-g fish has some-

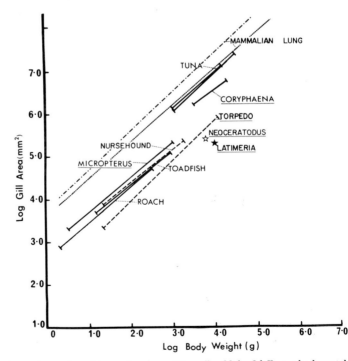

**Fig. 28.** Bilogarithmic plot of the surface area of the gills of fish of different body weight. The
dotted lines show the relationship for mammalian lungs of "free-living" and captive species.
(After Hughes, 1978b.)

times been used (Jager and Dekkers, 1975); however, where the size range is very great (e.g., 10 g–10 kg) it is perhaps better to compare fish over similar ranges, especially in cases where the regression coefficient (*b*) has not been determined and it is necessary to to utilize an average figure. The errors introduced by making such extrapolations may be quite large (Hughes, 1984b).

In fish that have accessory air-breathing organs the area of the gill surface is also small, and this is also true in some deep-water species that probably have a relatively sluggish existence. When assessing the importance of a respiratory surface for gas exchange it is important, however, to take into account not only the surface but also diffusion distances. This is because the diffusion of gas from the water to the blood, although directly proportional to surface area, is also inversely proportional to barrier thickness. Surfaces have sometimes been supposed to be important in gas exchange because of the degree of vascularization, although distances separating the blood and water may be very great, as in the cutaneous circulation of a number of fish, such as the icefish (Holeton, 1970). In comparison with the areas of lungs, no fish gills have such extensive surfaces as that of a mammal of comparable body mass. However, this is presumably compensated for by the greater ventilation of the gill surface and the lack of dead space and tidal ventilation.

Area measurements of lunglike air-breathing organs have not so far taken into account increases in surface due to the bulging capillaries. As has been suggested (Hughes and Munshi, 1978), the effective increase in area would be partly compensated by the nonrespiratory areas such as the "lanes" (Fig. 17). The need to take this heterogeneity into account when calculating diffusing capacity has been emphasized (Hughes, 1980d) and is especially important in fish (e.g., *Monopterus*) where the nonrespiratory regions are very extensive. In such cases, published values (Hughes *et al.*, 1974a) must be considerably reduced, but in fish with "lanes" (Table III) the effect may be compensated by projections of the respiratory islets (modified lamellae) above the air sac lining surface.

## 2. THICKNESS

As indicated earlier, the thickness of the barrier separating the water from the blood is an important factor in gas exchange. Measurements of these distances have been carried out in a number of fish gills, but unfortunately the methods adopted have varied quite considerably. In a number of cases the measurements have been confined to the "diffusion channel" that overlies the blood channels in the lamella. The minimum distances may be measured at regular intervals along the outside epithelium and an average value taken. In other cases only the minimum distance in the diffusion channel has been measured. One method that has been used for measuring the arithmetic mean thickness, particularly in lungs, is to obtain values for the surface area (intersection counting) of the air and capillary surfaces together with an estimate of the volume of the barrier using point counting. Thus, the arithmetic mean thickness of the barrier is the volume of the barrier divided by the mean area of the air and blood surfaces. Such a method has scarcely been applied to fish gills, but a method developed for measurement of the harmonic mean thickness in tetrapod lungs has given useful results. In this method, values for distances between the external surface and the nearest point on the blood capillary or the nearest blood corpuscle are measured using a special rule, and the numbers of occurrences between certain distances are calculated and, from their distribution, the harmonic mean thickness is calculated. For details of this method see Weibel (1971) and Perry (1983). In only a very few cases has the harmonic diffusion distance been measured for gills (Hughes, 1972c; Hughes and Perry, 1976). In this type of analysis the longer distances are of less importance, because in the estimation it is the reciprocal of the distance that is calculated. Where the harmonic diffusion distance has been used, then the appropriate area measurement is that of the total outer lamellar surface (Fig. 29). In cases where only the diffusion channel measurements have been used with a consequent smaller value, then the smaller area overlying the blood channels is the appropriate value for surface area (Hughes, 1982).

In all these discussions there has often been neglect of the water that is in immediate contact with lamellae. At least two different types of studies (Hills and Hughes, 1970; Scheid and Piiper, 1976) have indicated that this in fact forms an important part of the barrier to diffusion. Another problem relates to the thin layer of mucus that must overlie the lamellae during normal function, but the extent of this layer has so far proved impossible to determine in the living condition.

## 3. Diffusing Capacity

Respiratory physiologists introduced the term diffusing capacity to express the capability of a gas exchange organ to transmit gases, and it is the ratio of volume of gas transferred across the barrier in unit time to the mean difference in partial pressures of that gas on the two sides of the barrier. Diffusing capacities with respect to oxygen and also carbon monoxide have been determined in a number of cases. It is directly proportional to surface area and inversely proportional to the thickness. Gill diffusing capacity $(D_g)$ = $\dot{V}_{O_2}/\Delta P_{O_2}$. The diffusion constant is the Krogh permeation coefficient $(K)$, which has not been determined for many fish tissues. Such calculations give

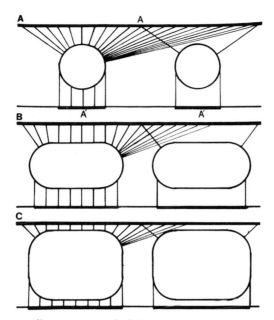

**Fig. 29.** Diagrams to illustrate two methods for measuring the thickness and surface area of a lamella for use in estimating diffusing capacity. The upper surface has an area (A), and the lines indicate distances for measurement of harmonic means. The lower surfaces show areas (A') over blood channels and lines for measurements used to obtain diffusion distances across the blood channels alone. The blood channels increase in relative size from **a** to **c**. (From Hughes, 1982.)

values for diffusing capacity of the tissue barrier alone $(D_t)$, but of course the transfer of oxygen from water to the hemoglobin molecules involves other stages, and these may be indicated by the following relationship:

$$1/D_g = 1/D_w + 1/D_m + 1/D_t + 1/D_p + 1/D_e.$$

where the subscripts w, m, t, p, and e indicate water, mucus layer, tissue barrier, plasma, and erythrocytes, respectively.

Although the term diffusing capacity is preferable because of its frequent usage in mammalian respiratory physiology and morphology, the term transfer factor introduced by Cotes (1965) for the mammalian lung has often been used for the fish gill. There are some drawbacks, however, as transfer factor was the term originally used in relation to blood platelet studies, and in mammalian respiratory physiology it usually refers to the transfer of carbon monoxide across the lung surface. A disadvantage of the term diffusing capacity, however, is that physiologically it concerns all the resistances between the oxygen contained in the water and its final combination with the hemoglobin molecules. It is thus greatly influenced by factors other than diffusion, notably convection within the water and plasma, and the kinetics of the reaction in the red blood corpuscles. Nevertheless, it is probably preferable to maintain the use of this term rather than confuse the literature still further. Accordingly, it would seem that for comparisons between different gill systems, data based on the diffusing capacity of the tissue barrier provide a useful guide to the systems' effectiveness (Table V). The greatest error probably arises because of the lack of information regarding the nature of the resistance in the water film. Once again this particular resistance will vary according to the degree of ventilation. Indeed, from a morphometric point of view, only diffusive resistances are considered with the exception of the reaction with hemoglobin.

Many of the different resistances to oxygen transfer between water and red cell are indicated in Fig. 27. Apart from the water, the main resistances are in the tissue, plasma, and erythrocytes, and these are the components included in morphometric estimates of diffusing capacity. Hence, $R_g = R_t + R_p + R_e$. Written as conductances, this becomes: $1/D_g = 1/D_t + 1/D_p + 1/D_e$, where $D_g$ is the gill diffusing capacity.

Estimates of these relative contributions in the lung of *Lepidosiren*, for example, suggest that the tissue component forms 75% of the morphometrically determined diffusing capacity, and 13% is made up by the resistance in the erythrocytes. However, at least two studies (Hills and Hughes, 1970; Scheid and Piiper, 1976) have concluded that a large contribution to the overall resistance is contained within the water. Consequently, for the total resistance, assuming 50% is in the water, this suggests that only 37% is contained within the tissue barrier and about 7% in the erythrocyte. That the major resistance is diffusive and not chemical has also been confirmed (Hills *et al.*, 1982) by analysis of in vivo measurements using isolated lamellae. There are of course many difficulties in estimating these resistances, and at present it would be wrong to place too much certainty on any of the figures, although they are a good guide. Perhaps the most certain are those for the morphometric part of the tissue barrier ($D_t = KA/t$), but even here there are difficulties. The need for caution is particularly great when discussing absolute values for diffusing capacity and using them in comparisons with other animals. On a relative basis, however, the values available have proved to be of considerable usefulness. For example, in studies of the effects of pollutants on the gills (Hughes *et al.*, 1979) it has been shown that the concept of relative diffusing capacity (Hughes and Perry, 1976) enables comparisons to be made between experimental and control fish without making absolute measurements of either gill area or barrier thickness. The assumption was made, however, that the permeation coefficient would be the same under the two conditions. Soivio and Tuurala (1981) have extended some earlier studies (Hughes *et al.*, 1978) to show that there are changes not only in the barrier thickness but also in other dimensions of the lamellae following exposure to hypoxia. These studies also indicated the value of simple stereological methods using point and intersection counting to assess the volume and surfaces to particular parts of a structure relative to other parts using sectioned material. Figure 30 illustrates the typical method, which makes it possible to compare the relative volumes of plasma and pillar cells within a lamella. The superimposed grid is rectilinear, which is preferred when the sections are at random to the surface of the lamellae. Despite the regular orientation of lamellae, it is difficult to ensure that sectioning is always at right angles to their surface. Careful control can make this possible, in which case a Merz grid can be used (Hughes and Perry, 1976).

To summarize, the whole study of fish gill morphometry has only just begun, and methods are becoming more standardized. In future this will lead to a situation where greater certainty can be laid on the results of comparisons of different types of material. Although the basic methodology is simple, it is advisable that anyone wishing to carry out such studies should take some of the precautions that have been shown to be necessary by these early exploratory studies.

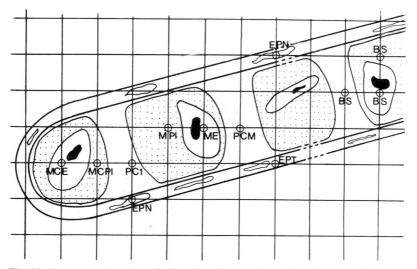

**Fig. 30.** Diagrammatic section of a lamella with superimposed rectilinear grid. Some of the points counted are indicated. E, erythrocyte cytoplasm; Nu, erythrocyte nucleus; Pl, plasma; PC, pillar cell; EPT, epithelial tissue; EPN, epithelial nontissue. (From Hughes 1979a.)

## C. Scaling

This term has become generally used in relation to the way in which changes in dimensions of a given structure or function are related to differences in body mass. It is thus a special form of allometry, which was first introduced by Huxley and Teissier (1936) and has provided a valuable synthesis of much morphological and physiological data. Stimulating studies by D'Arcy Thompson (1917) and Huxley (1932) as experimental biology began to expand must be given much credit for the early influence of allometry. Since then there has always been an interest in allometry, but this has been renewed considerably in recent years perhaps because of the greater interest in functional morphology in relation to physiology and the ready availability of facilities that greatly speed up the computation and analysis of data. Methods for processing morphometric measurements similar to those out-

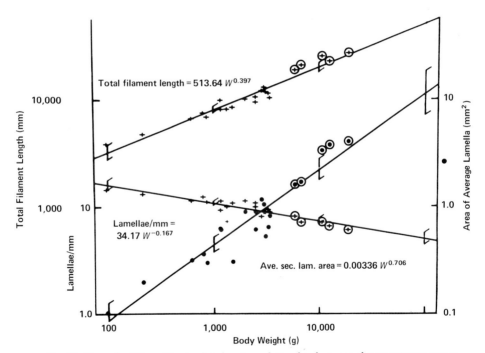

**Fig. 31.** *Torpedo.* Bilogarithmic plot showing relationship between the constituent parameters of the gill area and body mass for 22 specimens. The upper crosses represent total filament length (mm), the lower crosses show the number of lamellae per millimeter, and filled circles are for the area of an average lamella (mm²). Most points plotted refer to *T. marmorata*; five of the measurements (enclosed with circles) were made on large specimens of *T. nobiliana*. Bars indicating 95% confidence limits for each of the regression lines are shown at body weights of 100, 1000, and 10,000 g. (From Hughes, 1978c.)

lined earlier have also been greatly improved by these developments. For example, Gray (1954) had made some pioneering measurements of the gill area of many species of fish covering a range of body sizes, but they were not subjected to regression analysis, partly because of the time required. Later studies (Hughes, 1966b; Hughes and Gray, 1972) have greatly benefited by these measurements of Gray and his assistants. Scaling has been brought into prominence in more recent years by the analysis of data on the locomotion and respiration of vertebrates and by more detailed consideration of the relationships between body mass and oxygen consumption (Hughes, 1984a).

### 1. RELATIONSHIP OF GILL AREAS TO BODY MASS

In the early analysis of these measurements using logarithmic coordinates (Muir and Hughes, 1969; Hughes, 1970a), it was found that each of the constituent measurements of gill area (total filament length, lamellar frequency, and lamellar area) all showed straight lines when plotted bilogarithmically (Fig. 31). Furthermore, the slopes of these lines when added together gave the same slope as that for the plot of total area against body mass. Such relationships have now been confirmed for many species of fish. A summary of the $a$ and $b$ values in the relationship, gill area = $aW^b$ is given in Table VI. It is now apparent that no common slope $b$ can be used for all species of fish. However, a value of about 0.8 is fairly common, although $b$ may range from 0.5 to 1.0. Thus, it has been concluded that individual species show differences in the scaling of respiratory parameters, and some of the possible factors have been discussed but as yet are not understood.

The importance of emphasizing differences at an intraspecific level becomes apparent when comparisons are extended to relationships between oxygen consumption and body mass. Again the general relationship with a slope of 0.75 to 0.8 has been known for many years, and for a wide range of fish species Winberg (1956) suggested an average value of 0.82, which has often been adopted. Just as with the scaling of the respiratory surfaces, so there are variations between species with respect to the $\dot{V}_{O_2}/W$ relationship. Initially it was supposed (Muir and Hughes, 1969) that the two slopes would be approximately the same, suggesting that gill surface area was the main factor governing oxygen consumption. When measurements for more species were analyzed, however, differences between the slopes were recognized and it was further appreciated that the total gill area was not normally functional during resting conditions. Perhaps for this reason the slope of the gill area line should be considered more in relation to that for $\dot{V}_{O_2, \text{active}}$. As only two or three fish species have had their active metabolism measured over a range of body sizes, it was suggested that, in the absence of definite data, the slope of the gill area line might be taken to be indicative of the

## Table VI

### Summary of Results of Regression Analysis for Gill Dimensions of a Range of Fish Species[a]

| Species | Gill measurement | Body weight | | | Equation for total gill area $Y = aW^b$ | References |
|---|---|---|---|---|---|---|
| | | 10 g | 100 g | 1000 g | | |
| *Thunnus albacares, Thunnus thynnus* | $L$ | 13,481 | 32,488 | 78,293 | $3151W^{0.875}$ | Muir and Hughes (1969) |
| | $bl$ | 0.0176 | 0.0674 | 0.2581 | | |
| | $1/d'$ | 49.56 | 40.35 | 32.85 | | |
| *Coryphaena hippurus* | $L$ | 5069 | 13,675 | 36,369 | $5208W^{0.713}$ | Hughes (1970a) |
| | $bl$ | 0.0804 | 0.1707 | 0.3626 | | |
| | $1/d'$ | 31.12 | 28.65 | 26.37 | | |
| *Scomber scombrus* | $L$ | 4054 | 10,434 | 26,849 | $424.1W^{0.997}$ | Hughes (1970b) |
| | $bl$ | 0.0187 | 0.0672 | 0.2417 | | |
| | $1/d'$ | 28.56 | 30.14 | 31.81 | | |
| *Salmo gairdneri* | $L$ | 2434 | 6541 | 17,579 | $314.8W^{0.932}$ | Hughes (1980b) |
| | $bl$ | 0.0229 | 0.0853 | 0.3172 | | |
| | $1/d'$ | 23.78 | 20.53 | 17.72 | | |
| *Tinca tinca* | $L$ | 3012 | 7674 | 19,540 | $867.2W^{0.698}$ | Hughes (1970b) |
| | $bl$ | 0.0263 | 0.0601 | 0.1373 | | |
| | $1/d'$ | 23.76 | 22.17 | 20.69 | | |

| Species | | 10 | 100 | 1000 | Relationship | Reference |
|---|---|---|---|---|---|---|
| *Opsanus tau* | $L$ | 923.7 | 2825 | 8638 | $560.7W^{0.79}$ | Hughes (1970a) |
| | $bl$ | 0.1412 | 0.3347 | 0.7879 | | |
| | $1/d'$ | 13.37 | 11.27 | 9.51 | | |
| *Scyliorhinus canicula* | $L$ | 1623 | 3644 | 8181 | $262.3W^{0.961}$ | Hughes (1970b) |
| | $bl$ | 0.04998 | 0.2412 | 1.1655 | | |
| | $1/d'$ | 14.56 | 12.36 | 10.49 | | |
| *Torpedo marmorata* | $L$ | 1280.3 | 3191.1 | 79544 | $117.5W^{0.937}$ | Hughes (1978c) |
| | $bl$ | 0.0171 | 0.0869 | 0.4414 | | |
| | $1/d'$ | 23.29 | 15.87 | 10.82 | | |
| *Anabas testudineus* | $L$ | 559 | 1209 | 2615 | $556W^{0.615}$ | Hughes and Munshi (1973a) |
| | $bl$ | 0.0398 | 0.106 | 0.2828 | | |
| | $1/d'$ | 25.72 | 18.11 | 12.76 | | |
| *Saccobranchus fossilis* | $L$ | 832.2 | 2267 | 6172 | $186.1W^{0.746}$ | Hughes et al. (1974b) |
| | $bl$ | 1.02455 | 0.06276 | 0.1604 | | |
| | $1/d'$ | 25.45 | 20.46 | 16.46 | | |
| *Channa punctata* | $L$ | 1485.6 | 3955 | — | $470.4W^{0.592}$ | Hakim et al. (1978) |
| | $bl$ | 0.0236 | 0.0475 | — | | |
| | $1/d'$ | 26.24 | 19.12 | — | | |

[a]Values for total filament length ($L$, mm), bilateral area of an average lamella ($bl$, mm²), and frequency of lamellae on one side of a filament ($1/d'$, mm) are given for body weights of 10, 100, and 1000 g. The relationship for total gill area ($Y = aW^b$) is also given for each species.

slope for active metabolism (Hughes, 1972c, 1977). Under these circumstances, it was appreciated that in some species the lines relating to resting metabolism and gill area (= active metabolism) would diverge, and consequently, the difference between these two (= scope for activity; Fry, 1957) would increase with body mass. In fact such a relationship has been demonstrated for salmon (Brett, 1965). In some fish, however, the two regression lines would be parallel, so that there would be no increase in scope with body size, and in other instances they might converge, indicating a decrease in scope with increase in body mass. The size where the two lines cross would suggest that no activity was possible above such a size, and the respiratory surface could be considered the major factor governing growth of the fish. Until sufficient data are available comparing the active and resting metabolism of fish with morphometric measurements of gill area and diffusing capacity, these must represent tantalizing hypotheses that may or may not prove to be helpful in further generalizations.

## 2. DIMENSIONAL ANALYSIS

Engineers have long used a technique in which the dimensions of different features of a machine are analyzed in relation to the particular size of the whole structure and in this way to gain information regarding their functional relationships. A similar method has been applied to the gills of fish for which measurements during their development are available. The data for the small-mouthed bass, *Micropterus dolomieu* (Price 1931), proved invaluable for making such an analysis (Hills and Hughes, 1970), and a similar technique was applied later to measurements made for the rainbow trout by Morgan (1971). Exponents for the relationships between body mass and different parameters relating to the gill dimensions were established as shown in Table VII. In spite of differences in the detailed exponents for these two species, the final conclusions were almost identical and have given useful general information regarding the changes in the structures during development. From this information the following conclusions were reached:

1. Water flow past the lamellae is probably laminar.
2. Relative to the blood, countercurrent flow of water is more probable than cocurrent.
3. The perfusion of gill surfaces with blood is constant at all body sizes.
4. Resistance to overall oxygen transfer provided by the water is between 5 and 10 times greater than that due to the tissue barrier, and this remains almost constant at all body sizes.
5. The water velocity between secondary lamellae increases slightly with body size; in the case of the trout it is proportional to $W^{0.17}$, whereas in *Micropterus* it is proportional to $W^{0.10}$.

### Table VII

Exponents for the Relationships between Body Mass and Different
Parameters Relating to the Gill Dimensions of *Micropterus* and the
Rainbow Trout, *Salmo gairdneri*

| Parameter | *Micropterus* | *Salmo gairdneri* |
|---|---|---|
| Distance between secondary lamellae | 0.022 | 0.09 |
| Number of pores | 0.54 | 0.23 |
| Pore length | 0.21 | 0.24 |
| Ventilation volume | 0.93 | 0.73 |
| Gill area | 0.95 | 0.78 |

Although the whole gill sieve may be considered as a large number of pores of the dimensions just discussed, which are in parallel to one another, and consequently, the volume flow per unit time through each pore is relatively small, nevertheless there are other paths for water flow between the buccal and opercular cavity that may be considered as part of an anatomic and physiological dead space, the total forming a water shunt that has been estimated to be as much as 60% of the total ventilation volume in trout (Randall, 1970). Randall further subdivided the physiological dead space into diffusion and distribution components. There have, however, been very few studies on the resistance properties of the whole gill network in relation to differences in hydrostatic pressure across the gill resistance. In some early studies (Hughes and Shelton, 1962) using anesthetized fish and by changing the pressure difference, it was shown that flow increased with increasing pressure difference until a point came when the flow for a given increment of pressure suddenly increased. It was suggested that perhaps this indicated the instant at which the gill resistance falls as the tips of the filaments separate. Such studies have also indicated (Hughes and Umezawa, 1968) that the resistance from the opercular to buccal cavities is greater than that in the normal direction of water flow.

The conditions regarding blood flow through lamellae have been established by Muir and Brown (1971), and dimensional analysis suggests that blood viscosity must increase with body mass (Hughes, 1977).

## VII. CONCLUSIONS

It is apparent from investigations at a gross morphological level that there are many variations in the organization of fish gills. Nevertheless, the same basic structure is recognizable even when it has become modified to produce organs so diverse as the glandular pseudobranch of *Anabas* and the air tubes of *Saccobranchus*, which are so different from the highly evolved gills of tuna and other oceanic species. In all these situations the presence of pillar cells has proved to be an invaluable guide to the morphological nature of such highly modified organs. Despite much investigation, a great deal remains to be learned about the functioning of these interesting cells, which clearly can function not only in support but also may have important immunological functions (Chilmonczyk and Monge, 1980), and their similarity to some structures in the reticuloendothelial system of mammals is most striking (Hughes and Weibel, 1976). The filaments of fish gills also show a wide range in their structure and development, again related to the habits of the particular fishes.

The detailed nature of the vascular pathways at the filament level is still in course of investigation, and the control mechanisms involved remain to be elucidated. At the filament level it would appear that nervous mechanisms are involved (see Chapter 2, this volume), but for lamellae little information is available concerning either the sensory or the motor side of their functioning. It would appear that much of the motor activity of lamellae is regulated by blood-borne substances. These different control mechanisms must affect the overall balance between the respiratory and other functions of the gills and still presents a challenge to the fish physiologist. Much has been learned by the use of isolated gill preparations, but until more is known of the blood composition and a suitable perfusion medium is developed, it is important to compare constantly the condition of such preparations with those of the intact gill system (see Chapter 10, Volume XB, this series).

Comparison of the structure and function of fish gills with the vertebrate lung suggests that the gill is at least as well adapted to its respiratory function in water as is the lung to air. Their positions and roles in the circulation of the body have many similarities, although differences due to the single and double circulation have some influence. In particular, blood is supplied to the gill at a relatively high pressure compared with the lungs. The precise way in which this affects the filtration of plasma through the capillary beds is not yet fully understood, but the existence of low capillary permeability (Hargens et al., 1974) may be related to these circulatory differences. Indeed there are indications that the whole functioning of the capillary system in fish may be quite different from what is accepted as normal for mammals. Vogel (1981) and Vogel and Claviez (1981) have drawn attention to these differences and have identified a secondary circulation that differs from the lymphatic system of higher vertebrates. Thus, perhaps some of the most fascinating studies of fish gills concern the air-breathing fishes, where the gills are modified to a certain extent while a lung or other air-breathing organ takes over some of the functions that they normally perform. Investigation of the wide variety of these fishes continues to reveal important general principles as, for example, the presence of something equivalent to a double circulation in some species of *Channa* (Ishimatsu and Itazawa, 1983).

## REFERENCES

Acrivo, C. (1938). Sur l'organisation et la structure du corps caverneaux chez *Scyllium canicula* Cuv. *Bull. Histol. Appl. Tech. Microsc.* **12**, 362–372.

Adeney, R., and Hughes, G. M. (1977). Observations on the gills of the sunfish *Mola mola. J. Mar. biol. Assor. U.K.* **57**, 825–837.

Alexander, R. McN. (1970). Mechanics of the feeding action of various teleost fishes. *J. Zool.* **162**, 145–156.

Al-Kadhomiy, N. (1984). Gill development, growth, and respiration of the flounder, (*Platichthys flesus*). Thesis, Bristol University.

Ballintijn, C. M. (1968). The respiratory pumping mechanism of the carp, *Cyprinus carpio* (L). Thesis, University of Grøningen, Netherlands.

Berg, T., and Steen, J. B. (1965). Physiological mechanisms for aerial respiration in the eel. *Comp. Biochem. Physiol.* **15**, 469–484.

Bertin, L. (1958). Organes de la respiration aquatique. *In* "Traité de zoologie" (P. P. Grassé, ed.), vol. 13, pp. 1303–1341. Masson, Paris.

Bettex-Galland, M., and Hughes, G. M. (1973). Contractile filamentous material in the pillar cells of fish gills. *J. Cell Sci.* **13**, 359–366.

Bevelander, G. (1934). The gills of *Amia calva* specialised for respiration in an oxygen-deficient habitat. *Copeia* **3**, 123–127.

Bijtel, J. H. (1949). The structure and the mechanism of movements of the gill filaments in Teleostei. *Arch. Neerl. Zool.* **8**, 267–288.

Booth, J. H. (1978). The distribution of blood flow in the gills of fish: application of a new technique to rainbow trout (*Salmo gairdneri*). *J. Exp. Biol.* **73**, 119–129.

Booth, J. H. (1979). The effects of oxygen supply, epinephrine, and acetylcholine on the distribution of blood flow in trout gills. *J. Exp. Biol.* **83**, 31–39.

Brackenbury, J. H. (1972). Lung-air-sac anatomy and respiratory pressures in the bird. *J. Exp. Biol.* **57**, 543–550.

Brett, J. R. (1965). The relation of size to rate of oxygen consumption and sustained swimming speed of sockeye salmon (*Oncorhynchus nerka*). *J. Fish. Res. Board Can.* **22**, 1491–1501.

Brett, J. R. (1972). The metabolic demand for oxygen in fish, particularly salmonids, and a comparison with other vertebrates. *Respir. Physiol.* **14**, 151–170.

Brown, C. E., and Muir, B. S. (1970). Analysis of ram ventilation of fish gills with application to skipjack tuna (*Katsuwonus pelamis*). *J. Fish. Res. Board Can.* **27**, 1637–1652.

Byczkowska-Smyk, W. (1957). The respiratory surface of the gills in teleosts. 1. The respiratory surface of the gills in the flounder (*Pleuronectes platessa*) and perch (*Perca fluviatilis*). *Zool. Pol.* **8**, 91–111.

Byczkowska-Smyk, W. (1958). The respiratory surface of the gills in teleosts. 2. The respiratory surface of the gills in the eel (*Anguilla anguilla*), the loach (*Misgurnus fossilis* L.) and the perch-pike (*Lucioperca lucioperca* L.). *Acta Biol. Cracov. Ser. Zool.* **1**, 83–97.

Byczkowska-Smyk, W. (1959a). The respiratory surface of the gills in teleosts. 3. The respiratory surface of the gills in the tench (*Tinca tinca* L.), the silver bream (*Blicca bjoerkna*), and the chondrostomas (*Chondrostoma nasus* L.). *Acta Biol. Cracov.* **2**, 73–88.

Byczkowska-Smyk, W. (1959b). The respiratory surface of the gills in teleosts, 4. The respiratory surface of the gills in the pike (*Esox lucius*), stone-perch (*Acerina cernus*) and the burbot (*Lota lota* L.). *Acta Biol. Cracov.* **2**, 113–127.

Byczkowska-Smyk, W. (1962). Vascularisation and size of the respiratory surface in *Acipenser stellatus. Acta Biol. Cracov.* **5**, 304–315.

Carter, G. S. (1957). Air breathing. *In* "Physiology of Fishes" (M. E. Brown, ed.), Vol. 1, pp. 65–79. Academic Press, New York.

Carter, C. S., and Beadle, L. C. (1931). The fauna of the Paraguayan Chaco in relation to its environment. II. Respiratory adaptations in the fishes. *J. Linn. Soc. London, Zool.* **37**, 327–366.

Chilmonczyk, S., and Monge, D. (1980). Rainbow trout gill pillar cells: Demonstration of inert particle phagocytosis and involvement in viral infection. *J. Reticuloendothel. Soc.* **28**, 327–332.

Cooke, I. R. C. (1980). Functional aspects of the morphology and vascular anatomy of the gills of the Endeavour dogfish, *Centrophorus scalpratus* (McCulloch) (Elasmobranchii: Squalidae). *Zoomorphologie* **94**, 167–183.

Cotes, J. E. (1965). "Lung Function." Blackwell, Oxford.

D'Arcy, Thompson, W. (1917). "On Growth and Form." Cambridge Univ. Press, London and New York.

de Jager, S., and Dekkers, W. J. (1975). Relations between gill structure and activity in fish. *Neth. J. Zool.* **25**, 276–308.

Dejours, P. (1976). Water versus air as the respiratory media. *In* "Respiration of Amphibious Vertebrates" (G. M. Hughes, ed.), pp. 1–15. Academic Press, New York.

Dornescu, G. T., and Miscalencu, D. (1968a). Cele trei tipuri de branchii ale teleosteenilor. *Ann. Univ. Bucuresti, Ser. S. Nat. Biol.* **17**, 11–20.

Dornescu, G. T., and Miscalencu, D. (1968b). Etude comparative des branchies de quelques espèces de l'ordre clupéiformes. *Morphol. Jahrb.* **122**(H2), 261–276.

Farber, J., and Rahn, H. (1970). Gas exchange between air and water and the ventilation pattern in the electric eel. *Respir. Physiol.* **9**, 151–161.

Fromm, P. P. (1976). Circulation in trout gills: Presence of 'Blebs' in afferent filament vessels. *J. Fish. Res. Board Can.* **31**, 1793–1796.

Fry, F. E. J. (1957). The aquatic respiration of fish. *In* "Physiology of Fishes" (M. E. Brown, ed.), Vol. 1, pp. 1–63. Academic Press, New York.

Gans, C. (1976). Ventilatory mechanisms and problems in some amphibious aspiration breathers (*Chelydra, Caiman*—Reptilia). *In* "Respiration of Amphibious Vertebrates" (G. M. Hughes, ed.), pp. 357–374. Academic Press, New York.

Gans, C., DeJohngh, H. J., and Farber, J. (1969). Bullfrog (*Rana catesbeiana*) ventilation: how does the frog breathe? *Science* **163**, 1223–1225.

Goette, A. (1878). Zur Entwicklungsgeschichte der Teleosteirkieme. *Zool. Anz.* **1**, 52.

Goodrich, E. S. (1930). "Studies on the Structure and Development of Vertebrates." Macmillan, New York.

Graham, J. B. (1976). Respiratory adaptations of marine air-breathing fishes. *In* "Respiration of Amphibious Vertebrates" (G. M. Hughes, ed.), pp. 165–187. Academic Press, New York.

Gray, I. E. (1954). Comparative study of the gill area of marine fishes. *Biol. Bull. (Woods Hole, Mass.)* **107**, 219–225.

Grigg, G. C. (1970a). Water flow through the gills of Port Jackson sharks. *J. Exp. Biol.* **52**, 565–568.

Grigg, G. C. (1970b). The use of gill slits for water intake in a shark. *J. Exp. Biol.* **52**, 569–574.

Hakim, A., Munshi, J. S. D., and Hughes, G. M. (1978). Morphometrics of the respiratory organs of the Indian green snake-headed fish, *Channa punctata. J. Zool.* **184**, 519–543.

Harden-Jones, F. R., Arnold, G. P., Grier-Walker, M., and Scholes, P. (1979). Selective tidal stream transport and the migration of plaice (*Pleuronectes platessa* L.) in the southern North Sea. *J. Cons., Cons. Int. Explor. Mer.* **38**, 331–337.

Hargens, A. R., Millard, R. W., and Johansen, K. (1974). High capillary permeability in fishes. *Comp. Biochem. Physiol. A.* **48A**, 675–680.

Henschel, J. (1941). Neue Untersuchungen über den Atemmechanismus mariner Teleosteer. *Helgol. Wiss. Meeresunters.* **2**, 244–278.

Hills, B. A. (1974). "Gas Transfer in the Lung." Cambridge Univ. Press, London and New York.

Hills, B. A., and Hughes, G. M. (1970). A dimensional analysis of oxygen transfer in the fish gill. *Respir. Physiol.* **9**, 126–140.

Hills, B. A., Hughes, G. M., and Koyama, T. (1982). Oxygenation and deoxygenation kinetics of red cells in isolated lamellae of fish gills. *J. Exp. Biol.* **98**, 269–275.

Holeton, G. F. (1970). Fish respiration without haemoglobin. Ph.D. Thesis, Bristol University.

Holeton, G. F. (1971). Respiratory and circulatory responses of rainbow trout larvae to carbon monoxide and to hypoxia. *J. Exp. Biol.* **55**, 683–694.

Holeton, G. F. (1972). Gas exchange in fish with and without hemoglobin. *Respir. Physiol.* **14**, 142–150.

Holeton, G. F. (1980). Oxygen as an environmental factor of fishes. *In* "Environmental Physiology of Fishes" (M. A. Ali, ed.), pp. 7–32. Plenum, New York.

Holeton, G. F., and Jones, D. R. (1975). Water flow dynamics in the respiratory tract of the carp (*Cyprinus carpio* L.). *J. Exp. Biol.* **63**, 537–549.

Horstadius, S. (1950). "The Neural Crest. Its Properties and Derivatives in the Light of Experimental Research." Oxford Univ. Press, London and New York.

Hughes, G. M. (1960a). The mechanism of gill ventilation in the dogfish and skate. *J. Exp. Biol.* **37**, 11–27.

Hughes, G. M. (1960b). A comparative study of gill ventilation in marine teleosts. *J. Exp. Biol.* **37**, 28–45.

Hughes, G. M. (1963). "Comparative Physiology of Vertebrate Respiration," (Heinemann, London (2nd ed., 1974).

Hughes, G. M. (1966a). Evolution between air and water. *In* "Development of the Lung" (A.V.S. de Reuck and R. Porter, eds.), pp. 64–80. Churchill, London.

Hughes, G. M. (1966b). The dimensions of fish gills in relation to their function. *J. Exp. Biol.* **45**, 177–195.

Hughes, G. M. (1970a). Morphological measurements on the gills of fishes in relation to their respiratory function. *Folia Morphol. (Prague)* **18**, 78–95.

Hughes, G. M. (1970b). A comparative approach to fish respiration. *Experientia* **26**, 113–122.

Hughes, G. M. (1972a). Gills of a living coelacanth, *Latimeria chalumnae. Experientia* **28**, 1301–1302.

Hughes, G. M. (1972b). Distribution of oxygen tension in the blood and water along the secondary lamella of the icefish gill. *J. Exp. Biol.* **56**, 481–492.

Hughes, G. M. (1972c). Morphometrics of fish gills. *Respir. Physiol.* **14**, 1–25.

Hughes, G. M. (1973a). Respiratory responses to hypoxia in fish. *Am. Zool.* **13**, 475–489.

Hughes, G. M. (1973b). Comparative vertebrate ventilation and heterogeneity. *In* "Comparative Physiology" (L. Bolis, K. Schmidt-Nielsen, and S. H. P. Maddrell, eds.), pp. 187–220. North-Holland Publ., Amsterdam.

Hughes, G. M. (1976). Fish respiratory physiology. *In* "Perspectives in Experimental Biology" (P. Spencer Davies, ed.), Vol. 1, pp. 235–245. Pergamon Press, Oxford and New York.

Hughes, G. M. (1977). Dimensions and the respiration of lower vertebrates. *In* "Scale Effects in Animal Locomotion" (T. J. Pedley, ed.), pp. 57–81. Academic Press, New York.

Hughes, G. M. (1978a). A morphological and ultrastructural comparison of some vertebrate lungs. *Colloq. Sci. Fac. Med. Univ. Carol. [Pap.], 21st, 1978* pp. 393–405.

Hughes, G. M. (1978b). Some features of gas transfer in fish. *Bull. Inst. Math. Appl.* **14**, 39–43.

Hughes, G. M. (1978c). On the respiration of *Torpedo marmorata. J. Exp. Biol.* **73**, 85–105.

Hughes, G. M. (1979a). The path of blood flow through the gills of fishes—some morphometric observations. *Acta Morphol. Sofia* **2**, 52–58.

Hughes, G. M. (1979b). Scanning electron microscopy of the respiratory surface of trout gills. *J. Zool.* **188**, 443–453.

Hughes, G. M. (1980a). Structure of fish respiratory surfaces. *Int. Congr. Anat., 11th, 1980* Abstract 36.

Hughes, G. M. (1980b). Functional morphology of fish gills. *In* "Epithelial Transport in the Lower Vertebrates" (B. Lahlou, ed.), pp. 15–36. Cambridge Univ. Press, London and New York.

Hughes, G. M. (1980c). Ultrastructure and morphometry of the gills of *Latimeria chalumnae,* and a comparison with the gills of associated fishes. *Proc. R. Soc. London, Ser. B* **208,** 309–328.

Hughes, G. M. (1980d). Morphometry of fish gas exchange organs in relation to their respiratory function. *In* "Environmental Physiology of Fishes" (M. A. Ali, ed.), pp. 33–56. Plenum, New York.

Hughes, G. M. (1981). Fish gills—past, present and future. *Biol. Bull. India* **3**(2), 69–87.

Hughes, G. M. (1982). An introduction to the study of gills. *In* "Gills" (D. F. Houlihan, J. C. Rankin, and T. J. Shuttleworth, eds.), pp. 1–24. Cambridge Univ. Press, London and New York.

Hughes, G. M. (1984a). Scaling of respiratory areas in relation to oxygen consumption of vertebrates. *Experientia* (in press).

Hughes, G. M. (1984b). Measurement of gill area in fishes: Practices and Problems. *J. Mar. Biol. Assoc. U. K.* (in press).

Hughes, G. M., and Adeney, R. J. (1977). Variations in the pattern of coughing in rainbow trout. *J. Exp. Biol.* **68,** 109–122.

Hughes, G. M., and Ballintijn (1965). The muscular basis of the respiratory pumps in the dogfish (*Scyliorhinus canicula*). *J. Exp. Biol.* **43,** 363–383.

Hughes, G. M., and Gray, I. E. (1972). Dimensions and ultrastructure of toadfish gills. *Biol. Bull. (Woods Hole, Mass.)* **143,** 150–161.

Hughes, G. M., and Grimstone, A. V. (1965). The fine structure of the secondary lamellae of the gills of *Gadus pollachius. Q. J. Microsc. Sci.* **106,** 343–353.

Hughes, G. M., and Iwai, T. (1978). A morphometric study of the gills in some Pacific deep-sea fishes. *J. Zool.* **184,** 155–170.

Hughes, G. M., and Morgan, M. (1973). The structure of fish gills in relation to their respiratory function. *Biol. Rev. Cambridge Philos. Soc.* **48,** 419–475.

Hughes, G. M., and Munshi, J. S. D. (1968). Fine structure of the respiratory surfaces of an air-breathing fish, the climbing perch, *Anabas testudineus* (Bloch). *Nature (London)* **219,** 1382–1384.

Hughes, G. M., and Munshi, J. S. D. (1973a). Fine structure of the respiratory organs of the climbing perch, *Anabas testudineus* (Pisces: Anabantidae). *J. Zool.* **170,** 201–225.

Hughes, G. M., and Munshi, J. S. D. (1973b). Nature of the air-breathing organs of the Indian fishes, *Channa, Amphipnous, Clarias* and *Saccobranchus* as shown by electron microscopy. *J. Zool.* **170,** 245–270.

Hughes, G. M., and Munshi, J. S. D. (1978). Scanning electron microscopy of the respiratory surfaces of *Saccobranchus* (= *Heteropneustes*) *fossilis* (Bloch). *Cell Tissue Res.* **195,** 99–109.

Hughes, G. M., and Munshi, J. S. D. (1979). Fine structure of the gills of some Indian air-breathing fishes. *J. Morphol.* **160,** 169–194.

Hughes, G. M., and Perry, S. F. (1976). Morphometric study of trout gills: A light microscopic method suitable for the evaluation of pollutant action. *J. Exp. Biol.* **64,** 447–460.

Hughes, G. M., and Peters, H. M. (1984). Ventilation of the gills and labyrinthine organs with air and water in the gouramy (*Osphronemus goramy*). (In preparation.)

Hughes, G. M., and Shelton, G. (1958). The mechanism of gill ventilation in three fresh water teleosts. *J. Exp. Biol.* **35,** 807–823.

Hughes, G. M., and Shelton, G. (1962). Respiratory mechanisms and their nervous control in fish. *Adv. Comp. Physiol. Biochem.* **1,** 275–364.

Hughes, G. M., and Umezawa, S. (1968). On respiration in the dragonet, *Callionymus lyra* L. *J. Exp. Biol.* **49,** 565–582.

Hughes, G. M., and Weibel, E. R. (1972). Similarity of supporting tissue in fish gills and the mammalian reticulo-endothelium. *J. Ultrastruct. Res.* **39,** 106–114.

Hughes, G. M., and Weibel, E. R. (1976). Morphometry of fish lungs. *In* "Respiration of Amphibious Vertebrates" (G. M. Hughes, ed.), pp. 213–232. Academic Press, New York.

Hughes, G. M., and Weibel, E. R. (1978). Visualization of layers lining the lung of the South American lungfish (*Lepidosiren paradoxa*) and a comparison with the frog and rat. *Tissue Cell* **10,** 343–353.

Hughes, G. M., and Wright, D. E. (1970). A comparative study of the ultrastructure of the water-blood pathway in the secondary lamellae of teleost and elasmobranch fishes—benthic forms. *Z. Zellforsch. Mikrosk. Anat.* **104,** 478–493.

Hughes, G. M., Dube, S. C., and Munshi, J. S. D. (1973). Surface area of the respiratory organs of the climbing perch, *Anabas testudineus* (Pisces: Anabantidae). *J. Zool.* **170,** 227–243.

Hughes, G. M., Singh, B. R., Thakur, R. N., and Munshi, J. S. D. (1974a). Areas of the air-breathing surfaces of *Amphipnous cuchia* (Ham.). *Proc. Indian Natl. Sci. Acad., Part B* **40,** No. 4, 379–392.

Hughes, G. M., Singh, B. R., Guha, G., Dube, S. C., and Munshi, J. S. D. (1974b). Respiratory surface areas of an air-breathing siluroid fish, *Saccobranchus (Heteropneustes) fossilis,* in relation to body size. *J. Zool.* **172,** 215–232.

Hughes, G. M., Tuurala, H., and Soivio, A. (1978). Regional distribution of blood in the gills of rainbow trout in normoxia and hypoxia: A morphometric study with two fixatives. *Ann. Zool. Fenn.* **15,** 226–234.

Hughes, G. M., Perry, S. F., and Brown, V. M. (1979). A morphometric study of effects of nickel, chromium and cadmium on the secondary lamellae of rainbow trout gills. *Water Res.* **13,** 665–679.

Hughes, G. M., Horimoto, M., Kikuchi, Y., Kakiuchi, Y., and Koyama, T. (1981). Blood flow velocity in microvessels of the gill filaments of the goldfish (*Carassius auratus* L.). *J. Exp. Biol.* **90,** 327–331.

Hughes, G. M., Peyraud, C., Peyraud-Waitzenegger, M., and Soulier, P. (1982). Physiological evidence for the occurrence of pathways shunting blood away from the secondary lamellae of eel gills. *J. Exp. Biol.* **98,** 277–288.

Hughes, G. M., Perry, S. F., and Piiper, J. (1984). Quantitative anatomy of the gills of the larger spotted dogfish (*Scyliorhinus stellaris*). (In preparation.)

Huxley, J. S. (1932). "Problems of Relative Growth." Methuen, London.

Huxley, J. S., and Teissier, G. (1936). Terminology of relative growth. *Nature (London)* **137,** 780–781.

Ishimatsu, A., and Itazawa, Y. (1983). Difference in blood oxygen levels in the outflow vessels of the heart of an air-breathing fish, *Channa argus:* Do separate blood streams exist in a teleostean heart? *J. Comp. Physiol.* **149,** 435–440.

Iwai, T. (1963). Taste buds on the gill rakers and gill arches of the sea catfish *Plotosus anguillaris* (Lacapède). *Copeia* **2,** 271–274.

Iwai, T. (1964). A comparative study of the taste buds in gill rakers and gill arches of teleostean fishes. *Bull. Misaki Mar. Biol. Inst. Kyoto Univ.* **7,** 19–34.

Iwai, T., and Hughes, G. M. (1977). Preliminary morphometric study on gill development in Black Sea bream (*Acanthopagrus schlegeli*). *Bull. Jpn. Soc. Sci. Fish.* **43,** 929–934.

Jones, D. R., and Randall, D. J. (1978). The respiratory and circulatory systems. *In* "Fish

Physiology" (W. S. Hoar and D. J. Randall, eds.), Vol. 7, p. 425. Academic Press. New York.

Jones, D. R., and Schwazfeld, T. (1974). The oxygen cost to the metabolism and efficiency of breathing in trout (*Salmo gairdneri*). *Respir. Physiol.* **21**, 241–254.

Kazanski, V. I. (1964). Specific differences in the structure of gill rakers in Cyprinidae. *Vopr. Ikhtiol.* **4**, 45–60.

Kempton, R. T. (1969). Morphological features of functional significance in the gills of the spiny dogfish *Squalus acanthias*. *Biol. Bull. (Woods Hole, Mass.)* **136**, 226–240.

Krogh, A. (1919). The rate of diffusion of gases through animal tissue. *J. Physiol. (London)* **52**, 391–408.

Lauder, G. V. (1980). The suction feeding mechanism in sunfishes (*Lepomis*): An experimental analysis. *J. Exp. Biol.* **88**, 49–72.

Lauder, G. V. (1983). Prey capture hydrodynamics in fishes: experimental tests of two models. *J. Exp. Biol.* **104**, 1–13.

Leiner, M. (1938). "Die Physiologie der Fischatmung." Akad. Verlagsges., Leipzig.

Lewis, S. V., and Potter, I. A. (1976). A scanning electron microscope study of the gills of the lamprey *Lampetra fluviatilis* (L). *Micron* **7**, 205–211.

McMahon, B. R. (1969). A functional analysis of the aquatic and aerial respiratory movements of an African lungfish, *Protopterus aethiopicus*, with reference to the evolution of the lung-ventilation mechanism in vertebrates. *J. Exp. Biol.* **51**, 407–430.

Marshall, N. B. (1960). Swimbladder structure of deepsea fishes in relation to their systematics and biology. *'Discovery' Rep.* **31**, 1–122.

May, R. C. (1974). Larval mortality in marine fishes and the critical period concept. *In* "The Early Life History of Fish" (J. H. S. Blaxter, ed.), pp. 2–19. Springer-Verlag, Berlin and New York.

Miscalencu, D., and Dornescu, G. T. (1970). Etude comparative des branchies de quelques perciformes marins. *Zool. Anz.* **184**, 67–74.

Morgan, M. (1971). Gill development, growth and respiration in the trout, *Salmo gairdneri*. Ph.D. Thesis, Bristol University.

Morgan, M. (1974). Development of secondary lamellae of the gills of the trout, *Salmo gairdneri* (Richardson). *Cell Tissue Res.* **151**, 509–523.

Moroff, T. (1904). Uber die Entwicklung der Kiemen bei Fischen. *Arch. Mikrosk. Anat. Entwicklungsmech.* **64**, 189–213.

Muir, B. S. (1970). Contribution to the study of blood pathways in teleost gills. *Copeia* **1**, 19–28.

Muir, B. S., and Brown, C. E. (1971). Effects of blood pathway on the blood pressure drop in fish gills with special reference to tunas. *J. Fish. Res. Board Can.* **28**, 947–955.

Muir, B. S., and Hughes, G. M. (1969). Gill dimensions for three species of tunny. *J. Exp. Biol.* **51**, 271–285.

Muir, B. S., and Kendall, J. I. (1968). Structural modifications in the gills of tunas and some other oceanic fishes. *Copeia* **2**, 388–398.

Muller, J. (1839). Vergleichende Anatomie der Myxinoiden. III. Uber das Gefassystem. *Abh. Akad. Wiss., Berlin* pp. 175–303.

Munshi, J. S. D. (1962). On the accessory respiratory organs of *Heteropneustes fossilis* (BL.). *Proc. R. Soc. Edinburgh, Sect. B: Biol.* **68**, 128–146.

Munshi, J. S. D. (1968). The accessory respiratory organs of *Anabas testudineus* (Bloch) (Anabantidae, Pisces). *Proc. Linn. Soc. London* **170**, 107–126.

Munshi, J. S. D. (1976). Gross and fine structure of the respiratory organs of air-breathing fishes. *In* "Respiration of Amphibious Vertebrates" (G. M. Hughes, ed.), pp. 73–104. Academic Press, New York.

Munshi, J. S. D., and Hughes, G. M. (1981). Gross and fine structure of the pseudobranch of the climbing perch, *Anabas testudineus* (Bloch). *J. Fish Biol.* **19**, 427–438.

Needham, J. (1941). "Biochemistry and Morphogenesis." Cambridge Univ. Press, London and New York.

Nowogrodska, M. (1943). The respiratory surface of gills of Teleosts. VII. *Trachurus trachurus*. *Acta Biol. Cracov.* **6**, 147–158.

Oliva, O. (1960). The respiratory surface of the gills in teleosts. 5. The respiratory surface of the gills in the viviparous blenny (*Zoarces viviparous*). *Acta Biol. Cracov.* **3**, 71–89.

Osse, J. W. M. (1969). Functional morphology of the head of the perch (*Perca fluviatilis*)—an electromyographic study. *Neth. J. Zool.* **19**, 290–392.

Paling, J. E. (1968). A method of estimating the relative volumes of water flowing over the different gills of freshwater fish. *J. Exp. Biol.* **48**, 533–544.

Pasztor, V. M., and Kleerekoper, H. (1962). The role of the gill filament musculature in teleosts. *Can. J. Zool.* **40**, 785–802.

Pattle, R. E. (1976). The lung surfactant in the evolutionary tree. *In* "Respiration of Amphibious Vertebrates" (G. M. Hughes, ed.), pp. 233–255. Academic Press, New York.

Pattle, R. E. (1978). Lung surfactant and lung lining in birds. *In* "Respiratory Function in Birds, Adult and Embryonic" (J. Piiper, ed.), pp. 23–32. Springer-Verlag, Berlin and New York.

Perry, S. F. (1983). "Reptilian Lungs. Functional Anatomy and Evolution." Springer-Verlag, Berlin and New York.

Peters, H. M. (1978). On the mechanism of air ventilation in anabantoids (Pisces: Teleostei). *Zoomorphology* **89**, 93–124.

Piiper, J., and Scheid, P. (1982). Physical principles of respiratory gas exchange in fish gills. *In* "Gills" (D. F. Houlihan, J. C. Rankin, and T. J. Shuttleworth, eds.), pp. 45–61. Cambridge Univ. Press, London and New York.

Piiper, J., and Schumann, D. (1967). Efficiency of $O_2$ exchange in the gills of the dogfish, *Scyliorhinus stellaris*. *Respir. Physiol.* **2**, 135–148.

Price, J. W. (1931). Growth and gill development in the small-mouthed black bass, *Micropterus dolomieu* Lacapede. *Contrib. Stone Lab. Ohio Univ.* **4**, 1–46.

Rahn, H. (1966). Gas transport from the external environment to the cell. *In* "Development of the Lung" (A. V. S. de Reuck and R. Porter, eds.), pp. 3–23. Churchill, London.

Randall, D. J. (1970). Gas exchange in fish. *In* "Fish Physiology" (W. S. Hoar and D. J. Randall, eds.), Vol. 4, pp. 253–292. Academic Press, New York.

Randall, D. J., Burggren, W. W., Farrell, A. P., and Haswell, M. S. (1981). "The Evolution of Air Breathing in Vertebrates." Cambridge Univ. Press, London and New York.

Roberts, J. L. (1975). Active branchial and ram gill ventilation in fishes. *Biol. Bull. (Woods Hole, Mass.)* **148**, 85–105.

Scheid, P. (1979). Mechanisms of gas exchange in bird lungs. *Rev. Physiol., Biochem. Pharmacol.* **86**, 138–185.

Scheid, P., and Piiper, J. (1976). Quantitative functional analysis of branchial gas transfer: Theory and application to *Scyliorhinus stellaris* (Elasmobranchii). *In* "Respiration of Amphibious Vertebrates" (G. M. Hughes, ed.), pp. 17–38. Academic Press, New York.

Schlotte, E. (1932). Morphologie und Physiologie der Atmung bei wassers-schlamm-und landlebenden Gobiiformes. *Z. Wiss. Zool.* **140**, 1–113.

Schmidt, P. (1915). Respiratory adaptations of Pleuronectids. *Bull. Acad. Sci. St.-Petersbourg* [6] **9**, 421–444 (in Russian).

Sewertzoff, A. (1924). Die Entwicklung der Kiemen und Kiemenbogen-gefasse der Fische. *Z. Wiss. Zool.* **121**, 494–556.

Singh, B. N. (1976). Balance between aquatic and aerial respiration. *In* "Respiration of Amphibious Vertebrates" (G. M. Hughes, ed.), pp. 125–164. Academic Press, New York.

Smith, D. G., and Chamley-Campbell, J. (1981). Localisation of smooth-muscle myosin in branchial pillar cells of snapper (*Chrysophys auratus*) by immuno-fluorescence histochemistry. *J. Exp. Zool.* **215**, 121–124.

Soivio, A., and Tuurala, H. (1981). Structural and circulatory responses to hypoxia in the secondary lamellae of *Salmo gairdneri* gills at two temperatures. *J. Comp. Physiol.* **145**, 37–43.

Steen, J. B. (1971). "Comparative Physiology of Respiratory Mechanisms." Academic Press, New York.

Steen, J. B., and Kruysse, E. (1964). The respiratory function of teleostean gills. *Comp. Biochem. Physiol.* **12**, 127–142.

Tovell, P. W., Morgan, M., and Hughes, G. M. (1971). Ultrastructure of trout gills. *Electron Microsc., Proc. Int. Congr., 7th, 1970* Vol. III, p. 601.

Underwood, E. E. (1970). "Quantitative Stereology." Addison-Wesley, Reading, Massachusetts.

van Dam, L. (1938). On the utilisation of oxygen and regulation of breathing in some aquatic animals. Dissertation, University of Grøningen, Netherlands.

Vogel, W. O. P. (1981). Struktur und Organisationsprinzip im Gefassysten der Knochenfische. *Morphol. Jahrb.* **127**, 772–784.

Vogel, W. O. P., and Claviez, M. (1981). Vascular specialisation in fish, but no evidence for lymphatics. *Z. Naturforsch., C: Biosci.* **36C**, 490–492.

Wagner, H., Conte, F. P., and Fessler, J. (1969). Development of osmotic and ionic regulation in two races of juvenile chinook salmon, *O. tschawytscha. Comp. Biochem. Physiol.* **29**, 325–342.

Watson, D. M. S. (1951). "Palaeontology and Modern Biology," Chapter 1. Yale Univ. Press, New Haven, Connecticut.

Weibel, E. R. (1971). Morphometric estimation of pulmonary diffusion capacity. I. Model and method. *Respir. Physiol.* **11**, 54–75.

Weibel, E. R. (1979). "Stereological Methods," Vol. I. Academic Press, New York.

Winberg, G. G. (1956). (1) Rate of metabolism and food requirements of fishes. (2) New information on metabolic rate in fishes. *Fish. Res. Board Can.* (Transl. Ser. 194 and 362).

Wood, C. M. (1974). A critical examination of the physical and adrenergic factors affecting blood flow through the gills of the rainbow trout. *J. Exp. Biol.* **60**, 241–265.

Woskoboinikoff, M. M. (1932). Der Apparat der Kiemenatmung bei den Fischen. Ein Versuch der Synthese in der Morphologie. *Zool. Jahrb., Abt. Anat. Ontog. Tiere* **55**, 315–488.

Wright, D. E. (1973). The structure of the gills of the elasmobranch, *Scyliorhinus canicula* L. *Z. Zellforsch. Mikrosk. Anat.* **144**, 489–509.

Yazdani, G. M., and Alexander, R. McN. (1967). Respiratory currents of flatfish. *Nature (London)* **213**, 96–97.

# 2

---

# GILL INTERNAL MORPHOLOGY*

*PIERRE LAURENT*

Laboratoire de Morphologie Fonctionnelle et Ultrastructurale des Adaptations
Centre National de la Recherche Scientifique
Strasbourg, France

## I. INTRODUCTION

## A. Basic Concepts of the Gill Organization

It is now currently accepted that the gills fulfill several functions in fish, mainly dealing with respiration and osmoregulation (Smith, 1929). Long ago gill morphology was held to account for this dual role (Keys, 1931); however, subsequent studies have resulted in a complementary concept being proposed that considers the existence of two distinct sites for gill exchanges. These sites correspond to specialized epithelia with distinct blood compartments. This concept assumes that these sites selectively participate in transport or exchange of various materials, gas, ions, and water. Indeed a careful observation of the cellular components of gill epithelia as well as of the functional organization of their blood compartments brings evidence for separate functions. Those structural clues have received functional support

*Financially supported by Grant CNRS AI 03 4302.

FISH PHYSIOLOGY, VOL. XA

from measurement of ion fluxes, a question considered elsewhere in this volume. Thus, the modern studies on gill internal morphology have focused on two major subject areas: the perfusion patterns and the functioning of the ion-transporting cells. If some variation occurs in the perfusion pattern within the fish groups, these studies have revealed a basic arrangement that is one of the keystones of gill physiology. In contrast, variations in structure and distribution of transporting cells are now analyzed in terms of adaptation to environmental conditions. Those variations are revealing as to the responsiveness of the structures to the milieu and consequently of the gill exchange mechanisms.

The nervous system has different functions within the gills. It partly assumes the perfusion control by acting on specialized vascular areas present inside the gills. Other components are sensory in function, generating and channeling information that concerns internal and external physiological conditions. Little is known at this level, except that some mechanical receptors do exist. The discovery of abundant populations of neuroepithelial cells within the gill epithelia raises the problem of their activity as chemoreceptors and prompts new research (Dunel-Erb *et al.*, 1982a).

Within the gill and particularly the filaments, a muscular system presumably plays a crucial role. The geometry of the gill basket, the position of the filaments in the flow of respiratory water, and consequently, part of the respiratory efficiency of the gills, depend on its activity. The distribution and the action of striated muscles is well known. The discovery of smooth muscle bundles fixed on the skeleton of the filaments reopens the question of this control (S. Dunel-Erb and P. Laurent, unpublished).

Finally, when looking into the structural details of the gill, numerous cell types are not yet identified from complex cell populations, including so-called nondifferentiated cells, cells containing large vesicles. The richness in epithelial cells of different type calls for new investigations and predicts future evolution in the concepts of gill morphology and physiology.


## B. Historical Survey

### 1. THE GILL VASCULATURE

One of the first studies on gill vasculature was written in 1785 by Monro following the oldest paper of Duverney (1699). Probably the first comprehensive study is that of J. Müller (1839). In an extensive study, he was the first to emphasize the existence of two independent vascular systems. Arterio-arterial blood pathways assume the respiratory function of the gills by perfusing the finest subdivision of the branchial system. A second system, classically made up of arteries, capillaries, and veins, delivers nutrients to the different parts of the gill tissues. According to Müller's concept, the so-called nutritive arteries emanate from the efferent branchial vessel of the gill arch and irrigate abductor and adductor muscles. Other small arteries emanating from the efferent artery of the filaments and running along with the efferent arteries supply the inner regions of the filaments. J. Müller mentioned a venous network in the central core of the filament. He suggested that this system drains into the branchial veins. Thus, according to Müller, the gill vasculature consists of two independent systems: (1) arterio-arterial pathways assuming the exchanges of gas between the blood and the external milieu and supplying the systemic circulation and (2) arteriovenous pathways feeding the gill tissues.

Later on Riess, in his 1881 study, claimed that Müller was mistaken in his interpretation of the nutrient vasculature. The observations of Riess focused on the filaments of the pike, and he claimed that its central part contains interstitial spaces filled with fluid, lymphocytes, and erythrocytes exuding from the higher pressure respiratory capillaries. Some large vessels collect them and nutrient veins as well. Riess described a system of nutrient arteries that are bound mainly for gill muscles and are tributaries of afferent filament arteries. Thus, according to Riess, the vascular system of the gill should be composed of an arterial, a venous, and a lymphatic part. However, the relationships between lymphatic and arterio-arterial systems are not clear in the Riess study, and the presence of blood cells in the central "lymphatic compartment" is explained by a process of exudation.

The question of the lymphatic system in gills has been reopened recently in the eel (Rowing, 1981) and in the toadfish (Cooke and Campbell, 1980). The lymphatic system in fish has been the subject of several articles (Allen, 1907; Florkowski, 1930; Jossifov, 1906; Burne, 1927, 1929; Romer and Parsons, 1977; Vogel and Claviez, 1981). A compilation of these works leads one to question the concept of a venolymphatic system in fish, a concept that accounts for its morphology halfway between veins and lymphatics. This point is considered in detail in Section II, B, 5.

Another alternative when considering the vascular system of the filament core has been presented by Steen and Kruysse (1964). The central (lymphatic) compartment is considered a shunting pathway for the blood directly connecting the afferent and the efferent arteries without perfusing the lamellae. This arrangement might divert the blood from the respiratory surface when the fish is in a situation of low oxygen demand, for instance. This concept, widely accepted by some physiologists (Richards and Fromm, 1969, 1970; Rankin and Maetz, 1971; Kirschner, 1969), was questioned on a structural basis (Gannon et al., 1973; Morgan and Tovell, 1973; Laurent and Dunel, 1976).

The reader, incidentally, will find out that the modern (perhaps not definitive) concept is mostly in accordance with the description given by Müller. However, the central vascular compartment cannot be strictly considered a vein or a capillary but more a particular vascular structure proper to fish gill. This compartment has been sometimes equivocally called venolymphatic sinus, a terminology that is not used in this chapter.

Important progress has been made in gill vascularization studies by the use of vascular plastic casts and replicas. These methods have overcome the time-consuming stereological reconstruction from histological micrographs. Two kinds of materials are now used:

1. Silicone rubber is a nonaggressive, low-viscosity material that after polymerization gives flexible casts. As a consequence of its properties, this material fills the capillaries with minimum artifact. After curing the resin and clearing the tissues with glycerol, high-definition photographs can be taken. With this very simple technique it is still possible to identify the surrounding tissues that stay in place; however, the relatively superficial perception implies successive dissections in case of in-depth investigation. Observations are largely improved by using a stereomicroscope. The maximum possible magnification is about 100×.

2. Mechanically self-supporting replicas are obtained after perfusion of the vasculature with polyester plastic monomers (methyl methacrylate). The viscosity of the monomer must be adjusted in order to fill the smallest vessels. After polymerization, which often requires immersion of the specimen in hot saline, the surrounding tissues are removed with corrosive chemicals and the cast cleaned carefully. Thereafter, it is gold coated and mounted on the stage of a scanning microscope. The casts are brittle but the definition excellent, and the focus depth is only limited by the microscope performance. Some now commercially available resins are easier to use than the initial material (Murakami, 1971). The reader will find an excellent compilation of these methods in the review of Gannon (1978; see also Olson, 1980).

## 2. The Ion-Transporting Gill Structures

The history of the theory of an ion-transporting gill function might be subdivided into three steps: the physiological concept of an extrarenal excretion of electrolytes in fish, the structural concept of gill-specialized excreting cells, and, more recently, the morphophysiological concept of a specialized epithelium associated with a particular blood compartment. Smith (1929) was the first to claim that in some teleosts excretion of urea and ammonia occurs across the gill. In addition to that, the production of a hypotonic urine in marine as well as freshwater teleosts led Smith (1930) to consider other possible sources of salt elimination than renal. He postulated that gill should be the site of an extrarenal excretion. This last point was further demonstrated by Keys (1931), and in a series of articles (1931–1932) he presented physiological and histological evidence for an excretion of chloride. Keys and Willmer, in their 1932 study, established the presence of a secretory type of cells in large numbers in the gills of the eel and other teleosts. They were, however, unable to confirm the existence of such cells in the dogfish. In the same article they suggested a probable correlation of those cells with the chloride-secreting activity of the gill. It is interesting to note that the Keys and Willmer study makes a distinction between the role of chloride cells of saltwater fish and the chloride cells in freshwater teleosts. In the latter case, chloride cells are not involved in osmoregulatory excretion but in salt absorption since freshwater fish are facing the danger of salt depletion. Modern experiments suggest that these cells in freshwater fish are the site for a calcium uptake (Payan *et al.*, 1981). However, Keys and Willmer put into question the presence of chloride cells in all species. Another mistaken point was, for instance, Keys' assertion concerning the absence of chloride cells in elasmobranchs, which actually have gills provided with a rather different type of transporting cell.

The localization of chloride cells observed by Keys and Willmer on the afferent blood circulation in teleosts was certainly quite correct. Not so correct was the statement that the cells were irrigated by the blood flowing in the lamellae. The specific association of chloride cells within the filament epithelium and the central venous compartment has been emphasized in teleosts and elasmobranchs (Laurent and Dunel, 1978) and also proposed in cyclostomes (Nakao and Uchinomiya, 1978). These observations led to the concept of an osmoregulatory blood circulation distinct from the respiratory circuit. Several studies provide physiological evidence for an anatomic dichotomy of respiratory and osmoregulatory functions in the gill (Girard and Payan, 1980).

## II. THE GILL VASCULATURE

### A. Gill Arches

#### 1. BASIC ORGANIZATION

It is well known that the structure and arrangement of gill arches depend on the classes, orders, and families considered. In the past, numerous articles have focused on the phylogenetic aspects of the skeletal visceral arch distribution. Indeed the organization of gill slits is one of the most fundamental features of the vertebrate phylum. From a functional standpoint, gill slits serve as the conduit for the passage of a respiratory (and nutritive) current of water from the pharynx to the exterior. A system of gill slits is already present in the ancestors of fish (Tunicata, Enteropneusta, and Pterobranchia) and still occurs in Amphibia and in a more or less reduced condition in the embryonic stages of Amniota. Gill slits develop by a meeting of the endodermal wall of the pharynx with the ectoderm. Then the membranes so formed are pierced, and the limit between ectoderm and endoderm is no longer visible. As ectoderm invaginates above the endodermal pouches, the gill arches are predominantly covered by ectoderm. Thus, the gill lamellar epithelium develops from the ectoderm (Goette, 1901; Moroff, 1904). The gill slits develop backward, and their number is larger in the lower than in the higher fish. The first gill slit is always modified, is usually closed in Craniata, but remains open in some Pisces where it is called a spiracle. This slit, situated between the mandibular and the hyoid arches, is the site of the pseudobranch (see Chapter 9, Volume XB, this series). The number of slits in modern fish is never more than eight, and in the majority of fish (Pisces) there are five pairs of branchial slits (Fig. 1a). In cyclostomes, the number is variable: seven in *Petromyzontia*, six in *Myxine*, and more in *Bdellostoma* (Fig. 1b). Traces of vestigial posterior slits are visible in some elasmobranchs, and it is interesting to mention that the ultimobranchial body is considered to have been derived from a vestigial pouch behind the last branchial slit (Romer and Parsons, 1977). In Amphibia five pouches develop behind the spiracle. In amphioxus the number of gill slits approaches 180 pairs; this situation refers more to an adaptation to a mode of feeding than to a primitive organization that probably never has consisted of a very large number of slits. Another interesting point is that correspondence of gill clefts with the intersegments is afforded by the nerves disposed intersegmentally in amphioxus and cyclostomes.

The branchial slits are separated from each other by the skeletal arches, which contain branchial bars and associated muscles in their inner part and the septum in their outer part. The gill arch skeleton and the branchial muscles are studied in detail in Chapter 1 of this volume. The skeleton lies in the inner region of the gill arch and consists of four connected and articulated segments: pharyngo-, epi-, cerato-, and hypobranchial (Harder, 1964).

---

**Fig. 1.** Schematics of the vascular arrangement of the branchial arches in different groups of fish. (a) *Perca fluviatilis* (teleost), (b) *Lampetra* (cyclostomes), (c) *Amia* (Holostei), (d) *Acipenser* (Chondrostei), (e) *Scyliorhinus* (elasmobranch), (f) *Neoceratodus* (Dipnoi), (g) *Lepidosiren* (Dipnoi), (h) *Protopterus* (Dipnoi). Abbreviations: af.PA, afferent pseudobranchial artery; ef.PA, efferent pseudobranchial artery; MA, mandibular artery; IC, internal carotid; HA, hyoidean artery; DA, dorsal aorta; VA, ventral aorta; PA, pulmonary artery; HBA, hypobranchial artery. The branchial slits are numbered in arabic numerals: **1**, spiracle; **2**, hyoidean; **3–7**, aortic arches. Pseudobranchs are present in (a), (c), (d), and (e). Hyoidean hemibranchs are present in (d), (e), (g), and (h).

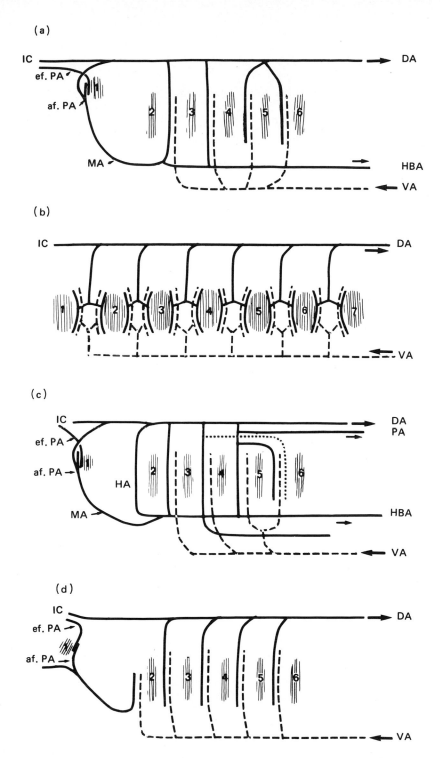

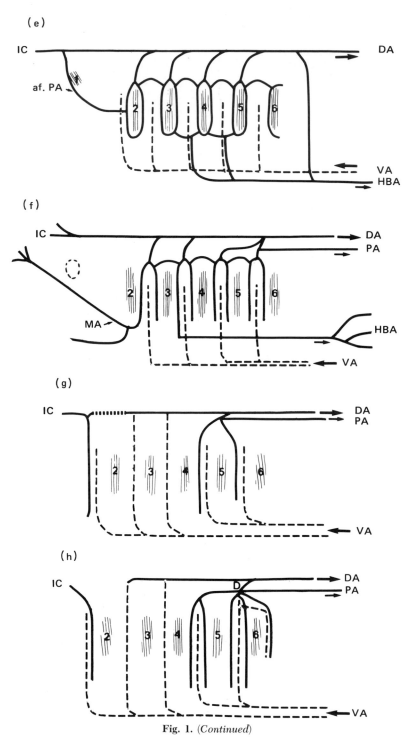

**Fig. 1.** (*Continued*)

Each gill arch skeleton is jointed with the posterior skull dorsally and with the copula ventrally. The septum contains nerves and blood vessels, and bears the filaments. Two rows of filaments are generally inserted on each gill arch. This whole forms the so-called holobranchs. The mode of insertion of the filaments on the gill arch depends on the morphology of the septum and varies with the group of fish. In some groups (Chondrostei and the holostean *Lepisosteus*), certain arches, the hyoidean and the mandibular only, bear a single row of filaments; they are called hemibranchs. In other groups or species, those arches are completely devoid of filaments or absent, as in *Amia* (Holostei), which has a mandibular but no hyoidean arch. The presence of lamellae on gill arches corresponds of course with the distribution of the true aortic arches. Aortic arches have been studied throughout the groups of fish with respect to the presence or the absence of mandibular and hyoidean arches (Kryzanovsky, 1934).

## 2. The Gill Arches in Teleosts

Teleostlike perch (Perciform) or trout (Clupeiform) have four arches bearing holobranchs and one bearing hemibranchs (Fig. 1a). Although this hemibranch is located close to the first branchial arch, it is actually vascularized by vessels derived from the ventral, the dorsal, or both sides of the mandibular aortic arch. This hemibranch, called the pseudobranch, has a structure and function different from holobranchs and will be considered in Chapter 9, Volume XB, this series. The pseudobranch is present in most teleosts. The afferent arteries, which emanate from the ventral aorta, and the efferent branchial arteries, which channel the blood from the branchial vasculature into the dorsal aorta, enter the gill arch by opposite sides. Indeed, as pointed out by Morgan (1974a), the branchial arteries at their earlier stages of development are continuous vessels connecting the ventral to the dorsal aorta. At its first stage of differentiation, the afferent branchial arteries develop from a branching of this primitive artery. This branch runs in parallel to the primitive one and forms the efferent arteries by constricting close to the branching point. Then afferent and efferent arteries become seperated. In some species, the afferent branchial artery divides into two anterior and posterior branches, giving rise to the tributaries supplying the filaments (Muir, 1970). In some others, the afferent artery runs straight through to its dorsal tip. In a second stage of differentiation, parallel vascular loops will connect the branchial arteries, giving rise to the filamental vasculature.

Thus, in a cross-sectioned arch (Fig. 2a,b) we find from the inside (gill arch properly) to the outside (gill septum) the gill bar and its associated gill rods (Bijtel, 1949). Muscles of two types are attached to the gill arch skeleton: striated muscles called abductor muscles (Bijtel, 1949) and smooth

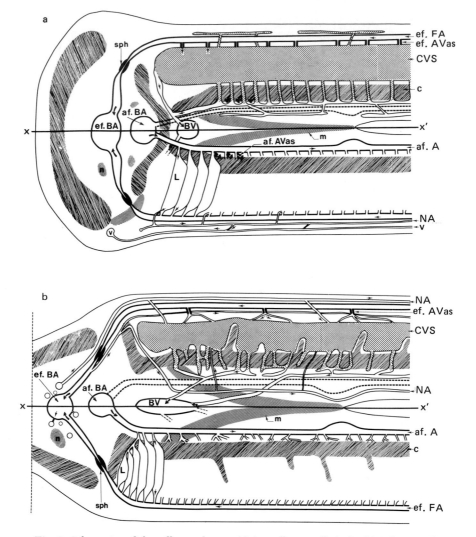

**Fig. 2.** Schematics of the gill vasculature. (a) *Anguilla anguilla* (eel), (b) *Salmo gairdneri* (trout), (c) *Scyliorhinus canicula* (dogfish), (d) *Acipenser baeri* (sturgeon), (e) *Amia calva* (bowfin), (f) *Lepisosteus osseus* (garpike), (g) *Protopterus aethiopicus* (lungfish), (h) *Raja erinacea* (skate). Abbreviations: af.FA, Afferent artery; af.BA, afferent branchial artery; af.AVas, afferent arteriovenous anastomosis; ef.AVas, efferent arteriovenous anastomosis; BV, branchial vein; c, cartilage; ci, contractile cisterna; cc, corpus cavernosum; cv, juxtacisternal vein; CVS, central venous sinus; ef.FA, efferent filament artery; ef.BA, efferent branchial artery; es, extra-cellular spaces; m, abductor muscle; n, nerve; NA, nutrient artery; F, filament; sh, arterio-arterial shunt; L, lamella; sph1, efferent artery sphincter; sph2, pre- and postlamellar sphincter; V, vein; VS, venous sinus. (Adapted from Dunel and Laurent, 1980.)

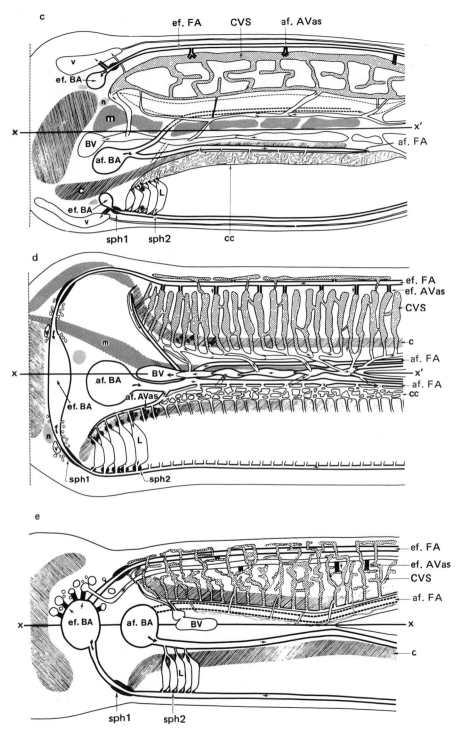

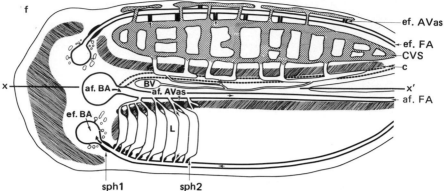

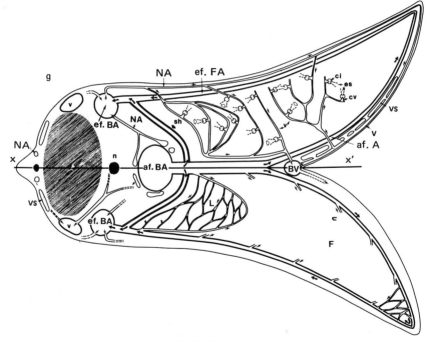

Fig. 2. (*Continued*)

muscles (Fig. 5a). The latter have been erroneously called ligaments (Dornesco and Miscalenco, 1967) but actually are formed by smooth muscle
fibers innervated by sympathetic endings (S. Dunel and P. Laurent, unpublished results). Branchial vessels include (*a*) one or two efferent arteries
(due to the splitting at a certain level of the gill arch). The presence of paired
efferent arteries is general and the point at which they separate varies with

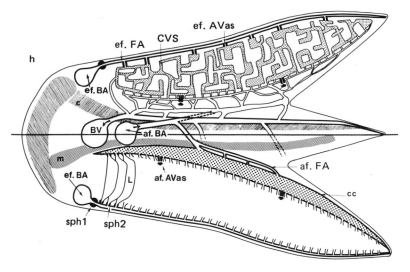

**Fig. 2.** (*Continued*)

the species (Muir, 1970). (*b*) An afferent artery, large walled, highly muscular, and in some species an afferent trunk. (*c*) One or two large veins, sometimes erroneously called *vaisseaux nourriciers centraux* (Dornesco and Miscalenco, 1967), though collecting the nutrient system (Laurent and Dunel, 1976). Nerves are located in the vicinity of the vessels. They are branches of the cranial nerves and are often called branchial nerves because of their relationship with the gill slits (Sewertzoff, 1911; Allis, 1920; Norris, 1925; Goodrich, 1930; Nilsson, 1983; Bailly, 1983). Gill arches are innervated by the metaotic group of nerves IX and X. Branches of these nerves enter the gill arch dorsally. The cranial nerve IX (or glossopharyngeal nerve) sends a protrematic branch to the mandibular arch, which bears the pseudobranch. The pseudobranch also receives fibers from the facial nerve. In addition, cranial nerve IX sends a posttrematic branch to the first branchial arch. Only part of this branch enters the arch itself by its motor fibers innervating the gill muscles. Cranial nerve X (the vagal nerve) sends a pretrematic branch to the first branchial arch, this branch mainly containing snsory fibers. The second to fourth arches are also supplied by the tenth cranial nerve, each arch receiving one or two posttrematic branches (visceral motor) and one or two pretrematic branches (visceral sensory). These nerves run along the arch and from there distribute the fibers within the filaments. For instance, in the Atlantic sea raven *Hemitripterus americanus*, three branches enter each gill arch. A small branch terminates in the musculature and is motoric (stimulation of the nerve causes filamental adduction); a second branch contains motor and sensory proprioceptive fibers. The third branch contains sensory fibers innervating the gill taste buds (Sutterlin and Saunders, 1969). The gill arch is covered by a multilayered epithelium rich in mucous cells in addition to pavement cells and taste buds (De Kock, 1963).

The mode of insertion of filaments depends on the configuration of the interbranchial septum. In teleosts the septum is reduced so that the two hemibranchs are fused only over a short distance. This disposition varies in lower groups.

The circulatory system of filaments develops as vascular loops joining the afferent with the efferent branchial arteries. The circulation becomes patent in the filament vascular loop as soon as the primitive aortic artery interrupts, leading to the separation of afferent and efferent vessels (Morgan, 1974a).

## 3. THE GILL ARCHES IN LOWER GROUPS

The pattern of branchial arch organization is relatively constant among teleosts but differs significantly in lower groups.

In Myxinoidei, the variable number of gill slits correspond to a series of pouches connected by narrower tubes with gut and body surface (see Grodzinski, 1926, 1932, for *Myxine glutinosa*; Jackson, 1901, for *Bdellostoma*; Marinelli and Strenger, 1956, for the microscopic anatomy of the gill).

In the lamprey (*Lampetra japonica*; Nakao and Uchinomiya, 1978), seven pouches open to the water tube at the inside and to the exterior via internal and external branchiopores, respectively (Fig. 1b). They delimit an archlike system contained with the esophagus within the branchial basket. The result is that the branchial bars are located in the inner side of the gill arch (Goodrich, 1930, p. 397). Afferent branchial arteries take their origin from each side of the ventral aorta, giving off ventral and dorsal branches in the medial portion of each septum; at their turn, these branches send off anterior and posterior filament arteries that, respectively, supply the anterior and the posterior hemibranchs lining each gill pouch cavity (Nakao and Uchinomiya, 1978). Within the septum one also finds extensive peribranchial venous sinuses collecting the lamellar vein and drained off by the vena cardinalis anterior and the vena jugularis impar.

In elasmobranchs, for example in *Scyliorhinus canicula*, the distribution of aortic arches is as shown in Fig. 1e according to Goodrich. There are five pairs of gill arches; the anterior arch, corresponding to the hyoid arch, bears a single posterior hemibranch. The organization of efferent branchial arteries is complicated by the presence of interarch connections. The interbranchial septum is well developed in elasmobranchs and bears gill filaments on its lateral borders. An afferent branchial artery supplies each gill arch, and paired efferent branchial arteries convey the blood to the dorsal aorta. In addition to these vessels, large venous sinuses and veins collect the venous blood from the filaments (Fig. 2c). Ceratobranchial cartilages are present in the inner side of the gill arch (Wright, 1973); they are part of the branchial arch skeleton, which is built on almost the same plan in the whole group of elasmobranchs (Goodrich, 1909).

In Chondrostei the distribution of aortic arches (Fig. 1d) shows that this group retains a hyoidean arch (hyoidean hemibranch) as well as a mandibular arch (pseudobranch) (Burggren et al., 1979). There are four branchial arches supporting holobranchs. In these holobranchs the interbranchial septum corresponds with the part of the filaments fused on half of their length (Fig. 2d). In this region, there is a dense system of blood sinuses and valved veins in relation with the venous system of the filaments and drained off in a large branchial vein (Dunel and Laurent, 1980).

Holostei are mainly represented by *Amia* and *Lepisosteus*. In *Amia* the distribution of the aortic arches is influenced by the bimodal pattern of respiration. There are four holobranchs (arches IV to VI) and a pseudobranch (arch I). Interesting is the distribution of efferent branchial arteries of arches V and VI (Wilder, 1877; Olson, 1981; Randall et al., 1981; Daxboeck et al., 1981) (Fig. 3a). The efferent branchial artery of arch V (ef.BA3) reaches the dorsal aorta or the celiacomesenteric artery. Arch VI efferent artery (ef.BA4) connects ef.BA3 via a short segment, the so-called commissural vessel (Olson, 1981). The air bladder artery stems from the commissural vessel. This pattern of circulation suggests that the blood supply to the gas bladder is independent, to some extent, of the systemic circulation. In *Lepisosteus*, the distribution of aortic arches includes a hyoidean hemibranch (arch II) in addition to a mandibular pseudobranch (arch I) (see Chapter 9, Volume XB, this series). This disposition is very close to *Acipenser*. In contrast to *Amia*, the air bladder receives its afferent blood supply from small bilateral tributaries of the dorsal aorta (Purser, 1926), (Fig. 3b).

Dipnoan fish display the most modified aortic arch distribution, which fits a bimodal respiration as in Holostei but presents a separation of the ventral aorta into two trunks, respectively fed by incompletely separated right and left parts of the heart. This organization occurs in the three species of dipnoan lungfish: *Protopterus*, *Lepidosiren*, and *Neoceratodus*. In the latter case (Fig. 1f), there are four holobranchs but no hyoidean hemibranchs in contrast with the two former species. In *Protopterus* (Fig. 1h) and *Lepidosiren* (Fig. 1g) the two anterior aortic arches are devoid of lamellae and the two posterior ones are equipped with very modified gills. In these species also the two posterior arches feed the dorsal aorta through a highly specialized vascular segment, the ductus arteriosus, which operates synchronously with a pulmonary artery vasomotor segment and drives blood either into the dorsal aorta or into the lung circulation (Laurent, 1981; Laurent et al., 1984a).

The gill arches of the South American and African lungfish, obligate air

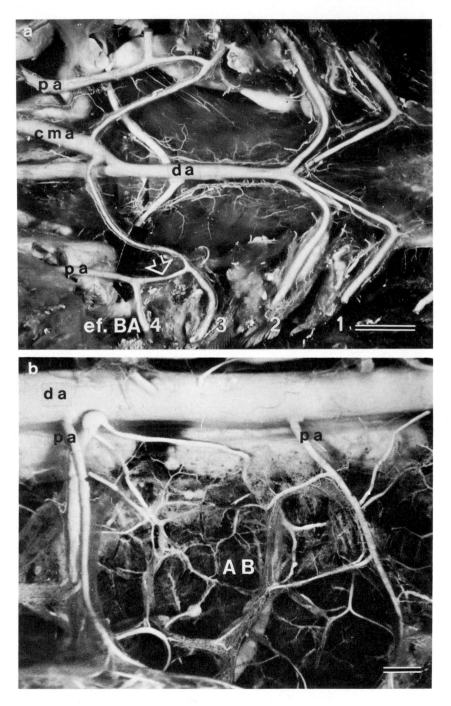

breathers, deserve particular mention. They are totally different from those of the Australian lungfish *Neoceratodus*, a facultative air breather. In the former species, gill arches give rise to a series of arborescent filaments (Fig. 4a), and in spite of their rather small number, they still constitute a holo-branch including, within the arch itself, a skeleton and afferent and efferent branchial arteries. There is still a rudiment of septum that includes, in addition to nutrient vessels and nerves, a large vein located at its posterior margin (Fig. 2g).

## B. The Filament

### 1. PRINCIPLE OF ORGANIZATION

The filaments are generally disposed on two rows except in hemibranchs. Lamellae are regularly spaced on these filaments. The filament is considered to form the gill unit, because it is provided with all functionally significant structures including the major sites of the vascular control, a proper motor system, and different types of epithelia. The role of the filament is of course to supply the lamellae, which in turn can be considered the respiratory unit, since in terms of morphometry the overall diffusive capacity of the gill is directly proportional to the number of these units (or their total surface area).

Each filament is supported by a barlike piece of skeleton, the gill rod, connected with the branchial bar. The gill rod increases the stiffness of the filament and participates in an appropriate relative positioning of each row of filaments on the same gill arch. The structure of the associated muscles and their typology, according to Duvernoy (1839) and Riess (1881), have been described in detail by Bijtel (1949). In a first type (e.g., *Esox, Umbra, Ameirus, Gasterosteus, Perca*), adductor muscles are cross-fixed on the rods of the two successive opposite filaments. In a second type (e.g., *Tinca, Leuciscus, Cyprinus, Salmo*) these muscles are inserted onto a tendon running longitudinally within the septum. The gill rod is made of a chondroid

---

**Fig. 3.** The postbranchial arterial circulation of holostean fish. (a) *Amia calva*. Microfil cast preparation of the efferent branchial arteries (numbered from 1 to 4). Note that efferent branchial arteries of arch V and VI, respectively, ef.BA3 and efBA4, are connected to each other by a short vessel (open arrow) that gives rise to air bladder artery (pa). Note also that ef.BA3 and ef.BA4 connect the celiacomesenteric artery (cma) and not the dorsal aorta (da). Bar = 10 mm. (b) *Lepisosteus osseus*. Microfil cast preparation of the dorsal aorta (da). A series of tributaries bilaterally disposed supply the air bladder (AB). Although it is a holostean fish like *Amia*, *Lepisosteus* has no specific air bladder artery (pa). The blood is collected from the air bladder by bilateral series of veins that connect the posterior cardinal veins. Bar = 1 mm. (Courtesy S. Dunel-Erb.)

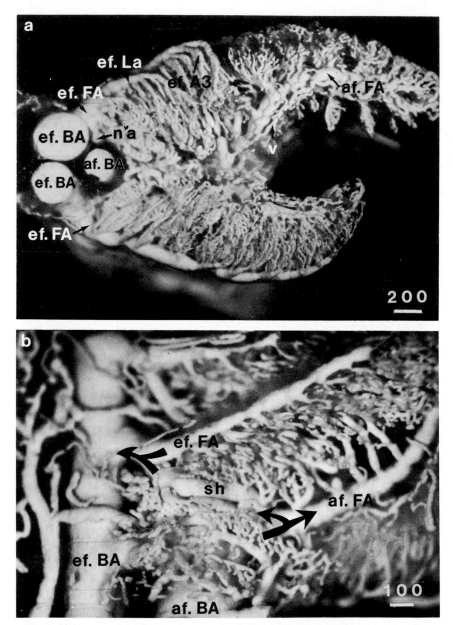

**Fig. 4.** (a) Cross section of a vascular cast of arch IV in *Protopterus aethiopicus*. The afferent branchial artery (af.BA) lies in the central position. From this vessel, a short segment (not visible) emanates and bifurcates to form the afferent filament arteries (af.FA) of the two opposite hemibranchs. In contrast, each efferent filament artery (ef.FA) is connected to an ipsilateral efferent branchial artery (ef.BA) located on both sides of the af.BA. Each filament artery gives

tissue with large cartilage-like cells and little intercellular substance. At its surface, the gill rod becomes slightly calcified. It is apparent from the histo-logical structure that the gill rod is rigid but still flexible. The shape of the gill rod is variable according to the species, round or flattened and crested.

It is now usual to distinguish two circulatory systems within the filament. The arterio-arterial or systemic vasculature, also called respiratory, actually connects the ventral to the dorsal aorta via the respiratory pillar capillaries of the lamellae. The second system is a complex arteriovenous circulatory sys-tem that takes origin from various parts of the former system, mainly from the efferent branchial and efferent filament arteries. It supplies the gill tissues and assumes a nutritional function. It also serves as a vascular com-partment for the osmoregulatory epithelium of the filaments.

## 2. THE ARTERIO-ARTERIAL VASCULATURE IN TELEOSTS

In a teleost holobranch, afferent filament arteries have their departure regularly spaced on the unique afferent branchial artery and alternatively enter the anterior and posterior rows of filaments. The filament artery gives off two branches: one straight forward to the tip of the filament supplying its medial and its distal part, the other one recurring and supplying its proximal part.

In the trout and other teleosts, a vascular enlargement forms a bleb on the afferent filament artery (Fig. 2b). Blebs are located in the most distal part of the so-called septal area: the part of the holobranch where the two rows of filaments start to diverge (Fromm, 1974). This enlargement has been in-terpreted as a distensible structure dampening the pulsative blood flow (Fromm, 1974). An alternative interpretation is that these blebs are the relict of the cavernous bodies, present within the gill of elasmobranchs. Indeed blebs by their lateral fusion make a communication between the filament arteries of alternate hemibranchs in the catfish (Boland and Olson, 1979), or of hemibranchs of the same row. This arrangement, called *canaux septaux longitudinaux* in the carp (Dornesco and Miscalenco, 1963) and other cypriniforms (Dornesco and Miscalenco, 1968c), is absent in clupe-iform (e.g., *Salmo, Alosa, Esox*) and in perciform (e.g., *Perca, Trachinus, Blennius, Gobius, Costus*). In the latter species blebs are reduced to simple enlargements if they are present (Dornesco and Miscalenco, 1967), and they

---

off lamellar arterioles (af. La or ef. La), and eventually each lamellar arteriole gives off tertiary arterioles (af. A3 or ef. A3). Note afferent nutritive arteries (na) emanating from the efferent filament artery. Veins (v) are connected to a branchial vein (not shown). Bar = 200 μm. (b) Lateral view of an arch IV filament in an estivating *Protopterus*. Note that the afferent–efferent shunt (sh) between af. FA and ef. FA is well dilated. Arrows indicate the direction of blood flow. Bar = 100 μm. (From Laurent *et al.*, 1978.)

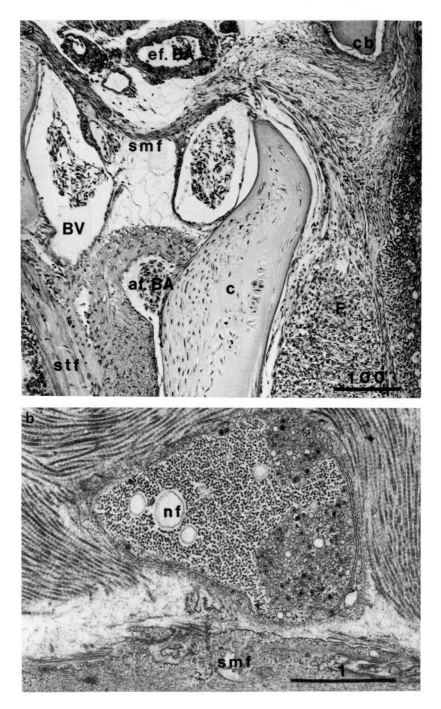

are absent in the toadfish *Torquiginer glaber* (Cooke and Campbell, 1980). The functional meaning of the arrangements seen in the carp or the catfish is not clear, but it can be suggested that a better blood distribution as well as mechanical effect might result.

Another point of interest regarding the afferent artery involves its relationships with the skeleton of the filament in some species. In the perch (Fig. 5a), the branchial artery is asymmetrical. The part of the arterial wall in contact with the filament rod has almost completely disappeared (Laurent and Dunel, 1976). This arrangement concerns the opposite part of the wall when the branchial artery comes into contact with the gill rod of the next hemibranch. Such an arrangement is also present in other species of perciforms (Dornesco and Miscalenco, 1967) as well as in the pike (Dornesco and Miscalenco, 1968b). In these species, the asymmetry of the arterial wall is still amplified by the presence of a longitudinal band of smooth muscle fibers closely associated both with the gill rod and the artery media. This muscular structure, previously considered purely elastic (Bijtel, 1943), is richly innervated by aminergi nerve fibers (Bailly, 1983).

From the afferent filament artery emanate short arterial segments, the lamellar afferent arteries, supplying the lamellae (Fig. 2b). These segments communicate with a variable number of lamellae generally not exceeding three; their shape and length depend on the course they have to follow before reaching the lamellae. The wall of these afferent lamellar arteries does not show any specific alteration suggesting a particular vasomotor function.

The shape and size of the lamellae vary considerably within the group of teleosts (Hughes and Wright, 1970; Hughes and Morgan, 1973). They are generally curved and consist mainly of a middle vascular layer constructed of pillar cells covered by a thin bilayered epithelium, the so-called respiratory epithelium (Fig. 6).

The teleost type of lamella is characterized by a polygonal distribution of pillar cells. This type is probably the most efficient system for exchanging gas

---

**Fig. 5.** (a) Branchial arch cross section and base of a filament (F) of *Perca fluviatilis*. Note the structure of the afferent branchial artery (af.BA), which is "stuck" on the cartilage rod (c). Part of the arterial wall is lacking. Compare the thickness of the afferent and the efferent branchial arteries (ef.BA). Several small arterioles take origin from the efferent artery. Two large veins (BV) are also seen on both sides of the gill arch. Smooth muscle bundles (smf) are fixed on the extremity of the gill rod. Their tonic contraction contributes to the positioning of hemibranchs. Striated muscle fibers (stf) belong to the adductor muscles. cb, cartilage bar. Bar = 100 μm. (From Laurent and Dunel, 1976.) (b) A motor nerve ending (nf) innervating the smooth muscle bundles (smf) shown in (a). Note the presence of small, clear, and also granular vesicles. This innervation is presumably adrenergic and displays formaldehyde-induced fluorescence. Bar = 1 μm.

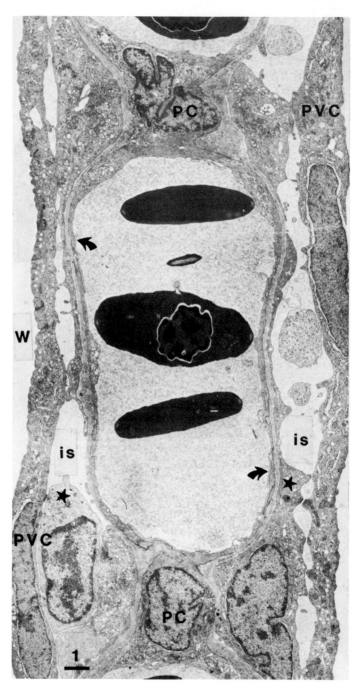

(Smith and Johnson, 1977), although its resistance to flow is higher than what could be expected from a linear distribution. In the lingcod the arrangement of pillar cells is regular, but that pattern varies with the region of the lamellae. The sheet flow theory applied to the lamellae perfusion suggests that some autoregulation occurs (Farrell *et al.*, 1980).

The lamellae have marginal outer vessels that embryologically differentiate from afferent and efferent filament arteries (Morgan, 1971). The marginal vessels are blood channels of a larger diameter than pillar capillaries and are currently considered as short-circuiting pathways for shunting the blood through the pillar capillaries. An alternative interpretation considers the marginal vessels as a feeder system whereby the distribution of blood flow across the lamellae is achieved (Smith and Johnson, 1977). Some authors describe inner basal vessels that have been also considered as a shunting pathway for the blood (Smith and Johnson, 1977). The basal vessels are often deeply buried within the epithelium of the filament and therefore not favorable to gas exchanges (Farrell *et al.*, 1980). However, in many cases there is clear evidence that basal vessels progressively taper off and that their diameters become negligible at the efferent tip of the lamellae (Dunel and Laurent, 1980).

The efferent lamellar artery is shorter than its afferent corresponding vessel, and does not display any significant anatomic characteristics in teleosts. This absence of reinforcement of the muscular media contrasts with that observed in lower groups (Dunel and Laurent, 1980) (see section II, B, 4). However, it has been often speculated that a particular vasomotivity occurs at this level in teleosts and controls the lamellar flow as does a sphincter (Holbert *et al.*, 1979). All efferent lamellar vessels connect the efferent filament artery in a way identical with the afferent side (Fig. 7).

The efferent filament artery is characterized mainly by the presence of a powerful sphincter located near its junction with the corresponding branchial artery (Laurent and Dunel, 1976; Dunel and Laurent, 1977) (Fig. 8). This sphincter is composed of a reinforcement of the muscle layer that is, as in the trout, twice as thick as the efferent artery (15 vs 6 $\mu$m). This sphincter is richly innervated by nerve endings filled with small, clear cholinergic-type vesicles (Dunel and Laurent, 1980) (Fig. 9a,b). Accordingly, the acetylcholinesterase histochemical reaction giving positive results suggests a cho-

---

**Fig. 6.** Cross section of a lamella of an eel. The lamellar epithelium consists of two layers of cells separated by intracellular spaces (is). Pavement cells (PVC) form the external layer. The internal layer consists of a thin cytoplasmic velum of poorly differentiated cells having their body (★) on both sides of the pillar cells (PC). The lamellar epithelium lies on a thick basal lamina. Thin pillar cell flanges line the blood compartment (arrows). W, external medium. Bar = 1 $\mu$m.

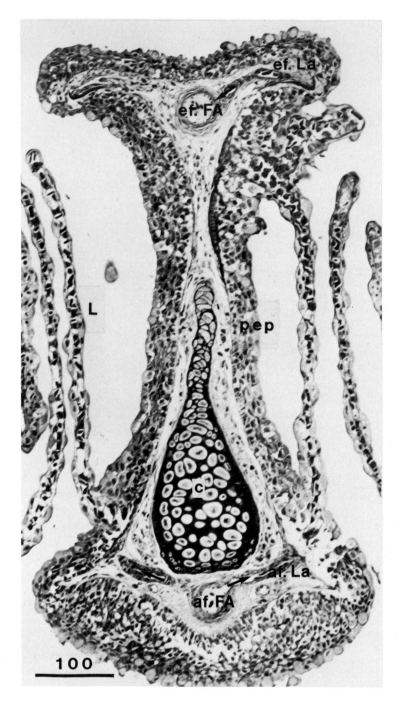

linergic innervation (Y. Bailly, personal communication). These anatomic evidences are supported by neuropharmacological experiments (Pettersson and Nilsson, 1979; Nilsson, 1983; Nilsson and Pettersson, 1981).

Pillar cells are specific structures common to gills of any fish and do not have an equivalent in any other vertebrate, whereas similar structures have been described in invertebrates (Dunel-Erb *et al.*, 1982b). Pillar cells were first described by Bietrix (1895). A pillar cell is made up of two parts. The nucleated body or perikaryon forms the pillar, properly supporting the two opposite epithelial layers as does a brace. Lateral processes forming flanges line the blood channel and meet the flanges of adjacent pillar cells (Fig. 6). Pillar cells are covered by a basement membrane underlying the lamellar epithelium. The basement membrane has been studied in detail (Newstead, 1967). It is rich in mucopolysaccharides (Bird and Eble, 1979). It is composed of the usual lucent zone and an electron-dense layer of packed filaments. This region is often infiltrated by fibrils of collagen. Collagen bundles anchored on the basement membrane traverse the lamella oriented normal to the lamellar surface and are deeply infolded by the pillar cells in extracellular channels (Bietrix, 1895; Hughes and Grimstone, 1965; Newstead, 1967; Hughes and Weibel, 1972; Dunel, 1975). These extracellular channels are open at either end but are closed from the blood space by junctional complexes. Collagenous fibrils of columns are apparently in continuity with the interstitial collagen. Six to ten columns traverse the pillar cell and indent its nucleus. Caveolae are rare in pillar cells, and Weibel–Palade bodies are absent. These characteristics make the identification of pillar cells with endothelial cells very doubtful (Hughes and Morgan, 1973), whereas pillar and endothelial cells could both originate from mesenchymal cells (Morgan, 1971). Another very important feature of pillar cells is the presence of fibrillar material running in bundles parallel to the columns. These fibrils are intracellular and continue into the cytoplasmic flanges of pillar cells. By their dimensions (50 Å in diameter) and their arrangement in tracts, these filaments look like myofilaments of the vascular smooth muscle more than tortuous tonofilaments (70 Å in diameter), so that the hypothesis has been advanced that pillar cells constitute a contractile system regulating the blood flow within the lamellae (Hughes and Grimstone, 1965; Newstead, 1967; Rankin and Maetz, 1971; Bettex-Galland and Hughes, 1973; Hughes and Bycskowska-Smyk, 1974; Smith, 1977; Ristori and Laurent, 1977). It has been concluded from experimental incubation of gill lamellar extracts with ATP that the thin cytoplasmic

**Fig. 7.** Histological cross section of a filament (trout). Afferent and efferent filament arteries and lamellar arterioles (af.FA, ef.FA, af.La, and ef.La, respectively are seen in cross section (filament) and longitudinal sections (lamellar). Arrows indicate the direction of the blood flow. Note the relative thickness of the epithelium lining the filament (pep) and the lamellae (L). Bar = 100 μm. (From Dunel-Erb *et al.*, 1982a.)

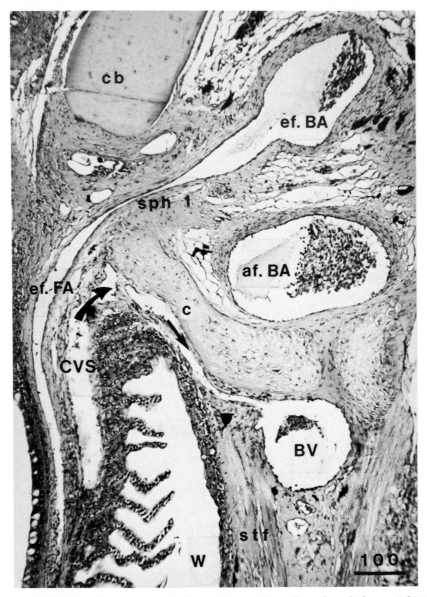

**Fig. 8.** Histological cross section of gill arch of the eel. The efferent branchial artery (ef. BA) collects the blood from the efferent primary arteries (ef. FA) via a well-muscled sphincter (sph1). Note that the central venous sinus (CVS) drains into the branchial veins (BV) via a connecting vessel (arrows). Note the position of the afferent branchial artery and its relationships with the cartilage rod (c). The arterial wall is thinner in this part. cb, cartilage bar (arch skeleton). Adductor muscles are seen in lower part of the micrograph (stf). W, external medium. Bar = 100 μm. (Adapted from Laurent, and Dunel, 1976.)

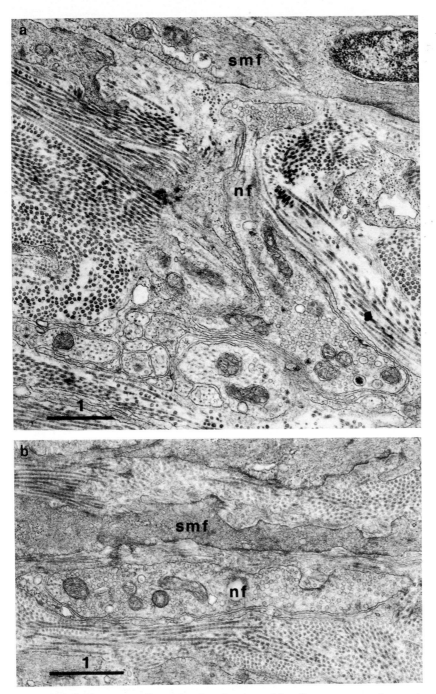

**Fig. 9.** (a) and (b) Nerve profiles within the adventitia of the efferent primary sphincter (see Fig. 8, sph1). Note vesiculated endings (nf) approaching smooth muscle fibers (smf). Bars = 1 μm.

filaments consist of an actomyosin-like contractile protein (Bettex-Galland and Hughes, 1973). Subsequently, specific immunofluorescence histochemistry has confirmed the presence of smooth muscle myosin in the pillar cells (Smith and Chamley-Campbell, 1981). However, it should be noted, in the hypothesis of a regulatory role of pillar cells in lamellar blood flow, that these cells are not innervated (Laurent and Dunel, 1980) and that their activation by cholinergic and adrenergic agonists is still under investigation. That leads to some uncertainty concerning the role of pillar cells. In the trout gills it has been shown, for instance, that changes in lamellae channel dimensions cannot explain the changes in response to adrenaline and acetylcholine in gill perfusion flow, since they vary in an opposite direction (Booth, 1979), but much better by an action on the efferent filament artery sphincters. In the same order, in the eel, microfil cast technique indicates that the lamellar vascular tonus, significant in saline-perfused preparations, is not affected by acetylcholine but greatly reduced like the whole gill vasculature by β-adrenergic agonists (Dunel and Laurent, 1977).

However, derivation of the sheet flow equations to the lamellar circulation suggests that a passive redistribution of lamellar flow occurs as a consequence of change in lamellar dimensions when pressure and flow are raised (Farrell *et al.*, 1980; Farrell, 1980b).

In conclusion, the functional meaning of pillar cells and of their "contractile" apparatus could be considered in the perspective of an autoregulative myogenic control according to which contractile myofilaments are more activated by rapid changes of blood pressure than by neurohumoral factors (Smith and Chamley-Campbell, 1981). Another possible function of pillar cells has been revealed by the demonstration of phagocytosis in the rainbow trout pillar cells and their involvement in viral infection (Chilmonczyk and Monge, 1980).

### 3. Variations of the Arterio-Arterial Vasculature among Teleosts

There are no wide variations in the arterio-arterial vasculature in teleosts.

Differences exist concerning the wall thickness of afferent filament arteries in the region of contact with the gill rods. From a large number of species examined, it appears that a lack or a reduction of the muscular media at this point has been observed in *Esox* (clupeiform), *Belone* (beloniform), *Gaidrospasarus*, and *Lota* (gadiform), *Pleuronectes* and *Solea* (pleuronectiform), and in all the species of perciforms examined (e.g., *Perca*, *Acerina*, *Trachinus*, *Blennius*, *Cottus*) (Dornesco and Miscalenco, 1963, 1967, 1968a,b; Laurent and Dunel, 1976).

In the icefishes (*Champsocephalus gunnari* and *Ch. aceratus, Pseudo-chaenichthys georgianus*), some modifications have been described that seem to be related to the peculiarity of these animals, which have no red blood cells. It is of interest to mention the uncommonly large diameter of afferent and efferent branchial arteries, poorly muscled in those species. However, the efferent filament artery has a particularly strong sphincter proximal to the efferent branchial artery (Vogel and Kock, 1981).

The polygonal distribution of pillar cells seen in most teleosts (Dunel and Laurent, 1980; Farrell *et al.*, 1980; Farrell, 1980a) suffers large alterations, some distributions being apparently at random but others having a more specific characteristic (Cooke and Campbell, 1980). For instance, in the bluefin tuna (*Thunnus thunnus*), the distribution of the pillar channels consists of two right-angled arrays (Muir, 1970). Some evidence has been provided for an alignment of pillar cells in air-breathing fish lamellae (Datta Munschi and Singh, 1968), an arrangement also seen in the holostean type (Acrivo, 1938; Dunel and Laurent, 1980) (Fig. 10b).

Another interesting peculiarity exists in the partial fusion of the lamellae by their margin. This arrangement has been observed in tunas and other fast-swimming fish that ram ventilate (Muir and Kendall, 1968).

Bimodal breathing teleosts, and particularly those that are obligate air breathers, are characterized by a reduction in the size and the perfusion system of their gills. This is particularly well demonstrated in the osteoglossid group, where two species represent respectively an obligate air breather (*Arapaima gigas*, the pirarucu) and an obligate water breather (*Osteoglossum bicirrhosum*, the aruana). The gill of aruana is similar to that of other freshwater teleosts, in contrast with that of *Arapaima*, which differs by several characteristics, here outlined in brief. First, the wall of the afferent as well as efferent filament arteries is particularly well muscled. Second, lamellar arteries give access successively to lamellae via sphincter-like openings. Finally, a large part of the lamellar vasculature is deeply buried below the surface of the filament epithelium. All these modifications cooperate to allow the blood to bypass the lamellae (Hulbert *et al.*, 1978). Another type of modification with respect to an adaptation to bimodal breathing consists of a suppression of the lamellae. A suppression or a reduction occurs in *Anabas*, afferent and efferent arteries being joined by broad spaces (Datta Munshi, 1968). In *Channa punctatus* the last two branchial vessels pass directly up to the dorsal aorta (Datta Munshi, 1968), whereas in *Channa argus* afferent and efferent arch vessels are joined by loops of small vessels (Wu and Chang, 1947; Ishimatsu *et al.*, 1979). In *Monopterus cuchia* (Datta Munshi and Singh, 1968) and *M. alba* (Liem, 1961), the fourth branchial arch is reduced to a single direct vessel linking the two aortas. In many cases of adaptation to air breathing, there is a reduction in number and size of the

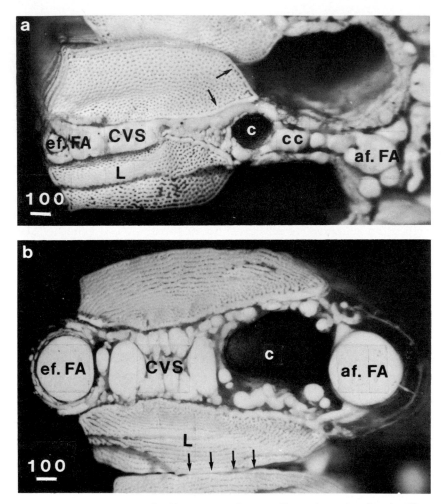

**Fig. 10.** (a) Cross section of a filament vascular microfil cast (*Acipenser baeri*). Afferent artery (af.FA) communicates with the corpus cavernosum (cc). The lamellae (L) have well-differentiated basal and marginal vessels (arrows). The polygonal disposition of pillar cells is clearly shown by the black dots representing their localization. The central venous compartment (CVS) is well injected; it surrounds the cartilaginous gill rod (c). Note that in this group (Chondrostei) the filaments are inserted by their trailing edge on the septum. (b) Cross section of a filament vascular cast (*Amia*). Compare with (a). In Holostei the septum is shorter and filaments are free. However, the lamellae are attached to their neighbors as seen on the lower part of the micrographs (arrows). Note the arrangement in parallel arrays of the pillar capillaries. The central venous sinus (CVS) extends processes that surround the efferent filament artery (ef.FA). On the afferent side these processes exist but are less injected. Bars = 100 μm.

filaments and lamellae and of the corresponding vasculature (Hughes, 1979). However, the poor development of gills in air breathing fish is not without some exceptions; for instance, the intestinal air breathing fish *Lepidocephalichthys guntea* has additional gill filaments and a greater exchange surface than other air breathing fish (Singh *et al.*, 1981).

It is interesting that the relict *Latimeria chalumnae*, a deep sea-dwelling fish and probably the closest living relative of the Australian lungfish, has its gills poorly developed and an interbranchial septum extending to the filament tip as in *Neoceratodus* and other ancient forms (Hughes, 1980).

## 4. VARIATIONS OF THE ARTERIO-ARTERIAL VASCULATURE IN LOWER GROUPS

There are some significant differences in the arterio-arterial vasculature in lower groups. They concern the presence of cavernous bodies in cyclostomes, elasmobranchs, and Chondrostei, the different forms of lamellae particularly in some lungfish, and the different patterns of vascular control. The presence of cavernous bodies interposed between the afferent filament artery and the afferent lamellar arteries is a common feature of lower groups of fish including larval and adult cyclostomes (Dröscher, 1882; Rauther, 1937; Acrivo, 1935a,b; Kempton, 1969; Wright, 1973; Dunel and Laurent, 1980; Lewis and Potter, 1982). Among other possible functions of the cavernous bodies in the gill, phagocytic activities have been suggested in relation to hemolysis of aging blood cells (Acrivo, 1935a) and defense mechanisms (Tomonaga *et al.*, 1975). However, it is likely that cavernous bodies are powerful blood pressure regulators in a manner similar to that of the conus arteriosus of the heart (Satchell and Jones, 1967).

The organization of the gill vasculature has been studied in *Myxine* with the help of corrosion casts (Pohla *et al.*, 1977). The afferent branchial artery of each pouch gives off a series of interconnected tributaries, the afferent radial arteries. On a strictly morphological basis, this complex system represents the cavernous body described in cyclostomes, elasmobranchs, Chondrostei, and Holostei. Radial arteries feed a system of secondary and tertiary gill folds, which fill the whole spherical branchial pouch. These folds or plica branchialis represent the site of respiratory exchange. An identical system of efferent radial arteries gathers the blood and sends it into the efferent branchial arteries (Fig. 11).

In the lamprey (*Lampetra japonica*), cavernous bodies (CB; or corpus cavernosum, cc) are located along the outer border of the axial plates within each filament. They are triangular structures in cross section, extending as far as the tips of the filaments. Internally, they consist of trabeculae of 3 to 6 μm in diameter delimiting blood cavities. The cavernous tissue freely com-

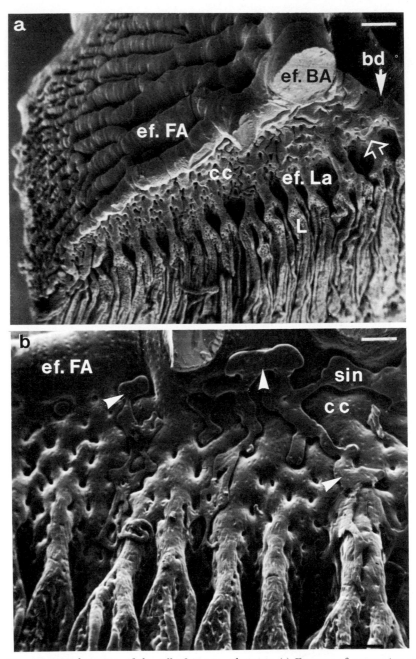

**Fig. 11.** Vascularization of the gill of *Myxine glutinosa*. (a) Fracture of a corrosion cast (circulatory efferent side). The wall of the branchial pouch lies on the left, the branchial water

municates with the afferent filament artery and connects the lamellar arteries feeding the lamellae. The whole is covered by several layers of smooth muscle and connective tissue. Trabeculae serve as pillars traversing the CB through and through. They consist of microfibrils and collagen fibrils. They also might contain smooth muscle fibers in continuity with the fibers surrounding the CB in its entirety (Sheldon *et al.*, 1962). A peculiar type of cell, the so-called cavernous body cells (Nakao, 1978), line the wall of the CB as well as the trabeculae. They are associated with collagen columns of 1 μm in diameter extending from one side to the other side of the cavernous body. These columns pass through extracellular channels made by infoldings of the CB cells. These cells have structural characteristics suggesting active functions: well-developed Golgi apparatus, mitochondria, and agranular endoplasmic reticulum. In addition, they display coated caveolae, vesicles, vacuoles, and cytoplasmic granules (Nakao, 1978). It has been noted that some characteristics of CB cells, including their close association with the collagenous columns, suggest that they could have a common origin with the pillar cells lining the blood axial plate lacunae and the lamellae (Nakao and Uchinomiya, 1978). However, other characteristics, including coated vacuolae, vesicles, and vacuoles, suggest an absorptive function, presumably of protein (Nakao, 1978).

In elasmobranchs (Kempton, 1969; Wright, 1973; Cooke, 1980; Olson and Kent, 1980), one or several filament arteries are supplied by a tributary of the afferent branchial arch, on both sides of the interbranchial septum (Fig. 12). These tributaries, which anatomically represent afferent filament arteries (for terminology, see Olson and Kent, 1980), give off a recurrent branch supplying the proximal third and a direct branch running all along the rest of the lamella almost to the tip. The cavernous body is connected to these recurrent and direct branches at various points (Wright, 1973; Dunel and Laurent, 1980) (Figs. 2c and 13b). Smooth muscle cells form a sphincter-

---

duct (bd) lies on the right. This fractured structure actually forms a sphere whose approximate axis is the branchial duct. A fairly similar arrangement forms the vascular afferent pole. Blood flow (open arrow) and water flow (white arrow) are arranged in a countercurrent system. The efferent branchial artery (ef.BA) forms a ring at the upper pole. This artery gives off efferent radial arteries (ef.FA), which are equivalent to the efferent filament arteries in the higher groups of fish. Efferent filament arteries give rise to anastomotic trabeculae, which are equivalent to the cavernous tissue in higher fish (corpus cavernosum, cc). Finally, from cc emanate efferent lamellar arterioles (ef.La) that feed lamellar capillaries of several dichotomic orders (L). The afferent pole has a fairly similar symmetrical organization. Bar = 0.35 mm. (b) Higher magnification showing the connection of lamellae with the radial arteries through the cavernous tissue (cc). Arrows indicate a complex system of sinuses (sin) fed by the radial arteries. This system probably represents the equivalent of the venous sinuses of higher groups. Bar = 0.14 mm. (From Pohla *et al.*, 1977.)

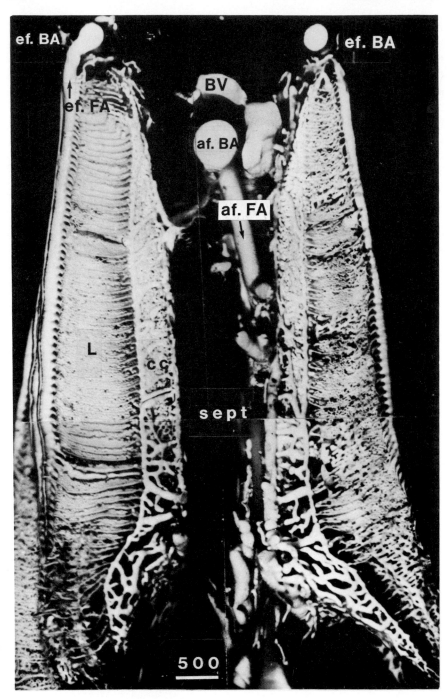

like arrangement around each communication. It is of interest that no sphincter has been described in the corresponding regions of cyclostomes. From the description given in the lamprey (Nakao, 1978) and those given in elasmobranchs (Cooke, 1980), it appears that cavernous bodies display a much more complicated arrangement in the latter animals.

In *Scyliorhinus* the cavernous body develops within each filament. It parallels the afferent filament artery and forms a T structure, the filament artery being at the intercept of the horizontal and the vertical bars. At the base of the filament, lateral cavernous bodies (horizontal bar of the T) are little developed, if present. Near the tip of the filament, they develop and come into contact with the cavernous bodies of the neighboring filaments. The artery is no longer visible, losing its identity within the cavernous body (Wright, 1973; Cooke, 1980). Actually, the so-called septal cavernous body (Cooke, 1980) develops as bilateral aisles from the afferent filament artery within the septum, and the so-called filament corpus cavernosum develops in an orthogonal plan from the same artery and corresponds with the part of the corpus cavernosum feeding the lamellae (Fig. 14b). In a certain way, this disposition accounts for the given description of three corpora cavernosa (Cooke, 1980), but more simply they could be considered as forming a single structure.

A similar description has been given in *Squalus acanthias* and *Raja erinacea* (Olson and Kent, 1980); the so-called lateral afferent sinuses and the medial afferent sinuses correspond with the preceding description in spite of the unusual terminology used by these authors and leading to a confusion with the venous sinuses. Briefly, it might be said that the lateral extension of the cavernous tissue assumes the connection with the corresponding region of the neighboring filaments, and that the medial extension assumes the communication with the lamellae. Thus, it is of interest to compare the cavernous body of elasmobranchs with the structure described by Dornesco and Miscalenco (1963) in the carp and already mentioned (Section II,B,2). Structurally, the corpus cavernosum forms a large cavity spanned by columns forming a loose network of intermingled channels of various sizes (Fig. 15). The shape and the arrangement of the columns differ among the species (compare Figs. 14 and 15). The wall of the corpus cavernosum has a structure of artery associating smooth muscle fibers and connec-

---

**Fig. 12.** Cross section of vascularly injected gill arch in an elasmobranch (*Raja clavata*). Afferent (af.BA) and efferent (ef.BA) arteries, and branchial vein (BV) are located within the gill arch. The septum (sept) fills the gap between the hemibranchs. An afferent filament artery (af.FA) emanating from the branchial artery runs within the septum, supplying the corpora cavernosum (cc) and from there the lamellae (L). The blood is collected into two efferent filament arteries (ef.FA) via sphincters (largely open on the left, closed on the right). Bar = 500 μm.

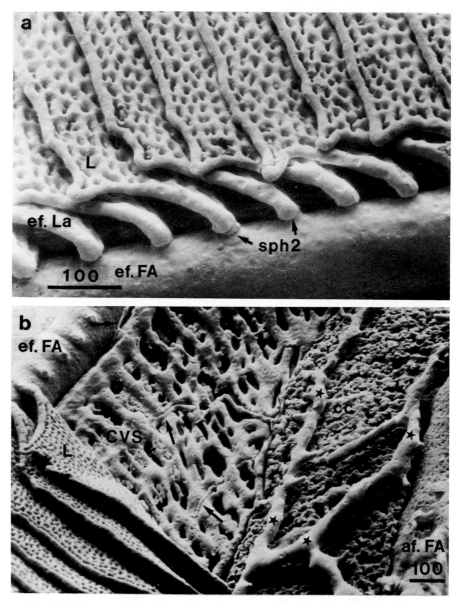

**Fig. 13.** (a) Corrosion cast replica of the efferent part of lamellae (L) and of their connection with the efferent filament artery (ef. FA) in *Scyliorhinus canicula*. Arrows indicate the location of postlamellar sphincters (sph2) (these sphincters are morphologically well characterized in lower groups of fish). Note the arrangement of the lamellar capillaries and the existence of a well-delineated marginal vessel. (b) Corrosion cast replica of a filament, *Scyliorhinus canicula*. Lamellae, still visible on the bottom left, are removed, revealing the central venous sinus (CVS)

tive tissue. Columns consist of smooth muscle fibers having some continuity with those of the wall. Columns also contain connective tissue. The wall and the columns are lined by closely packed large endothelial cells (Kempton, 1969). These cells contain densely stained granules and vacuoles (Wright, 1973), which have been considered to have phagocytic activity. Sphincters have been observed to exist on the afferent lamellar artery (Wright, 1973). Comparative ultrastructural study reveals that they are noninnervated and consists of a local muscular reinforcement of the lamellar afferent artery with a circular arrangement of smooth muscle that contrasts with the spirally oriented fibers in the lamellar artery itself (Dunel and Laurent, 1980).

In elasmobranchs the external shape of the lamella roughly represents a trapezium lying on the septum at a small angle of about 60°. Such a disposition allows an appropriate channeling of the respiratory water, which is finally collected in a water canal alongside the septum (Kempton, 1969). Lamellae often display outfoldings of the outer marginal channel on the efferent side (Olson and Kent, 1980), which leads one to think that the outer marginal channel is probably not a means of fast blood flow but is perfectly designed for distributing and collecting the lamellar blood flow. However, the pillar cells near the outer marginal vessels are more or less organized in one or several parallel rows that probably favor a faster bloodstream (Fig. 13a). The efferent lamellar artery on cast preparations (Fig. 13a) displays characteristic constrictions or blebs (Dunel and Laurent, 1980), and ultrastructural observations confirm the presence of a sphincter-like structure identical to that of the afferent side. In the same way the basal marginal vessel does not seem to form a preferential pathway, since its small diameter and its tortuous course probably preclude a large stream (Olson and Kent, 1980). The distribution of pillar cells as well as their structure do not present any significant difference in comparison with teleosts. Characteristic sinuous aspects of the collagen fibers in the columns in parallel with straight bundles of the intracellular microfilaments encountered in the dogfish favor the hypothesis of contractility of pillar cells (Wright, 1973). Efferent lamellar arteries (postlamellar arteries) are collected by an efferent filament artery. As in teleosts, a sphincter is located at the base of this artery near its junction with the efferent branchial artery (Dunel and Laurent, 1980). These sphincters, provided with seven to eight layers of smooth muscle fibers, have their most external layer closely associated with large nerve endings containing

---

interposed between the efferent filament (ef. FA) and the cavernous tissue (cc). The cavernous tissue is supplied from the afferent artery (af. FA) by connections not visible here and feeds, in turn, the lamellae. Note that processes of the CVS (stars) overlap the cc. The CVS is fed by long, narrow vessels (arrows) arising from ef. FA or by anastomoses, not visible here. Bars = 100 μm. (Dunel-Erb, unpublished.)

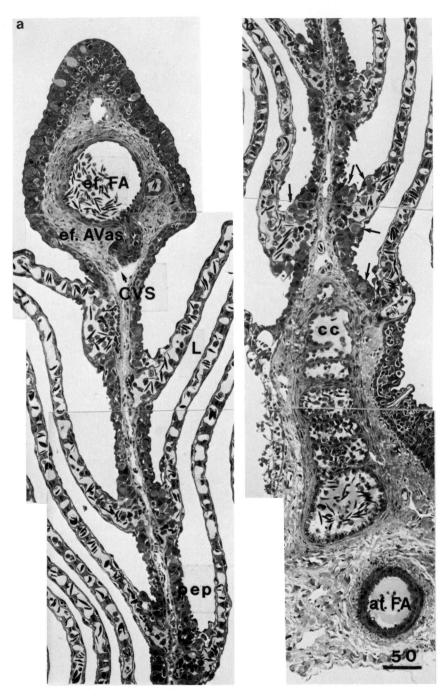

numerous large granular and small agranular vesicles. A part of this innervation displays a formaldehyde-induced histofluorescence (Y. Bailly, personal communication).

Chondrostei show remarkable affinities in terms of gill arterio-arterial vascularization both with teleosts and elasmobranchs. They retain a cavernous body that is structurally identical with that of elasmobranchs but less developed (Fig. 16a). They have a shorter septum than elasmobranchs, and there is no lateral fusion of the cavernous tissue from adjacent filaments in the septum, as occurs in elasmobranchs (Fig. 2d). Sphincters are present both on efferent filament arteries (innervated) and on afferent and efferent lamellar arteries (noninnervated) (Fig. 17) (Dunel and Laurent, 1980). The lamellae have their pillar cells arranged in a polygonal disposition (Fig. 10a). The outer marginal vessel is well defined, and the inner marginal vessel progressively tapers off from the afferent to the efferent side. Lamellar circulation and profiles have been studied in *Acipenser transmontanus* (Burggren *et al.*, 1979).

Holostei with their two representatives, *Lepisosteus* (Fig. 2f) and *Amia* (Fig. 2e), represent a further step. The cavernous bodies have completely disappeared (Fig. 19). The arterio-arterial vasculature is equipped both with innervated sphincters on the efferent filament arteries, and pre- and post-lamellar noninnervated sphincters on the lamellar arteries (Dunel and Laurent, 1980). Holostei also have some peculiarities related to their bimodal respiration. These fish are able to estivate in moist mud during the dry season; at that time they rely on their gas bladder for aerial respiration. First, in *Amia* lamellae, pillar cells are arranged in parallel arrays and are fused side by side (Dunel and Laurent, 1980; Olson, 1981) (Fig. 10b). Probably this arrangement is a compromise between respiratory efficiency and low hemodynamic resistance. Such an arrangement could explain how gill resistance decreases when the fish breathes air. Cast preparations of *Lepisosteus* also show the same arrangement of pillar cells (Dunel and Laurent, 1980). Second, and as in tunas, in *Amia* each lamella of each adjacent filament is joined to its neighbor at its distal edge by an interlamellar tissue (Bevelander, 1934; Olson, 1981; Daxboeck *et al.*, 1981). This tissue consists of a stratified epithelium. No other constituents such as smooth muscle or collagen fibrils are present (S. Dunel-Erb, personal communication). This

---

**Fig. 14.** The leading (a) and the trailing (b) edges of a cross-sectioned filament in an elasmobranch (*Scyliorhinus canicula*). The lower part of (b) shows the insertion of the filament on the septum. This part contains the afferent filament artery (af.FA) connected to the corpus cavernosum (cc) that in turn supplies the lamellae (L). Note the difference in thickness of the filament and the lamellar epithelium and the presence within the filament epithelium (pep) of chloride cells (arrows). The upper part of (a) shows the efferent filament artery (ef.FA) within the leading edge of the filament. Note the presence of an efferent arteriovenous anastomosis (ef.AVas) that communicates with the central venous sinus (CVS). Bar = 50 μm.

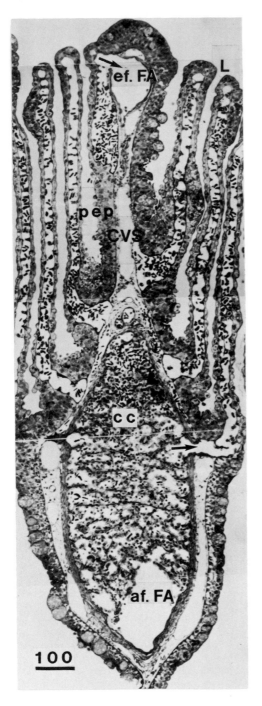

**Fig. 15.** Cross section of a filament in *Raja erinacea* (elasmobranch). At this level, the filament is no longer attached to the septum. The corpus cavernosum (cc) is well developed (compare with Fig. 14b); it almost entirely surrounds the afferent filament artery (af. FA). Several small vessels (nutrient?) are seen in cross section within the core of the filament. Note the thickness of the filament epithelium (pep), which is provided with large cells. Mucous cells are visible on the filament epithelium. Large sinuses are located around the afferent filament artery. They communicate with the central venous sinus (CVS). Bar = 100 μm.

arrangement, which is not present in *Lepisosteus*, transforms the organization of parallel arrays of lamellae in *Amia*, into a sort of sieve that supposedly gives more rigidity and avoids lamellar collapse during estivation (Fig. 20b). (see Daxboeck *et al.*, 1981, for estivation of *Amia*). Finally, it is interesting to note the presence of blebs or bellows-like ampullae on the afferent filament arteries, a structure that has been presumably retained from ancestral forms (Olson, 1981).

It is obvious that among the different groups, Dipnoi display the most altered gills. This alteration is of course in keeping with the aerial respiration. This is particularly significant in the African (*Protopterus*) and the South American (*Lepidosiren*) lungfish, which are obligatory air breathers. Alterations are not so evident in the facultative Australian lung breather *Neoceratodus*. In *Lepidosiren* and *Protopterus*, as already mentioned, archs IV to VI are supplied by afferent arteries emanating from the right part of the bulbus, and consequently they receive deoxygenated blood. Arch VI is supplied by collaterals of the arch V afferent branchial artery. Efferent arteries of arches IV to VI (efferent VI flows into the efferent V) join to form the pulmonary artery (Fig. 1g,h). In the gill each lamella is provided with a blood supply by a series of parallel afferent arteries arising from the branchial artery (Fig. 2g). After a short course, the afferent arteries bifurcate into two branches supplying opposite lamellae. By successive dichotomies, each filament artery then gives off secondary and tertiary arteries to lamellae of corresponding orders (Fig. 4a). These tributaries form loops around the lamellae and connect with the corresponding efferent artery (Fig. 21a). The arrangement of the blood vessels on secondary and tertiary lamellae is the same as in filaments. The afferent arteries along the edge of the lamellae are continuous with the efferent artery; no specialized structures are identified in the transition zone. The wall of the afferent artery thins progressively as the efferent side is approached. In cast preparation, the injected microfil tapers but does not always stop (Fig. 21a)—implying an increasing resistance along the afferent vessel (Laurent *et al.*, 1978).

There are two points of interest in addition to the arborescent structure of this type of gill. The first point concerns the presence of a large and direct channel that connects the afferent to the efferent filament artery (Fig. 4b). This short connection represents a shunt pathway that might be formed by the external vascular loop of an abortive lamella (Fig. 2g). In histological sections the wall of this vessel appears to have a thick muscular layer, suggesting a potent role in vasomotor control (Laurent *et al.*, 1978). The second point of interest consists in the structure of the "respiratory" lamellae. A very complex network is present inside the lamellar loop (Fig. 21a). It consists of a system of interconnecting large capillaries that originate from the afferent side and terminate on the efferent side of the loop. It is worthy of

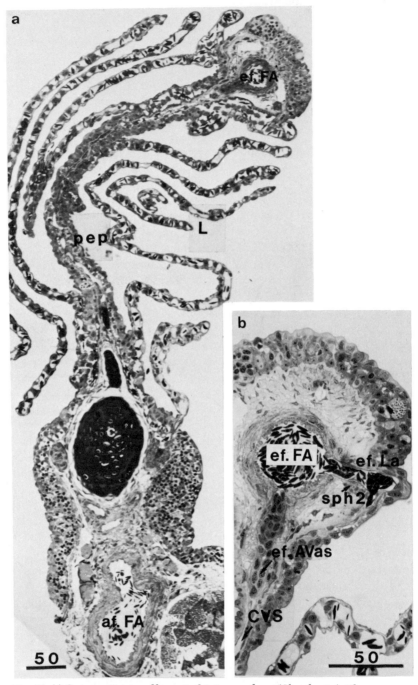

**Fig. 16.** (a) Cross section in a filament of *Acipenser baeri* (Chondrostei). The corpus cavernosum that emanates from the afferent filament artery is poorly developed (compare with Fig. 15). (b) Enlarged view of the leading edge in cross section from the same material as in (a). Note

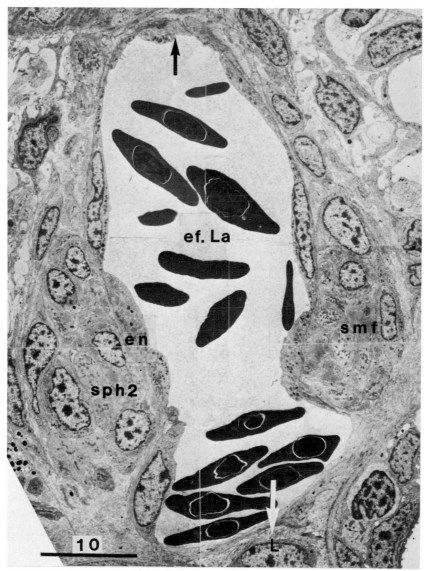

Fig. 17. Electron-microscopic photomontage of a lamellar sphincter (sph2) of *Acipenser baeri*. A symmetrical thickening of the efferent lamellar arteriole (ef.La) wall is attributable to concentric piles of smooth muscle fibers (smf). Note the special arrangement of endothelial cells (en). Arrows indicate the connections of lamellar arteriole to filament artery (↑) and to pillar capillary (↓). Bar = 10 μm. (From Dunel and Laurent, 1980.)

the efferent lamellar arteriole (ef.La) that communicates with the efferent filament artery (ef.FA) through a sphincter (sph2). In addition, note that an efferent arteriovenous anastomosis (ef.AVas) connects the artery to the central venous sinus (CVS). Bars = 50 μm.

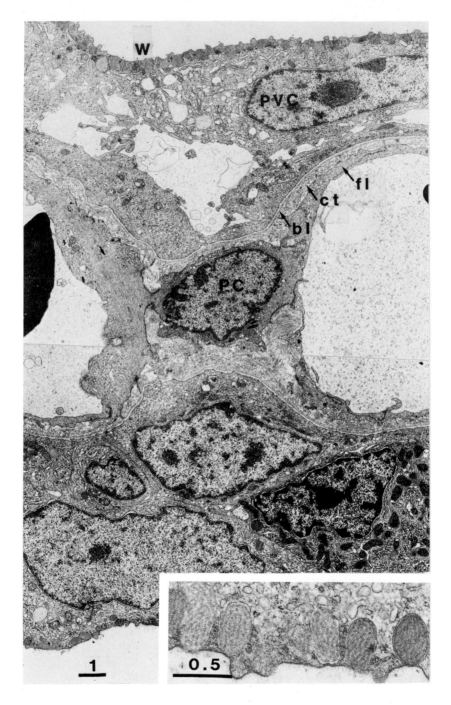

note that the so-formed network is entirely different from the pillar cells system of the lamellae encountered in the other groups of fish and resembles the arrangement found in amphibian external gills (see Laurent, 1982). The walls of the arterio-arterial capillaries consist of an endothelium with frequent cells protruding into the lumen. These cells show numerous pinocytotic vesicles and large, opaque inclusions, possibly similar to Weibel–Palade granules. The tunica media consists of a single layer of circular smooth muscle fibers associated with elastic tissue and resting on a basal lamina that separates it from the endothelium. In many places nexus are formed by direct apposition of endothelial and sarcolemmal membranes. Arterio-arterial capillaries show large and abrupt variations in diameter, suggesting local constriction of minute sphincter-like structures. Moreover, they often branch and form an irregular two-dimensional network. At various points the arterio-arterial vessels approach the external milieu, from which they are separated by a rather thick epithelium (Laurent et al., 1978).

## 5. The Arteriovenous Vasculature in Teleosts

Except for the mechanisms of its vasomotor control, the arterio-arterial vasculature does not raise any particular problem at structure or functional levels. With regard to the arteriovenous vasculature, on the contrary, the structural bases have been only recently understood thanks to sophisticated methods of vascular casting. Nevertheless, some points remain obscure, and discussions now focus on the real nature of these compartments, some authors considering them venous and others lymphatic. In addition, from a physiological standpoint, the role of the different parts of the arteriovenous vasculature is not completely understood; for instance, the functional meaning of the large and polymorphic venous compartment within the core of the filaments is still a matter of debate. Detailed and careful description based on corrosion cast subdivides the arteriovenous vasculature into different parts distinguished by terminology depending on the species and the authors (Farrell, 1979, in the cod Ophiodon; Cooke and Campbell, 1980, in the toadfish Torquiginer; Boland and Olson, 1979, in the catfish Ictalurus). These distinctions are also of interest in lower groups (Cooke, 1980, in the

---

**Fig. 18.** Lamella of Acipenser baeri. Note the pillar cell body (PC) and its cytoplasmic flanges (fl) surrounded by connective tissue (ct) lying on the basal lamina (bl) of the lamellar epithelium. Note also different cell types within the epithelium. In addition to pavement cells, a mitochondria-rich cell is also visible. Pavement cells are characterized in Chondrostei and other lower groups by the constant presence of numerous large vesicles attached to the apical membrane and apparently releasing some material outside (inset). The function of these vesicles is still unknown. Pavement cells are also rich in small membraned or coated vesicles, suggesting an important pinocytotic activity. W, external medium. Bars = 1 μm (inset, 0.5 μm).

endeavor dogfish *Centrophorus*; Olson, 1981, in the bowfin *Amia*; Olson and Kent, 1980, in the spiny dogfish shark *Squalus acanthias* and the skate *Raja erinacea*). Other studies based on electron-microscopic observations concern the relationships between vasculature and epithelia or other tissues (Laurent and Dunel, 1976; Dunel and Laurent, 1980; Vogel *et al.*, 1973). The concept of Riess (1881) tended to establish the existence of two non-respiratory systems, the so-called lymphatic and nutritive systems, both situated in the core of the filament. Contrary to the concept of Müller (1839), which has presented the nonrespiratory system as exclusively belonging to the venous circulation, the concept of Riess led to more complication and confusion.

To bring more clarity, an objective description of the arteriovenous vasculature is first given. Some elements for a discussion will be added thereafter. The arteriovenous vasculature schematically includes (1) a system of vessels that supply the venous compartments, (2) the venous compartment consisting mainly of one large or several small interconnected sinuses, and (3) a system of collecting venous pathways. This basic pattern of organization is common to all the groups of fish except Dipnoi.

The relationship of the arteriovenous vasculature with the systemic circulation has been described presumably for the first time by Müller (1839), who called them pores. Actually, connections are achieved by two sorts of vessels: (1) short vessels often called arteriovenous anastomoses on account of their morphology and (2) long arterioles often called nutrient vessels. In both cases, these vessels might connect any point of the systemic circulation to any point of the venous gill circulation.

Anastomoses have been described in teleosts in several publications (Dunel, 1975; Laurent and Dunel, 1976; Vogel *et al.*, 1973, 1974, 1976). Basically anastomoses are rather short channels ($< 100$ μm), but they may be extended by capillaries that connect the venous side. They have a wall thicker than an artery of the same size, a character that makes anastomoses easily recognizable on histological sections. Their thin lumen is lined by numerous and particular endothelial cells (Fig. 22d). Detailed studies of the arterio-venous anastomoses in *Tilapia mossambica* (Vogel *et al.*, 1974) reveal that endothelial cells are of two types according to their location on the

---

**Fig. 19.** Cross section in a filament of *Amia* (Holostei). The central venous sinus (CVS) is made up of numerous loosely interconnected lacunae, (some of them indicated by stars). Its "spongy" structure is also visible on cast preparation (Fig. 10b). The massive organization of the filament in *Amia* (compare with Figs. 14 and 16) is presumably related to the respiration pattern of this bimodal breather, which can survive burrowed within the mud during the dry season. Note the arrangement in parallel arrays of the lamellar capillaries (arrow), and compare with the cast of Fig. 10b. Bar = 100 μm.

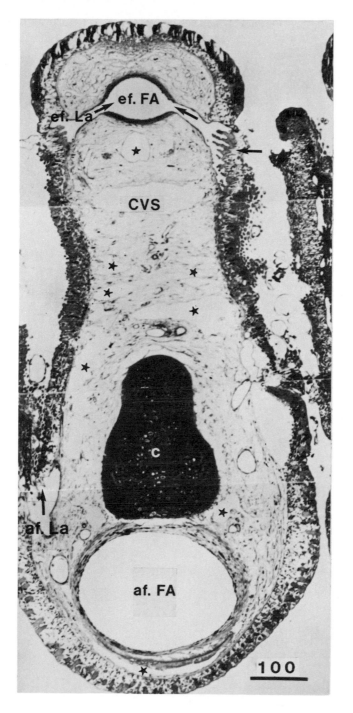

arterial side or on the venous side of the anastomosis. In both cases they fundamentally differ from the endothelial cells of an artery or a vein. For instance, they do not display Weibel–Palade bodies. Type I endothelial cells, which are preferably located on the arterial side, are large and frequently bulge into the arterial lumen. They have microvilli or deep indentations. They are immediately linked to arterial endothelial cells by a junctional complex and filled with concentric whorls of filaments presumably forming irregularly oriented tubes throughout the cytoplasm. Few mitochondria and membraned electron-dense bodies are packed between the filament whorls. Granules of glycogen are located within and outside the tubes. Type II endothelial cells are located near the anastomose apertures on the venous side. They are more cuboïdal and smaller than type I and in most cases devoid of filamentous whorls. Membrane caveolae are numerous. They contrast with the flat endothelial cells lining the venous compartment. In the rainbow trout only one type of endothelial cell is observed. This unique type does not display the characteristic filamentous whorls of type I endothelial cells of *Tilapia*. Endothelial cell microvilli are well demonstrated by scanning electron micrographs of the arterial side (Vogel *et al.*, 1976). Endothelial cells of the anastomoses in icefish have identical characteristics (Vogel and Kock, 1981).

Another point of interest is the presence around anastomoses of layers of more or less circular smooth muscle fibers resting on a basal lamina. The number of layers is variable, one to four according to the species; for instance, in *Anguilla anguilla*, two layers are commonly seen (Dunel and Laurent, 1980). In *Tilapia* anastomoses are not muscular; instead of muscle fibers there is a sheet of particular cells, the so-called cover cells with an electron-lucent cytoplasm, numerous mitochondria, a few caveolae, and rare intracytoplasmic filaments. (Vogel *et al.*, 1974). Nexus have been described between endothelial and cover cells across the basal lamina as well as between endothelial and smooth muscle cells when present. Anastomoses have been also found in various species of teleosts: *Ictalurus melas, Silurus glanis, Ciliata mustella, Solea solea, Blennius pholis, Platichthys flesus* (Dunel and Laurent, 1980). In all species studied so far the anastomosis lumen is very crowded and narrowed by endothelial cells bulging out into the lumen (Fig.

---

**Fig. 20.** (a) Cast preparation of the filament vasculature in *Amia*. Note the efferent arteries (ef. FA) surrounded with central venous sinus (CVS) processes. The lamellar capillaries (L) are almost in contact with those of the next filaments. Bar = 500 μm. (b) Scanning electron micrograph (critical point dried preparation) of the same region as in (a). Note that tips of lamellae (L) are buried within a longitudinal septum (arrow) running parallel to the filament. Bar = 100 μm.

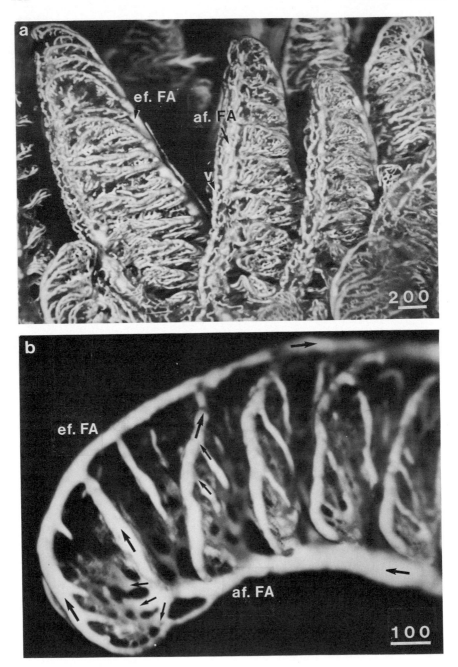

23). However, perikarya are often bent in a way that suggests the existence of blood flow from the arterial toward the venous side (Fig. 22d). On fixed sagittal sections erythrocytes are often seen journeying among the endothelial cell perikarya. Often epithelial cells are tightly packed so that the lumen is no longer visible, a situation that suggests a state of closure (Fig. 23).

In teleosts, anastomoses depart from the filament artery and in most cases from the efferent artery. In other words, the blood that flows into the anastomosis has passed through the lamellae and is oxygenated. This general rule suffers exception. In Siluroidei, *Ictalurus melas* and *Silurus glanis* display anastomoses from both afferent and efferent filament arteries (Dunel and Laurent, 1980). That is also the case with the eel (Laurent and Dunel, 1976) and the trout (Vogel, 1978b). In this latter species, the number of anastomoses on the afferent artery is not as large as the number on the efferent one (which is almost as high as the number of lamellar arteries). In *Tilapia* anastomoses are also present on both sides, whereas the number on the afferent side is very small in comparison to the efferent, the minimum ratio of the efferent to the afferent side number being 17.8 (Vogel *et al.*, 1973). Thus, except for these few exceptions, anastomoses connect the postlamellar vasculature in teleosts.

It has been claimed that anastomoses are innervated (Vogel *et al.*, 1974). Nerve fiber bundles, mainly amyelinated, run perpendicularly to the anastomoses axis, and axons are observed in the vicinity of anastomoses cover cells in *Tilapia* (Vogel *et al.*, 1974). However, this assumption should be questioned for two reasons. First, it is unusual to consider that nonmuscular vessels (as is the case of *Tilapia* anastomoses) are under neural control. Second, all the species examined so far by neurofibrillar staining methods, as well as by TEM, display along the efferent filament arteries nerve bundles that innervate different parts of the filament including the neuroepithelial cells and the central venous sinus wall (Fig. 26a). These fibers often surround the anastomoses, but at least in teleosts, careful examinations have failed to detect any functional relationships between nerve and anastomoses muscle (Fig. 23).

The second type of vessel that connects the arterio-arterial vasculature

---

**Fig. 21.** (a) A vascular cast preparation of gill in the lungfish *Protopterus aethiopicus*. This arborescent organization totally differs from that of obligate water-breathing fish and is reminiscent of the external gill of amphibians. ef.FA, efferent filament artery; af.FA, afferent artery; v, vein. Bar = 200 μm. (b) An enlarged view of the tip of a lungfish gill (*Protopterus aethiopicus*). Note the dichotomized pattern giving rise successively to lamellae of first, second, and tertiary order. There are direct connections between the afferent and the efferent side, thus making some sort of a vascular loop. The direction of blood flow is indicated by arrows. Bar = 100 μm. (From Laurent *et al.*, 1978.)

with the venous compartments consists of long, thin arteries. They travel within the filaments following various routes depending on the species. Various and complex topologies have been so far reported under such terms as venolymphatic vessels (Farrell, 1980a), nutritive pathways (Boland and Olson, 1979), nonrespiratory vessels (Cooke and Campbell, 1980), nutritional vessels (Morgan and Tovell, 1973), central filament artery (Vogel and Kock, 1981), and Fromm's arteries (Vogel, 1978a). It has been shown that these arteries are tributaries of the efferent branchial arteries and also in some cases emanate from the efferent filament arteries (Laurent and Dunel, 1976). Later on, this assumption was confirmed by analogous studies from various species (Dunel and Laurent, 1980; Boland and Olson, 1979; Cooke and Campbell, 1980; Vogel, 1978a). In such species as the trout and the perch these arteries take their origin from a group of short, narrow tributaries departing from the efferent branchial arteries and forming a sort of periarterial vasculature network (Laurent and Dunel, 1976). Each of them has a sphincter at its origin and gathers into a single artery (Fig. 2b). Apparently this arrangement is similar to the secondary arterial system (Burne, 1927, 1929; Vogel and Claviez, 1981; Vogel, 1981). Electron microscopy in the trout reveals the presence of numerous nerve endings in their vicinity. The so-formed arteries, which have also been mentioned by Fromm (1974), run bilaterally alongside the efferent filament arteries. These arteries are often interconnected with those of neighboring lamellae, particularly at their base. Moreover, other small arteries of identical origin run deeply within the core along the gill rod in the perch and along the efferent filament artery in the trout. The efferent origin of all these vessels has been clarified by the use of casting methods (Laurent and Dunel, 1976; Vogel, 1978a).

An identical system of arteries has also been observed in the icefish; these so-called central filament arteries (Vogel and Kock, 1981) originate from the efferent branchial artery without the interposed network of small periarterial vessels that exists in the trout. In the catfish a very complex system of small arteries has been described (Boland and Olson, 1979). Among them, filament nutrient vessels originate from the efferent filament arteries. These small vessels anastomose and form a vascular web around the efferent filament artery as they do in other species (Laurent and Dunel,

---

**Fig. 22.** Different types of arteriovenous anastomoses. (a) In *Scyliorhinus canicula*, anastomosis between the efferent filament artery (ef. FA) and the central venous sinus (CVS). Bar = 50 μm. (b) In *Raja clavata*, an anastomosis between the corpus cavernosum (cc) (afferent side) and the central venous sinus (not visible here). Bar = 50 μm. (c) In *Acipenser baeri*. Note the presence of several myelinated fibers seen in cross section. They have no relationships with the anastomosis. Bar = 10 μm. (d) In *Anguilla*. Note the disposition of endothelial cells that are arranged like valves and an erythrocyte traveling through the anastomosis. Bar = 10 μm. (Adapted from Dunel and Laurent, 1980.)

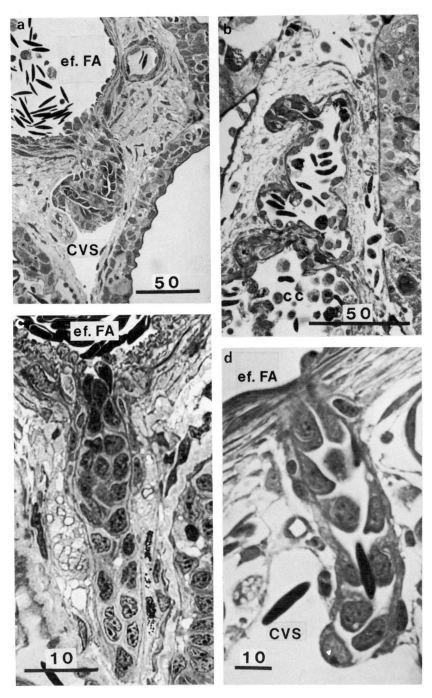

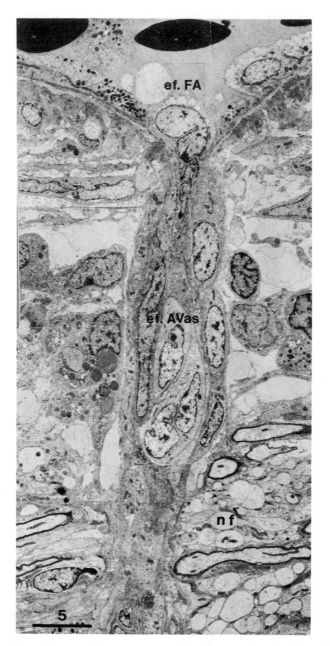

**Fig. 23.** Photomontage of electron micrographs of an anastomosis (ef.AVas) (*Ictalurus melas*). The section plane is not quite sagittal. Note the endothelial cells (clear) and the circular smooth muscle cells (darker). Note also large bundles of myelinated fibers (nf) that pass around the anastomosis but have no relationship with it. Bar = 5 μm.

1976). Other larger arteries of the same origin traverse the filament to provide a nutrient blood flow to the tissues around the afferent filament artery. In the toadfish, tributaries of the efferent branchial artery give off nutritive arteries within each filament where they feed adductor muscles. Nutritive capillaries also arise from efferent lamellar arteries or even from the basal channel of the lamellae themselves. However, most of the nutritive capillaries arise from the inner side of the efferent filament artery. In addition, some capillaries arise directly from the afferent filament artery (Cooke and Campbell, 1980).

It is now well accepted that the central core of each filament consists in a venous network that is fed by the arterio-arterial vasculature and that finally drains into the branchial vein (Laurent and Dunel, 1976). This venous compartment has been variously termed filament-lymph vessel (Morgan and Tovell, 1973), central or filament space venous sinus (Vogel *et al.*, 1973), interlamellar network (Olson, 1981), central venolymphatic sinus (Daxboeck *et al.*, 1981), filament sinus (Cooke and Campbell, 1980), collateral and medial sinuses (Olson and Kent, 1980), lymph spaces (Kempton, 1969), and central compartment (Wright, 1973). For reasons of unity and clarity, the term central venous sinus of filament will be adopted throughout this chapter, with the restriction that the adjective "central" does not limit the compartment to a central location and also includes the processes of the venous sinus.

The main part of the central venous sinus occupies the core of the filament, and its morphology is variable. In the eel the sinus is a simple saclike structure (Fig. 24a) with numerous extensions forming vessel-like sinuses mostly around the afferent filament artery, but fewer around the efferent artery (Fig. 2a). In the trout and the perch the sinus forms an anastomotic network of branches that extend in the direction of the afferent artery between the base of the lamellae. These extensions pass around the gill rod and partially surround the afferent filament artery, whereas on the opposite side shorter prolongations form a vascular network around the efferent filament artery (Laurent and Dunel, 1976) (Fig. 2b). In the catfish the sinus consists of a series of parallel finger-like cavities fitting the interlamellar space, which suggest functional relationships with the interlamellar (filament) epithelium (Dunel, 1975; Boland and Olson, 1979). In *Tilapia mossambica* (Vogel *et al.*, 1973) and in the toadfish *Torquigiker* (Cooke and Campbell, 1980), the morphology of the sinus and its pattern of ramification resemble the arrangement seen in the trout.

From an ultrastructural standpoint, venous sinuses have a veinlike structure with a low ratio of wall thickness to lumen width. Their walls are lined by a thin endothelium resting on a discrete basal lamina. Some smooth muscle fibers lie on the basal lamina, and a loose connective tissue surrounds

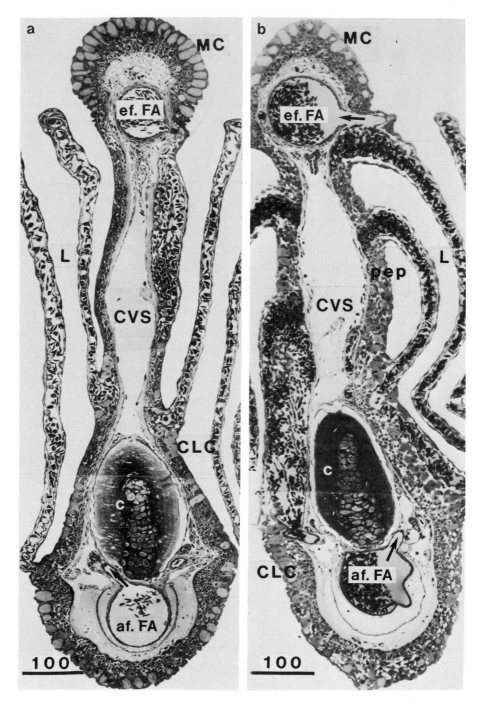

these elements externally. In some regions, a dense population of varicose amyelinated nerve fibers is seen within the connective tissue (Fig. 25). Formaldehyde-induced fluorescence had demonstrated the existence of a perisinusal network (Fig. 26b) (Bailly, 1983).

The junctions between the sinus and the branchial veins are made by veins running parallel to the afferent filament arteries in the trout (Fig. 2b) and in the perch. In the eel the efferent blood pathways differ by the existence of a direct connection between the sinus and the branchial vein through a short vessel (Fig. 2a). In addition, a group of small veins connect the sinus to collaterals of the branchial vein. These collaterals also communicate with a large sinus located at the posterior part of the gill arches (Laurent and Dunel, 1976).

The problem of lymphatics in the head region has been reevaluated in the eel (Rowing, 1981) with a particular emphasis on their distribution within the gills, in order to determine the precise nature of the filament sinuses. Some authors call these sinuses lymphatic (Riess, 1881; Florkowski, 1930; Steen and Kruysse, 1964; Morgan and Tovell, 1973). Others speak of a venolymphatic system (Cooke and Campbell, 1980; Farrell, 1980a). A third contingent consider them as venous (Vogel *et al.*, 1976; Laurent and Dunel, 1976). Actually this problem has several facets. Discussion about the ontogeny of the gill sinuses and their relationship with the lymphatic system per se continues a long spell of discussion on the lymphatic pathways in lower vertebrates (cf. Jossifov, 1906), but the role of these sinuses depends presumably less on their ontogeny than on their vascular connections

On a strictly structural basis, the walls of sinuses look like lymphatic vessels at some places, whereas at other places (there is no clear topology so far) the presence of a structured wall including basal lamina, smooth muscle fibers, and connective tissue leads to the opposite conclusion (Laurent and Dunel, 1976; P. Laurent and S. Dunel, unpublished data). However, from some evidence given, filament central sinuses are claimed to be connected dorsally to the systemic lymphatic system and from these to the internal jugular vein. Ventrally, sinuses of arches I to III connect to a periarterial sinus located around the ventral aorta and from there to a posterior lymphatic compartment draining the sinus of arch IV, and in its turn drained into the external jugular vein (Rowing, 1981). Nevertheless, antidromic ink perfusion of the so-called branchial vein does inject the sinus of the filament

Fig. 24. Cross section of a filament of the eel: (a) freshwater; (b) seawater adapted. For abbreviations, see Fig. 2 legend. Note the presence of more numerous chloride cells (CLC) within the filament epithelium in seawater, especially on the trailing edge (bottom); conversely, the mucous cells (MC) are less abundant. Another interesting observation concerns the dilated extracellular space in lamellar (L) epithelium in fresh water (the medium is hypo-osmotic). Bars = 100 μm. (Adapted from Laurent and Dunel, 1980.)

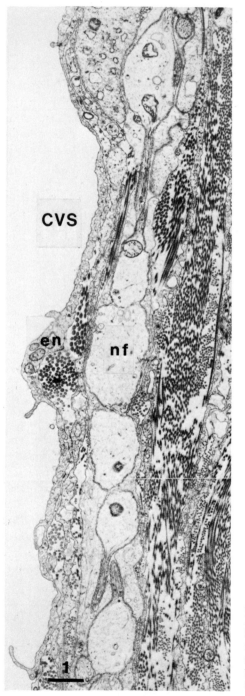

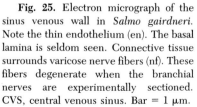

**Fig. 25.** Electron micrograph of the sinus venous wall in *Salmo gairdneri*. Note the thin endothelium (en). The basal lamina is seldom seen. Connective tissue surrounds varicose nerve fibers (nf). These fibers degenerate when the branchial nerves are experimentally sectioned. CVS, central venous sinus. Bar = 1 μm.

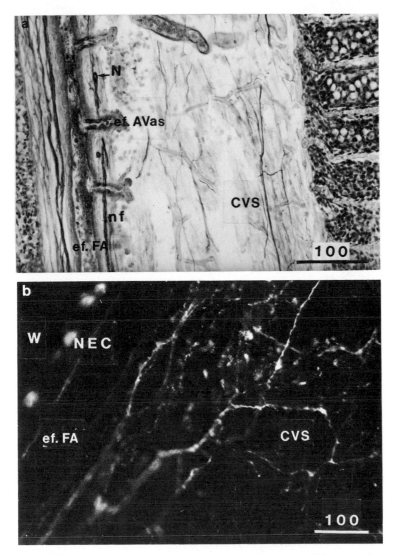

Fig. 26. (a) Parasagittal plane of section in a filament (*Micropterus dolomieui*, silver stained (Cajal method). The section plane passes within the wall of the central venous sinus (CVS). Several anastomoses are seen (ef.AVas) connecting the efferent filament artery (ef.FA) with the central venous sinus. Several nerves are running together underneath the filament epithelium, both sides of the artery (nf). Neurons are visible (N). A nonmyelinated network is located within the wall of the central venous sinus. Note that nerve fibers pass straight through on both sides of the anastomoses. On the right, the lamellae are sectioned obliquely. (b) Formaldehyde-induced fluorescence in a filament (*Ictalurus melas*). Parasagittal plane of section. The filament epithelium is located on the left side of the micrograph (W). Neuroepithelial fluorescing cells (NEC) are seen on the innermost part of the epithelium. A faint fluorescence of nerves is seen within the subepithelial region. The right side of the micrograph represents the sinus wall and its network of varicose nerve fibers. Bars = 100 μm.

and of the arch itself, and orthodromic injection of microfil in the afferent branchial artery shows a direct connection between filament sinuses and the branchial vein (Laurent and Dunel, 1976). Thus, whatever the distal connections of the branchial vein, sinuses collect blood or a mixture of blood and interstitial fluid, sending it into the branchial vein (which contains erythrocytes), and from there into the venous systemic circulation. Possibly the red lymphatics of *Myxine* (Cole, 1923) resort to the same type of vascular pathways (see Section II,B,6).

The collection of extracellular fluid from the epithelial interstitial compartment is a function of veins as well as lymphatics. On a strictly structural basis, it has been shown that exchange of substances might occur across the gill epithelium between the external medium and the filament sinuses via the chloride cells (Laurent and Dunel, 1978). This route has been functionally demonstrated in saltwater-adapted trout (Girard and Payan, 1977a,b). In addition, the admixture of blood through arteriovenous anastomoses actively contributes to the washout of the venous sinuses and secondarily supplies oxygen to the filament epithelium and its chloride cells. Finally, the subepithelial micropumps within the lungfish gill, obviously designed to inject interstitial fluid forcibly into veins, is also suggestive of a similar function of the venous vasculature (see Section II,B,6).

It is therefore not necessary to include a lymphatic apparatus. Only local conditions of oncotic and hydrostatic pressures need be considered. These considerations lead us to reject the hybrid terminology of venolymphatic.

## 6. THE ARTERIOVENOUS VASCULATURE IN LOWER GROUPS

In the lamprey, the base of each hemibranch, on both sides of the interbranchial septum, rests on a peribranchial venous sinus that surrounds the afferent filament artery. Some proportion of the venous blood is diverted via arteriovenous anastomoses into veins. These filament veins run along both sides of the afferent filament artery and finally unite in a single vein near the posterior end of the filament. They are drained into peribranchial sinuses via anastomosing canals. Peribranchial sinuses are continuous with sinuses extending around the ventral aorta, the water tube, and the esophagus, and finally are drained out by the vena jugularis impar and the vena cardinalis anterior (Nakao and Uchinomiya, 1978). An almost identical system has also been described in *Myxine* (Pohla *et al.*, 1977). It is interesting to note that this circulatory pattern is quite similar to the venous drainage described in the eel (Rowing, 1981), but no attempt has been made to relate it to some kind of lymphatics. For Nakao and Uchinomiya, there is no doubt about the venous nature of the systems described hitherto.

Two types of arteriovenous anastomoses have been described in the lamprey. Type I connects afferent filament arteries to filament veins, and type II connects efferent filament arteries to peribranchial venous sinuses. Both types of anastomoses consist of an inner endothelial layer and an outer smooth muscle layer. The endothelial cells are distinguished by their dark cytoplasm both in semithin toluidine blue-stained light micrographs and in ultrathin uranyl–lead citrate-stained electron micrographs. These cells have the numerous short microvilli similar to their counterparts in teleosts (Vogel et al., 1976). The presence of a basal lamina between endothelial cells and smooth muscle cells is still questioned, since only fine filamentous material and an accumulation of dense material are seen in the region that is currently occupied by the basal lamina. Five to eight layers of smooth muscle fibers envelop the endothelium concentrically. Some close contacts between endothelial processes and sarcolemma are seen, but no specialization of membranes equivalent to a nexus has been identified (Nakao and Uchinomiya, 1978). The walls of the filament veins are extremely thin with an endothelial layer displaying numerous gaps. Endothelial cells have caveolae associated with both luminal and basal surfaces. The presence of a basal lamina is problematic, although accumulation of dense material was noted on the basal cell surface, which in turn consists of a few adventitial cells but apparently no muscle cells. Among other interesting points of similarity is a spatial relationship between chloride cells and the venous compartments in the lamprey. Such a relationship yields to a general rule, since it has been also observed in other lower groups as well as in teleosts. Thus, the lamprey gill venous vasculature reveals certain structural characteristics in common with teleosts. The general design is, however, quite different.

In elasmobranchs, the first description of the gill venous system in *Scyliorhinus canicula* (O'Donoghue, 1914) mention the presence of central venous sinuses, supplied by small arteries and drained via the inferior jugular veins. The central venous sinuses are lined by a very thin endothelium (0.2–0.6 μm) containing flattened, elongated nuclei, few mitochondria, and numerous pinocytotic vesicles. They are surrounded by connective tissue (Wright, 1973). They have a structure very similar to that of intestinal veins and vesicles described in *Squalus acanthias* (Rhodin and Silversmith, 1972). The arteriovenous (or nonrespiratory) vasculature forms an extensive vascular bed within the gill filaments in the dogfish (Dunel and Laurent, 1980; Cooke, 1980; Olson and Kent, 1980) and the skate (Dunel and Laurent, 1980; Olson and Kent, 1980). The distribution of this vasculature is quite comparable to that found in teleosts. It consists of arteriovenous anastomoses and arteries of small diameter leading to large sinuses drained into venous branchial vessels and from there into the inferior jugular vein (Fig. 2c).

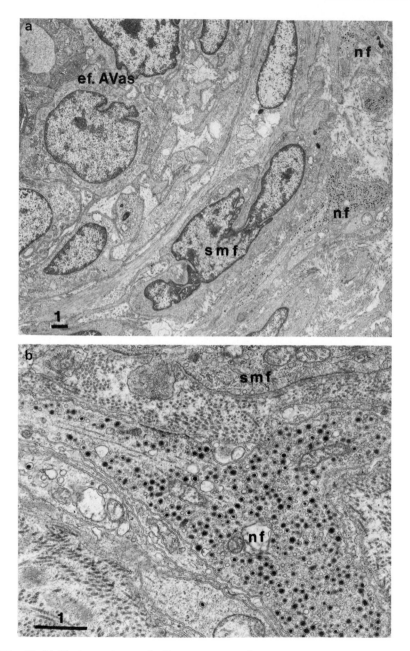

**Fig. 27.** (a) Electron micrograph of a cross-sectioned arteriovenous anastomosis (*Scyliorhinus canicula*). Note the endothelial cells tightly packed within the lumen, indicating a state of
closure. They are surrounded by smooth muscle fibers (smf). Amyelinated nerve profiles and

The location of the anastomoses is different according to the considered species. In *Scyliorhinus canicula,* anastomoses are probably exclusively located on the efferent primary arteries (Figs. 14 and 22a) (Dunel and Laurent, 1980). Anastomoses are often bifurcated, giving two openings in the central venous compartment. In the endeavor dogfish (*Centrophorus scalpatus*), anastomoses are seen connecting the corpus cavernosum to the central sinus (Cooke, 1980). Anastomoses of *Scyliorhinus* have a thicker muscular layer than in teleosts (Fig. 27a), and the presence of cilium-like structures must be mentioned in the endothelial cells. Large nerve endings of about 5 μm filled with dense-cored vesicles have been found within 0.5 μm of muscle fibers (Fig. 27b) (Dunel and Laurent, 1980). In *Squalus acanthias* the central venous sinus is supplied both from pre- and post-lamellar sources. In this species arteriovenous anastomoses are located on the cavernous tissue and connect the sinus network (Olson and Kent, 1980). In *Raja clavata,* anastomoses are located on the afferent as well as the efferent side. In the former case, anastomoses directly connect the corpus cavernosum with the sinuses at irregular intervals (Fig. 22b), whereas in some other lower groups anastomoses connect the lamellar arteries with the sinuses.

In elasmobranchs the central venous sinus is complex. It consists of a central part, relatively compact in the endeavor dogfish, and forming a wide, flat, saclike sinus in the core of the filament between the corpus cavernosum and the efferent artery (the so-called central canal; Cooke, 1980). In *Scyliorhinus canicula,* the central sinus consists of a complex system of digitations (Fig. 13b) that becomes a dense network of finger-like processes in *Raja clavata* (Fig. 12). Similar arrangements have been found in *Squalus acanthias* and *Raja erinacea* (Olson and Kent, 1980). In all species studied so far, two systems of sinus processes are distributed respectively around the afferent and the efferent filament arteries. They envelop them with an entanglement of vessel-like sinuses. Different and complex terminologies have been proposed by other authors; therefore, we have tried to keep this description as simple as possible unless details are of some importance. That is particularly the case when we consider the parallel sinuses running bilaterally alongside the primary arteries sometimes called "companion" vessels (Cooke, 1980; Cooke and Campbell, 1980). This appellation might cause confusion with arteries and must be avoided. It is claimed indeed that the central venous sinuses as well as sinuses located within the septum communicate with these so-called companion vessels in the endeavor dogfish (Cooke, 1980). A similar arrangement seen also in *Scyliorhinus canicula* is

---

large endings are seen within the adventitia (nf). (b) Enlarged view of a nerve ending from the preparation shown in (a). Note the numerous dense-cored vesicles and their homogeneous size beside smaller clear vesicles with a dark membrane. Note that muscle fibers (smf) are within 1 μm from the nerve. Bars = 1 μm.

still more complex in *Raja clavata*. In the latter species, an important system of septal sinuses develops from the tip of the septum down to the branchial vein. In *Scyliorhinus* the central venous system also communicates with large anterolateral sinuses within the arch via veinlike vessels running alongside the efferent arteries. Finally, all sinuses drain off into the branchial veins (Fig. 2c).

Data concerning the arteriovenous vasculature of chondrostean fish is seemingly limited to one article (Dunel and Laurent, 1980) despite the great interest of this old stem group from which originates the modern actinopterygians (Romer, 1959). *Acipenser baeri* displays anastomoses on both sides of the secondary lamellae (Fig. 2d). The afferent side anastomoses are short channels connecting the afferent secondary arteries to the pericartilaginous processes of the central venous sinuses. Efferent side anastomoses are often much longer and directly connect the efferent arteries to the venous sinus (Figs. 16b and 22c). *Acipenser*, like *Scyliorhinus*, has a thick muscular layer around anastomoses and cilium-like structures in the endothelial cells.

The central venous sinus is made up of numerous parallel interconnected sacs. Each of them underlies an interlamellar distribution. This arrangement again suggests the existence of precise relationships between the sinus and the primary epithelium lining the interlamellar areas. On the afferent side, these sacs alternating with the afferent lamellae open in veinlike vessels. Those vessels communicate with a series of septal sinuses communicating to each other through openings equipped with valves preventing any backflow. Septal sinuses are drained into a branchial vein. On the efferent side a network of sinusal processes surrounds the efferent filament artery. They do not have a particular drainage in contrast with *Scyliorhinus* (compare Fig. 2c and 2d).

Holostei with their two representatives *Lepisosteus* and *Amia* constitute a closer step to teleosts. In *Lepisosteus* the distribution of anastomoses is not different from what is observed in *Acipenser* (compare Fig. 2d and f); they are located on afferent and efferent filament arteries as well (Dunel and Laurent, 1980). A series of small arteries emanate from the efferent branchial artery as they do in *Salmo* (Laurent and Dunel, 1976; Vogel, 1978a). These arteries supply the filament tissues. In *Amia*, anastomoses between arteries and the central sinuses are sparsely distributed and of very small size, and thus probably only minimally effective. Actually in this species numerous large sphincters are located on the branchial efferent artery itself. These sphincters communicate with an entanglement of vessels that come together and form a large single channel running along the filament and successively supplying finger-like subdivisions of the central sinus (Fig. 2e). This channel is generally in a median position just below the efferent filament artery (Dunel and Laurent, 1980). In other specimens small vessels arise from individual hillocks of the efferent filament artery and, after a brief tortuous course, connect to each other, generally forming two large vessels. They generally anastomose before entering the filament and supplying the central sinus. Frequently also narrow arteries depart from the efferent artery at irregular intervals, distribute within the filamental tissues, and are finally collected by the venous sinus. The central sinus consists in two rows of parallel and anastomotic irregular channels (one per interlamellar space) passing on both sides of the gill rod and collected by longitudinal channels. From there an irregular network of anastomotic processes envelops the afferent primary artery. The whole system of sinuses drains into a branchial vein. A valvelike structure is located on the branchial vein at the entrance of each filament vein.

Among the dipnoan lungfish only the arteriovenous vasculature of Protopteridei has been studied (Laurent *et al.*, 1978). In *Protopterus annectens*, *P. aethiopicus*, and *Lepidosiren paradoxa* as well, the gill arteriovenous system originates from nutrient vessels (or vessels considered as such). These nutrient vessels are collateral branches of efferent arteries of different orders (filament, lamellar, or even tertiary; see Section II,B,4). The venous part of the system consists of small vessels having the structural characteristics of a vein, that is, a thin endothelium with flat and rather rare endothelial cells. The elastic layer is also thin and occasionally contains smooth muscle fibers. The lumens of these veins are broad. The system drains into a branchial vein that runs parallel to the afferent branchial artery (Fig. 2g). Interspersed in the arteriovenous system are specialized cisternae. Each cisterna communicates with a vein and large subepithelial extracellular spaces. The cisternal lumens vary in size and shape from a sphere of up to 60 μm to an ellipsoid of up to 50 by 80 μm. The wall of the cisterna is made up of three layers: endothelium, smooth muscle, and elastin. The endothelium is of the continuous type according to the terminology of Rhodin (1968) and displays few caveolae; it is underlaid by a double basal lamina and completely surrounded by a network of smooth muscle fibers. A thick layer of elastin forms an external envelope. The communication with the vein is made up by a narrow slit through the wall. The slit is lined by cells, the perikaryon of which protrudes into the adjacent vein. A symmetrical arrangement forms the communication with the extracellular spaces. In this case cells lining the slits protrude inward. The slit lining cells differ from the endothelial cells, in that they have an opaque cytoplasm that at high magnification resolves into parallel packed microfilaments. Both types of apertures are consistently present in each cisterna. Thus, cisternae are interposed between the intracellular spaces and the venous system.

Cisternae are also present in the venous circulation of other vascular beds in Protopteridei (P. Laurent, unpublished observations) and probably represent a structural characteristic of this group. They are particularly abundant in the gill lamellae, where they probably play an important role. Their structure indeed suggests that they act as one-way valved micro-pumps, collecting extracellular fluids from the intercellular epithelial chan-nels and injecting this fluid into the "juxtacisternal" veins by contracting their musculature. By the disposition of the slit cells, the first type of com-munication (slit cells inside) allows an inward movement of fluid from the intercellular spaces when the musculature relaxes and a depression occurs in the cisterna. The second type of communication (slit cells outside) lets the fluid flow out when the musculature contracts. By this time the inward slit is due to the inner pressure in contrast with the outward slit that opens.

The function of these cisternae appears to be similar to that of lymphatic microhearts and is of potential importance in maintaining the hydromineral balance across the gill epithelium. Certain functional analogies between these cisternae and the venous sinus of the filament should be supposed in respect to gill epithelial physiology. Indeed blood pulsatile pressure and respiratory water movements possibly favor the blood circulation within the central sinus and help encourage an epithelial intercellular drainage.

In concluding this structural analysis of the arteriovenous vasculature among the different groups of fish, there are essentially two points that call for some consideration.

1. Arteriovenous anastomoses are present in all groups of fish except in Dipnoi and probably are locally controlled.
2. The nutrient vasculature, if it exists *sensu stricto,* has unusual struc-tural characteristics; except in the striated gill muscle, the existence of an efficient exchanging system of arterioles provided with capillaries is not obvious.

Thus, the distinction often made between nutrient vessels and a system of sinuses is possibly too artificial; for example, it is not clear that functional differences exist between long, narrow arteries and short anastomoses. In contrast, the development of venous sinuses and their associated more or less superficial complex of processes suggests precise functional relationships with the epithelium. Direct anastomoses, which are equivalent to ar-teriovenous shunts, possibly represent a powerful accelerator of the venous washout, a function that in Dipnoi might be assumed by the cisternae. In addition, a supply of oxygen to gill tissue should result.

## III. THE GILL EPITHELIA

The functions of the gills consist of gas and ion transport. These trans-ports are performed by and across specialized epithelia. The morphology of these epithelia described in this section therefore constitutes the structural ground of exchange mechanisms. As already stated, the fish gill is a complex organ that cannot be considered as a simple tissue like the epithelium of the gut, the gallbladder, or the urinary bladder. It consists of no less than two kinds of epithelia including several different cell types supplied by complex systems of vessels under neural and humoral control, systems that have been described earlier in the present chapter. At first glance, it is easy when looking at a cross section of a gill filament (Fig. 7) to see that the epithelium covering the core of the filament differs from that which surrounds the lamellae. The latter, called lamellar or respiratory epithelium, is thin and therefore adapted to gas exchange. This observation is consistent whatever the group of fish with the exception of some lungfishes. *Protopterus* and *Lepidosiren* have their lamellae covered with a thick epithelium. This observation is in good accordance with the fact that these fish use their gills little for gas exchange. In contrast, the filament epithelium in nondipnoan fish extends between the lamellae, so that the filament and lamellar epithelia alternate in register with their respective vascular compartments, the arteriovenous and the arterio-arterial.

## A. The Filament Epithelium

### 1. GENERAL ORGANIZATION

This multilayered epithelium surrounds all the filaments including the anterior and the lateral surfaces of the branchial arches with the exception of the lamellae themselves (Fig. 28a,b). Threading its way between the regularly spaced lamellae, it is also often called interlamellar filamental epithelium (Conte, 1969). This term should be avoided because it tends to restrict the physiologically active area to the interlamellar spaces, whereas functionally important cells spread over the entire filament epithelium. The term filament epithelium will consequently be used throughout this chapter unless an interlamellar location is strictly implicated. Nevertheless, the vascular compartments closely associated with the filament epithelium are the venous sinuses of the filament. In its interlamellar parts, the filament epithelium is also surrounded by vascular elements belonging to the arterio-arterial vasculature. They consist of the basal blood channel of the lamellae and one to three rows of pillar capillaries that are buried within the filament epithelium.

The relationship with the arteriovenous compartment is more clearly seen on the trailing or the leading edge of the filament, where extensions of the venous sinus represent the single vascular compartment underlying this part of the filament epithelium.

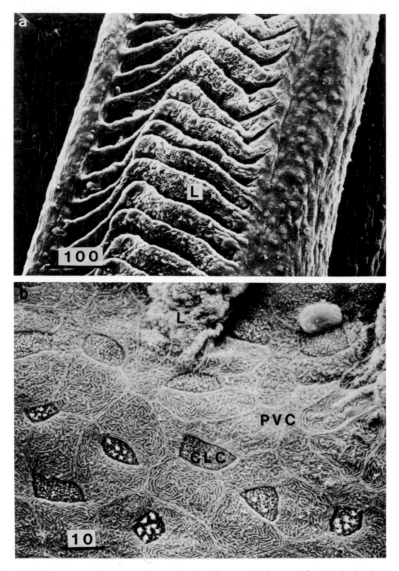

**Fig. 28.** (a) Scanning electron micrograph of a filament (*Salmo gairdneri*), the leading edge on the left. Note the asymmetrical shape of lamellae. Bar = 100 μm. (b) Enlarged view of the trailing edge. Note the pavement cells (PVC) ornamented by a complex network of microridges. Between pavement cells are intercalated apical surfaces of chloride cells (CLC). Note that chloride cells are spreading on the interlamellar epithelium. Bar = 10 μm.

This epithelium consists of five major cell types, including squamous, mucous (goblet), nondifferentiated, and chloride cells (Conte, 1969). The fifth type, consisting of numerous neuroepithelial cells, has been discovered within the filament epithelium (Dunel-Erb et al., 1982a).

## 2. SQUAMOUS CELLS

The external boundary is lined by squamous pavement cells, the surface of which has been extensively studied with the scanning electron microscope (Hughes, 1979; Hossler et al., 1979a,b; Rasjbanshi, 1977; Olson and Fromm, 1973; Lewis, 1979a,b; Hughes and Datta Munshi, 1978; Lewis and Potter, 1976b; Kendall and Dale, 1979; Kimura and Kudo, 1979; Dunel and Laurent, 1980). The surface of the filament epithelium displays a complex system of microridges. These ridges vary in length from 5 to 15 μm and in width from 0.15 to 0.20 μm, and they differ slightly in shape and size between species. Boundaries between pavement cell surfaces are generally not always discerned from the external side, although in some cases cell boundaries are clearly defined by microridges running parallel to the cell limits (Fig. 28b). Epithelial cells are polygonal in shape and measure 3 to 10 μm in diameter. They have a conspicuous Golgi apparatus. Another pavement cell that is not distinguished externally from the other squamous cells is columnar and is provided with a large number of cytoplasmic vesicles and deeply infolded parietal membranes. Squamous cells also cover the gill arch and the gill raker surfaces, and their surface ridges often form whorls.

## 3. MUCOUS (GOBLET) CELLS

The assumption that mucous cells are mainly located on the leading edge of the filaments is far from being a general rule. Mucous cells are predominant on the leading and the trailing edges as well of the filament, where they are often disposed side by side in many species (Fig. 24). A few additional cells are scattered on the interlamellar areas, where they stay close to the chloride cells (Fig. 30).

It is obvious, on account of long series of observations bearing on different species and various experimental conditions (S. Dunel-Erb and P. Laurent, unpublished), that the populations of gill mucous cells are quite variable in numbers and locations. For instance, in the white sucker, the main population of mucous cells is concentrated on the primary lamellar trailing edge. Nevertheless, from a conclusive standpoint, it is obvious that the leading edge should represent a more efficient localization in terms of distribution of the mucous sheet over the gill.

Mucous cell populations might also be affected by changes in salinity. In

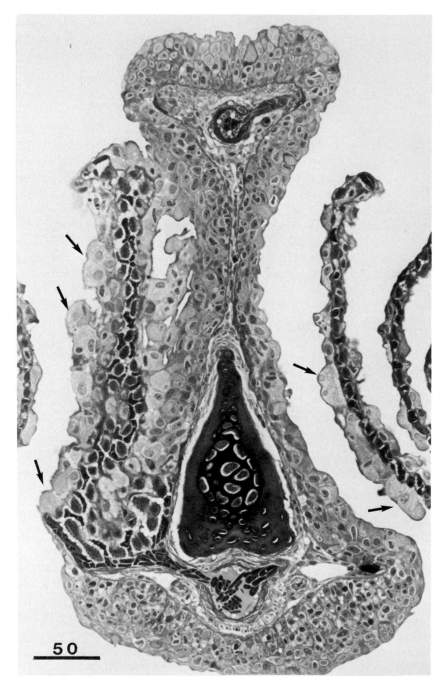

50

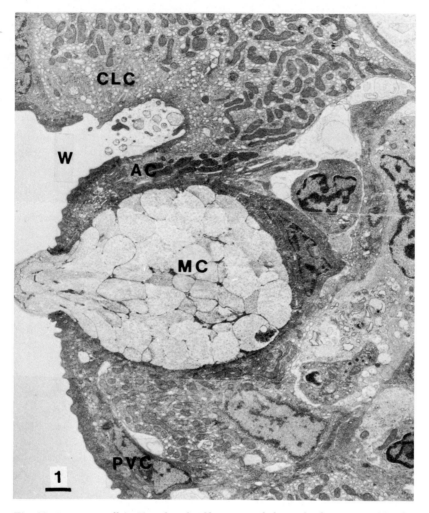

**Fig. 30.** A mucous cell (MC) within the filament epithelium of *Solea*. Note a chloride cell (CLC) and its accessory cell (AC). W, external medium. Bar = 1 μm.

**Fig. 29.** Cross section in a filament of a trout (*Salmo fario*) caught from a mountain lake (Vosges). Water of this lake has a low $Ca^{2+}$ concentration (< 0.1 mEq liter$^{-1}$). Note the proliferation of chloride cells on the lamellae and their very large apical surface in contact with the external milieu. Bar = 50 μm.

the freshwater teleosts *Gambusia* and *Calta*, mucous cells disappear completely in chloride-containing media, even at low concentration (Ahuja, 1970). In *Anguilla japonica*, it has been shown that the number of mucous cells decreases after its adaptation to seawater for a month. A euryhaline fish *Etroplus maculatus* adapted step by step from fresh water to 100% salt water has its mucous cells increasing in size but decreasing in number. At 100% salt water, mucous cells have been reduced to negligible numbers (Virabhadrachari, 1961). In *Barbus filamentosus*, goblet cells are found mostly in the epithelium of the gill arches and to a lesser extent in the filament epithelium between the lamellae. In the same article, it was also reported, on the basis of light microscopy, that mucous cells are able to transform into chloride cells after transfer from fresh water into 8% salt water (Zacone, 1981). Such observations, which need further evidence from the electron microscope, have already been reported in the freshwater, teleosts gill (Das and Srivastava, 1978) or in the abdominal epidermis of the guppy (Schwerdtfeger and Bereiter-Hahn, 1978). In the seawater-adapted eel (S. Dunel-Erb and P. Laurent, unpublished), mucous cells are less numerous on the leading and the trailing edge as well (Fig. 24a), than in the freshwater eel (Fig. 24b). In the former, however, chloride cells develop mainly on the trailing edge, where they seem to replace the mucous cells. A transformation of one type into the other has never been observed with the electron microscope.

Gill mucous cells have their development stimulated by prolactin (Ball, 1969; Bern, 1975; Mattheij and Sprangers, 1969), which reduces the gill water permeability (Ogawa, 1974). In hypophysectomized *Fundulus heteroclitus*, mucous cells degenerate (Burden, 1956), a situation that could be counteracted by injection of pituitary homogenate. A correlation exists between number of gill mucous and adenohypophyseal prolactin cells in *Gasterosteus aculeatus* (Leatherland and Lam, 1968) and *Anoptichthys jordani* (Mattheij and Sprangers, 1969). However, in the teleost *Cichlasoma biocellatum* the mucous cells of the gills are not affected by osmotic conditions (Mattheij and Stroband, 1971).

Thus, the role of mucous cells in osmoregulation is still unclear. However, the fact that mucous cells are most abundant in fresh water suggests that they might control the loss of ions or the water influx.

## 4. Chloride Cells

One of the main features of the gill filament epithelium is the presence of the so-called chloride cells, the description of which first appeared in the original work of Keys and Willmer (1932) under the name chloride-secreting cells. The name chloride cell, probably attributable to Copeland (1948), is now widely used, although such cells or cells close to this type are now often called ionocytes or more generally mitochondria-rich cells. Since the first studies with the electron microscope (Doyle, 1960; Doyle and Gorecki, 1961; Kessel and Beams, 1962; Philpott, 1962), numerous morphological and physiological studies have been devoted to chloride cells. In addition to their localization within the filament epithelium, chloride cells have been observed in the fish skin (Korte, 1979; Lasker and Threadgold, 1968), in the pseudobranch (Harb and Copeland, 1969; Dunel and Laurent, 1973), and in the opercular epithelium (Karnaky and Kinter, 1977; Foskett et al., 1981). In the gill, chloride cells are located within the filament epithelium of the interlamellar region and on the trailing edge of the filaments (Fig. 28b). They are irregularly spaced or form clusters of nonadjacent cells. In some instances, they are seen spreading along the base of lamellae resting on the pillar capillary basal lamina. In some species of coldwater fish, chloride cells are also located on the lamella (Boyd et al., 1980). More often chloride cells are lying on the basal lamina of the filament epithelium exclusively.

Chloride cells have quite specific ultrastructural characteristics, that is, a densely branched network of tubules associated with a large number of mitochondria (Fig. 31a).

The tubular system forms a three-dimensional network more or less evenly distributed within the cell except in the apical part where round and elongated vesicles appear. The distance between tubules and the outer mitochondrial membrane is often less than 10 nm. The tubule structure has been studied by different methods including freeze-fracture (Sardet, 1980). By this method, the tubules show a repetitive organization of intramembranous particles helicoidally disposed, as was already shown in *Fundulus* chloride cells in lanthanum-treated tissues (Ritch and Philpott, 1969). From these latter studies, it follows that particles are closely associated with the luminal surface of the tubular membranes. It has been suggested that these repeating particles represent a part of the $(Na^+, K^+)$-ATPase transport complex (Ritch and Philpott, 1969). There is clear morphological evidence for the continuity of the tubular lumen with the basolateral extracellular spaces (Doyle and Gorecki, 1961; Strauss, 1963; Threadgold and Houston, 1964; Dunel, 1975). The use of horseradish peroxidase as a tracer injected into the gill vasculature shows that large quantities of this protein penetrate the entire tubular system within a 15-min interval (Philpott, 1966), a result confirmed by using fixative containing lanthanum salt (Ritch and Philpott, 1969). However, these methods were unable to demonstrate a functional continuity of the tubules with the apical membrane via the vesicular complex. Nevertheless, it has been repeatedly suggested that the mechanism of ion transfer from the tubular system toward the apical lumen involves those vesicles acting as shuttles in the direction of the apical surface (Philpott and

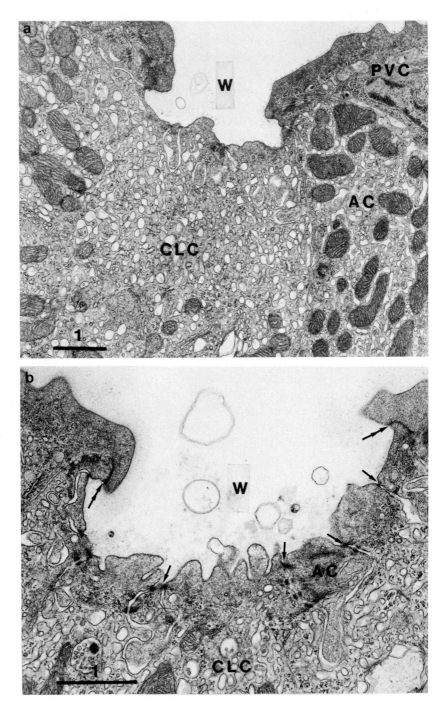

Copeland, 1963; Threadgold and Houston, 1964; Sardet, 1980). The movement of large tracer molecules across the gill seemed to support this idea (Masoni and Garcia-Romeu, 1972). However, neither the needed rate of vesicle movement accounting for the measured rate of transfer (Potts, 1977) nor the paucity of pinocytotic images (P. Laurent, unpublished) fits this hypothesis. However, the use of osmium quick-fix technique has resulted in some evidence for the existence of direct connections between the tubular and the apical membranes through a rapid vesicular shuttle. Interestingly, the same technique revealed vesicle populations of two kinds: noncoated (supposed to be exocytotic) and coated (supposedly endocytotic), the latter carrying a less concentrated solution (Bradley, 1981). It should be noted that the existence of coated and noncoated apical vesicles was previously reported from freeze-fracture observations (Sardet, 1977).

Another study points out that the intracytoplasmic membranous system of chloride cells is distributed among three independent components: (1) the tubular system, in continuity with the basolateral plasma membrane, (2) the vesiculotubular system, and (3) the endoplasmic reticulum (Pisam, 1981). The absence of relationships, particularly between the two former forms as they are shown by the lanthanum and peroxidase experiments, weakens the speculation drawn from the quick-fix experiments. In contrast, the use of periodic acid–chromic acid–silver methenamine, colloidal thorium, and [³H]glucosamine autoradiography has shown that a large amount of polysaccharides present in the apical region originates from the Golgi area, accumulates within the vesiculotubular system, and is finally released in the apical cavity (Pisam et al., 1980, 1983). In contrast, the tubular system is well colored with the lead technique of Thiery and Rambourg (1976; see also Sardet et al., 1979).

Chloride cells, which are present in both freshwater and marine teleosts, display significant differences in relation to the milieu where the fish live. These differences have been estimated both from direct comparisons be-

---

**Fig. 31.** (a) Electron micrograph of the apical surface of a chloride cell from a saltwater fish (sole). Note the presence of the accessory cells (AC) beside the chloride cells (CLC). Pavement cells (PVC) surround the chloride cell complex. Processes of the accessory cells spread within the apical part of the chloride cell. Note the richness in mitochondria within accessory and chloride cells as well. W, external medium. (b) Enlarged view of the chloride cell pit (flounder). Note the processes of the accessory cells sharing the apical surface with the chloride cells. Simple arrows show the junction between the accessory and chloride cells. This junction has a short occluding zone. Compare with the junction between chloride cells and pavement cells (double arrow). Note also the electron opacity of the apical region in both types of cell due to an accumulation of microfibrils. Different cytoplasmic organelles are visible within the chloride cells: convoluted tubules, vesicles, multivesicular bodies. The apical cytoplasm of the accessory cell seems more compact. Bars = 1 μm.

tween individuals of each group (Liu, 1942; Doyle and Gorecki, 1961; Philpott and Copeland, 1963; Karnaky *et al.*, 1976b; Dunel and Laurent, 1973; Dunel, 1975) and from the structural changes the chloride cells display when a euryhaline species is transferred from fresh water to seawater, and vice versa (Copeland, 1948; Pettengill and Copeland, 1948; Getman, 1950; Threadgold and Houston, 1964; Conte and Lin, 1967; Olivereau, 1970; Shirai and Utida, 1970; Dunel, 1975; Doyle, 1977; Laurent and Dunel, 1980; Pisam *et al.*, 1980; Hossler, 1980; Zaccone, 1981). These studies have shown that chloride cells increase in size and number and exhibit darkening of cytoplasm and alterations in the branching pattern of the tubular system as well as formations of an apical crypt in marine or seawater-adapted fish when compared with chloride cells in freshwater fish (compare Fig. 24a and b). However, these modifications are generally difficult to interpret on a functional basis. They could reflect either a more active state (size and number) or an osmotic reaction (formation of concave apical crypt in seawater versus a convex shape in fresh water). The observation of an increased density of the tubular system as well as the number of chloride cells might correlate with the increasing $(Na^+, K^+)$-ATPase activity measured in saltwater conditions (Philpott, 1980). However, the most specific and reliable structural change consists of the presence, or the development during adaptation to seawater, of an accessory cell beside each chloride cell. Accessory cells were first described in the pseudobranch and the gill of saltwater fishes (Dunel and Laurent, 1973). Later on, the development of accessory cells was observed in the eel gill during its transfer from fresh water to seawater and, in contrast, their disappearance from the mugil gill during the transfer from seawater to fresh water (Dunel, 1975).

The accessory cells form an oblong structure much smaller than chloride cells and indenting the lateral surface of the adjacent chloride cells (Fig. 31a). It is usually wedged and inserted between the chloride cell and the pavement cell layers. The accessory cells extend cytoplasmic processes that are embedded and sometimes burrowed into the apical cytoplasm of chloride cells (Fig. 31b). These processes are extensively developed in length, but are narrow and form with the proper chloride cell cytoplasm some sort of mosaic, as revealed by the scanning microscope (Dunel and Laurent, 1980) (Fig. 32). There, the plasma membranes of both types are fastened together by short zonulae occludentes of 100 to 200 Å (Dunel, 1975) that contrast with the long junctions linking chloride cells to neighboring pavement cells (Fig. 31b). Freeze-fracture reveals that those short junctions are formed by a single strand contrasting with the multistrand anastomosing network featuring the junctions linking pavement cells to each other or to chloride cells (Sardet *et al.*, 1979; Ernst *et al.*, 1980). Consequently, according to Claude and Goodenough (1973), such chloride–accessory cell junctions are thought

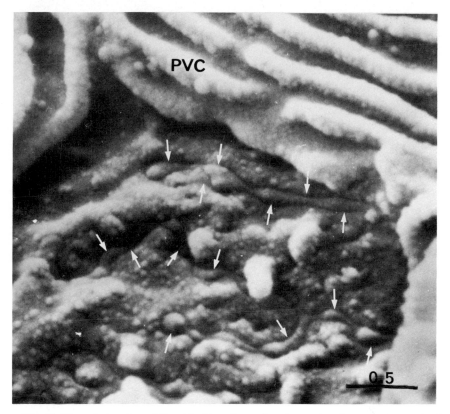

**Fig. 32.** The apical surface of the chloride cell complex. A scanning electron micrograph of the pit of a saltwater chloride cell (sole). Note the accessory cell processes (white arrows), which clearly stand out on the chloride cell surface. Note also a cluster of long microvilli. PVC, pavement cells. Bar = 0.5 μm.

to be leaky, and from that a hypothesis has been put forward that considers that transepithelial movement of $Na^+$ occurs via the intercellular junctions (paracellular pathways), whereas $Cl^-$ is actively transported across the chloride cells (transcellular pathways) (Sardet *et al.*, 1979).

The dynamics of development of accessory cells during the transfer to seawater means that these cells are by no means a developing (juvenile) chloride cell but really represent a quite distinct cell type. The evidence is as follows:

1. Examination of a very large number of chloride cells in long term-adapted fish shows that the shape and the size of the accessory cells are quite constant but smaller than chloride cells.

2. During adapataion, no form of transition corresponding to cells in the process of maturing to the size of chloride cells is observed.

3. When fish are returned to fresh water, accessory cells disappear abruptly in contrast to chloride cells.

Nevertheless the origin of accessory cells remains obscure, and the question arises whether accessory cells develop in parallel with chloride cells from some stem cells, or accessory cells develop beside preexisting chloride cells. Some observations give estimations of the duration of saltwater adaptation required to discern accessory cells and the threshold concentration of salt. Adaptation of freshwater eel for 1 week to 50% salt water causes accessory cell development, whereas none were found after several weeks at 10% salt water. *Fundulus heteroclitus* when transferred from seawater to freshwater loses its accessory cells within 5 days (Laurent and Dunel, 1980). It has also been reported that in the pinfish, *Lagodon rhomboides*, accessory cells are more numerous at 100% than 33% salt water (after 14 days) (Hootman and Philpott, 1980).

Besides the appearance of a chloride cell–accessory cell complex, there are other ultrastructural modifications within the filament epithelium that have long been associated with the transfer from fresh water to seawater. The first report of chloride cell proliferation during transfer to seawater goes back to 1942 (Liu, 1942). In *Anguilla rostrata* (Getman, 1950), it has been observed that during adaptation to seawater the number of chloride cell apical pits increases during the first 15 hr after the transfer. This phenomenon is reversible, but it is difficult to return to the initial stage. These conclusions have been subsequently confirmed by scanning electron microscopy in that changes in size of apical pits, depth, and cellular surface are already evident 6 hr after transfer to seawater and completed by 24 hr of salinity change (Hossler, 1980). It has been reported that acclimation to 8% salt water of the freshwater teleost *Barbus filamentosus* is accompanied by a thickening of the lamellar epithelium and a marked proliferation of chloride cells at the base of the lamellae (Zacone, 1981). Other observations concern the tubular system, which in the pupfish *Cyprinodon variegatus* abruptly increases when fish are transferred from 100 to 200% salt water. These observations suggest a relationship between this system and the degree of salinity below and even above the saltwater concentration—in other words, more or less proportional to the osmotic work and also as has been shown, to $(Na^+, K^+)$-ATPase gill activity (Karnaky *et al.*, 1976a,b). The meshes of the vesiculotubular system are smaller and the system more dense and regular in seawater. The endoplasmic reticulum, which is formed by dilated cisternae in fresh water, forms well-developed and anastomosed smooth sheets interdigitating with the tubular system in seawater (Pisam, 1981).

Cell renewal in the epithelium has been studied in freshwater fish, *Oncorhynchus* (Conte and Lin, 1967) and *Barbus conchonius* (McKinnon and Enesco, 1980) by [³H]thymidine autoradiography. This amino acid, rapidly incorporated into the newly synthesized DNA of dividing cells, migrates with the newly formed cells. Because at 1 hr intervals labeled nuclei were seen only at the base of gill filaments, and because at later stages these cells reach the lamellae, there is some evidence that gill epithelial cells, including chloride cells, stem from the base of the filament close to the arch and that there is no division within the filament epithelium itself. It appears from these experiments that cell migration rather than cell proliferation characterizes the gill cell renewal. These results are of some interest when considering the proliferation of chloride cells and the development of accessory cells. Unfortunately, these results concern a freshwater species and should be extended to euryhaline species during their transfer to seawater. The development of chloride cells in relation to the ionic composition of the medium seems to be controlled by unknown humoral factor(s) isolated from the corpuscles of Stannius of same teleosts (Wendelaar-Bonga *et al.*, 1976).

## 5. NEUROEPITHELIAL CELLS

A large population of cells has been described (Dunel-Erb *et al.*, 1982a) within the gill filament epithelium of teleosts and elasmobranchs as well. They have common characteristics with the neuroepithelial cells described within the wall of the respiratory tract of mammals and submammalian vertebrates (cf. Lauweryns *et al.*, 1972). Gill neuroepithelial cells are mainly featured by their capacity to fluoresce after formaldehyde treatment according to Falck *et al.* (1962), a histochemical reaction that indicates the presence of biogenic amines. The color of the emitted fluorescence, the conditions of cyclization, and the characteristics of photodecomposition of the fluorophore suggest that 5-hydroxytryptamine (5-HT) is the major monamine contained in the gill cells as well as in the mammalian neuroepithelial cells (Lauweryns *et al.*, 1973). Further evidence concerning the nature of the amine is derived from the action of parachlorophenylalanine (PCPA), a chemical that inhibits 5-HT synthesis. The presence of 5-HT has been confirmed by immunocytochemical methods (Bailly, 1983). Combined fluorescence and toluidine blue-stained semithin sections reveal that these cells are located on the serosal side of the filament epithelium (Fig. 33a,b). They do not reach the mucosal surface (Fig. 33a). They are seen clustered or isolated along the part of the epithelium facing the buccopharyngeal cavity and the inhalant water flow (Fig. 34). They are found over the full length of the filaments without exception, and no particular distribution has been seen so far in relation to the filaments or the arch position. Calculation of the number of

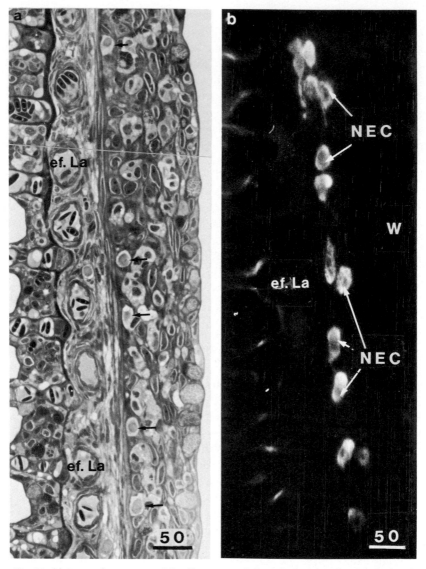

**Fig. 33.** (a) A semithin section of the filament epithelium (*Salmo gairdneri*), toluidine blue staining. The epithelium is multilayered and about 50 μm thick. The section plane is there oblique. Neuroepithelial cells are indicated by arrows. They lie directly on the basal lamina. Note the large diversity in the cellular population, suggesting that this epithelium might have several functions. (b) With the same plan as in (a), the Falck method reveals the fluorescence of neuroepithelial cells (NEC). Note that some of them have long fluorescent processes. Bars = 50 μm. (Adapted from Dunel-Erb *et al.*, 1982a.)

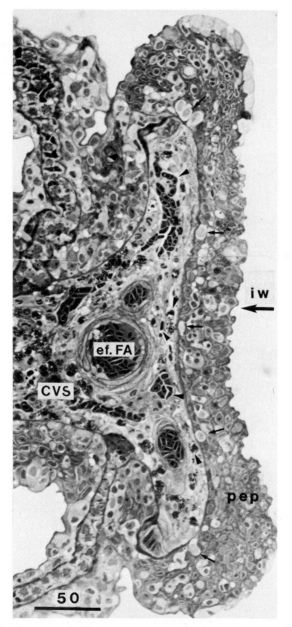

**Fig. 34.** The leading edge of a filament. Same method as for Fig. 33. Several neuroepithelial cells are indicated (arrows) within the filament epithelium facing the inhalant water flow (iw). Blood compartments of various sizes are evidenced by erythrocytes. The arteriovenous vasculature is also observed (arrowhead). Bar = 50 μm.

cells per filament in *Salmo gairdneri* is within the range of 781 to 3252 (mean value 2774), or more than $4 \times 10^6$ neuroepithelial cells per gill apparatus (S. Dunel-Erb, unpublished). With the electron microscope these cells are seen resting on the epithelial basal lamina (Fig. 35a). They are mainly characterized by dense-cored vesicles (DCV) of 80 to 100 nm. Often these vesicles appear to be empty (Fig. 35b). However, when using a procedure for specific preservation of biogenic amines (Tranzer and Richards, 1976), vesicles are seen completely filled with dense material. The DCV are generally scattered within the cells, but in many cases, the accumulation of vesicles in a part of the cell close to the basal lamina suggests that those cells are polarized. Exocytosis of DCV occurs, and sometimes isolated cores are seen just outside the part of the cell membrane resting on the basal lamina. The ultrastructure of the neuroepithelial cells depends somewhat on the species. In the eel an enlarged basal process lies on the basal lamina, and dense-cored vesicles are mostly located in this part of the cell. The spheroid nucleus, situated in the apical part, is surrounded by densely packed microfilaments. Mitochondria are grouped in two separate localizations, perinuclear and apical. In addition, a large population of clear vesicles of different sizes, coated or noncoated, are scattered within the cytosol. In the perch, the nucleus is bilobed and surrounded by a characteristic accumulation of microfilaments. Dense-cored vesicles are scattered without apparent pattern of distribution. In the trout, microfilaments are few and the Golgi apparatus and the reticulum are particularly well developed. In the catfish there are no microfilaments, but dense-cored vesicles are very numerous and packed in different parts of the cell. Those characteristics seem to indicate a high rate of cellular function presumably associated with the formation of cored vesicles.

The most significant characteristic of this cell is in its innervation. First, fluorescent nerve profiles that are located beneath the filament epithelium extend all around the whole filament. Its formaldehyde-induced fluorescence is green, presumably involving the presence of secondary amines. The electron microscope likewise reveals a dense subepithelial nervous network that consists mostly of amyelinated nerve fibers and a few myelinated fibers. From that amyelinated nerve, fibers separate and approach the basal lamina near the neuroepithelial cells (Fig. 35b), pass across the basal lamina, and join the cells (Fig. 37a). Only occasionally the reverse situation is seen:

---

**Fig. 35.** (a) A neuroepithelial cell (NEC) within the filament epithelium is seen lying on the basal lamina (bl). Note the dense-cored vesicles and small mitochondria unusually poorly preserved. Note also a nerve profile indenting in a groove of the NEC (nf) (*Salmo gairdneri*). (b) A neuroepithelial cell (NEC) separated from a vesiculated nerve profile by the basal lamina (bl). Note the presence of a smooth muscle fiber (smf) surrounding the central venous sinus (*Salmo gairdneri*). Bars = 1 μm. (From Dunel-Erb *et al.*, 1982a).

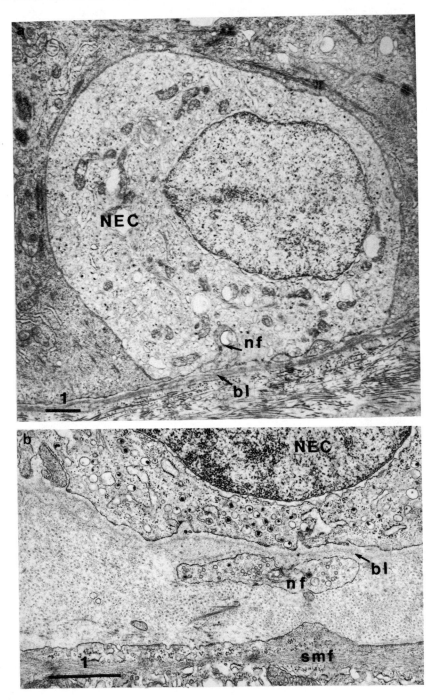

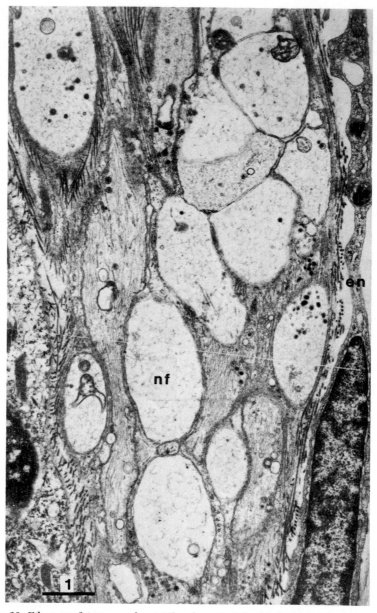

**Fig. 36.** Filament of *Acipenser baeri* (Chondrostei). Subepithelial network of nerve fibers (nf). These fibers are located between the filament epithelium (on the left) and the endothelium (en) of the central venous sinus. Note that nerve fibers form large varicosities, some of them containing dense-cored vesicles; other nerve profiles are rich in neurofilaments. The Schwann tissue is scarce. Bar = 1 μm.

neuroepithelial cells send vesiculated processes across the basal lamina and approach subepithelial nerve fibers.

Nerve profiles of the subepithelial network are varicose, and their swollen regions display a clear cytoplasm with scattered neurotubules and a few DCV (Fig. 36). In the narrower parts, neurotubules and neurofilaments are densely packed. Some profiles containing mitochondria, small clear vesicles, and dense-cored vesicles of small size might be considered preterminal. But it is not definitively settled whether or not they take a part of the observed fluorescence (Fig. 26b). After they have crossed the basal lamina, some fibers come into contact with the neuroepithelial cells. Usually more than one profile is seen apposed on the same cell, indicating that several nerve fibers are related to a single cell or that the same fiber gives several contacts *en passant* (Fig. 37a). The former pattern seems to be probable, since two types of profiles are observed. In the first type they are small and contain a dense population of clear vesicles and a few cored vesicles. In the second type, they are large and contain few organelles and large mitochondria (Fig. 37b). In both cases, however, accumulation within the neuroepithelial cells of dense-cored vesicles on the cell membrane facing the nerve profile indicates synaptic relationships (from the cell to the nerve) in a manner already described in the carotid body or in the so-called intensely fluorescent cells (see Taxi, 1971).

As has already been pointed out, the subepithelial region near the cells, which consists mostly of connective tissue (besides the nerves), surrounds a large and complex vascular compartment (the central venous sinus and its processes lined by a thin endothelium). In this region, there are also the basal part of the lamellae, their basal blood channels, and the efferent lamellar arteries that collect the arterialized blood. Due to the presence of this complex system of vascular compartments, the minimum distance between the blood and the neuroepithelial cells does not exceed 20 μm at some places.

Neuroepithelial cells from hypoxic trout have been studied, and several ultrastructural changes have been observed. The degranulation of the vesicles suggests that gill neuroepithelial cells might release some substances, presumably serotonin, into the surrounding tissues in response to hypoxia. The released serotonin might have some effects on epithelial transport (Legris *et al.*, 1981); alternatively, they might activate (through the synapses) the afferent nerve fibers, a mechanism similar to that already postulated for the carotid body.

Finally, it should be mentioned that neuroepithelial cells have been described in all species of teleosts considered so far, including the rainbow trout (*Salmo gairdneri*), the catfish (*Ictalurus melas*), the black bass (*Micropterus dolomieui*), the eel (*Anguilla anguilla*), and the perch (*Perca perca*).

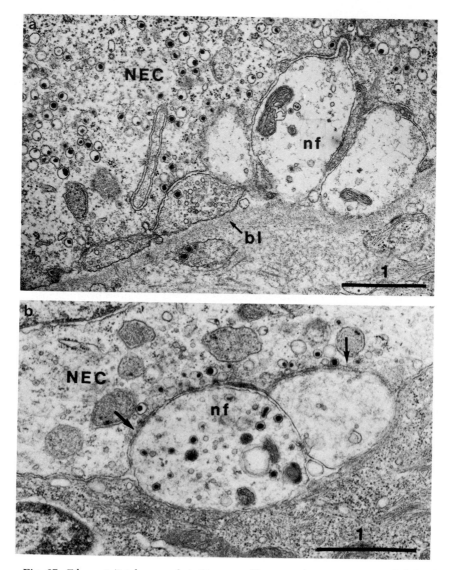

**Fig. 37.** Filament (*Ictalurus melas*). Nerve profiles are indenting a neuroepithelial cell (NEC). Note that three profiles are poor in organelles, one is completely encased within the cell, another profile is provided with clear vesicles, and a profile is seen on the other side of the basal lamina (bl). (From Dunel-Erb *et al.*, 1982a.) (b) Filament (*Ictalurus melas*). Two nerve profiles synapse with a neuroepithelial cell (NEC). Note the accumulation in the NEC of dense-cored vesicles close to the part of the membrane delimited by the arrows and the presence of dense material associated with the membrane. Such an organization is currently considered as efferent (centripetal) synapse. Note also the various types of vesicles within the nerve profiles. Bars = 1 μm.

## 6. Other Cell Types

In addition to the neuroepithelial cells, several other types of granulated cells have been found within the filament epithelium. Cells of 8 to 10 μm in diameter, with short, large processes containing dense-cored vesicles of 1500 to 1800 μm, have been found in some species (e.g., *Ictalurus melas*). These cells apparently have no contact with nerve endings and are located in the middle part of the filament epithelium. For those reasons, they must be considered to be different from neuroepithelial cells. Another component of this epithelium consists of cuboidal cells devoid of any granular vesicles but rich in microfilaments anchored on desmosomes. These cells have many mitochondria, free ribosomes, and a well-developed rough endoplasmic reticulum. They are probably actively involved in protein synthesis and presumably serve as a mechanical strengthening. Cells containing large vacuoles filled with osmiophilic material are also seen in large clusters surrounding the afferent and efferent filament arteries in the sucker and the brown bullhead (P. Laurent, unpublished). Their function is still unknown. Finally, nondifferentiated cells are abundant, particularly within the innermost epithelial layer.

## 7. Variations of the Filament Epithelium in Lower Groups

Chloride cells are also present in the filament epithelium of the Holostei, *Lepisosteus* and *Amia* (Laurent and Dunel, 1980). Their morphology is the same as in teleosts. In Chondrostei, the freshwater species *Acipenser baeri* has chloride cells that display a poorly branched tubular system and a well-developed vesicular apical system. The euryhaline forms of Chondrostei possess chloride cells whose number increases in response to an increased salinity (Chusovitina, 1963). Unfortunately, no data are yet available concerning the presence of accessory cells in marine forms of Chondrostei.

In elasmobranchs, the first description of chloride cells from *Scyliorhinus canicula* concerns a pear-shaped cell type having a basolateral membrane deeply infolded similar to many other transporting cells. It has a narrow apical end in contact with the external milieu. Those cells are rich in mitochondria and vesicles (Wright, 1973).

Subsequent comparative studies in the skate *Raja clavata* and the dogfish *Scyliorhinus canicula* actually revealed the existence of two types of chloride cells in these species (Laurent and Dunel, 1980). In the first type the apical membrane is deeply buried in a cul-de-sac connected to the external milieu by a narrow opening (Fig. 38a); in the second type, the apical membrane protrudes outward (Fig. 38b). Both types have long microvilli; however, the most evident characteristic of both cells is a lack of the tubular system found

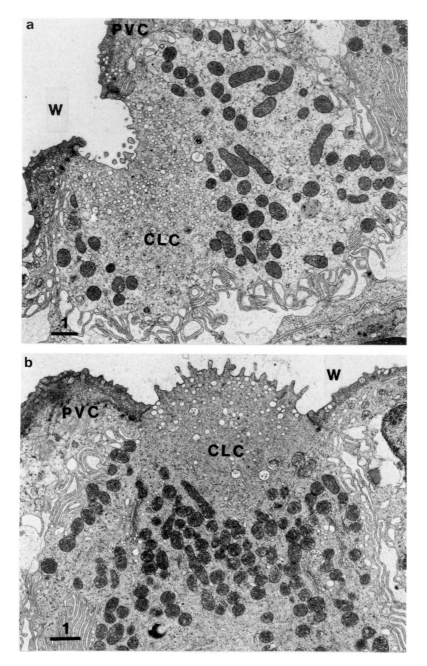

in teleosts—a system that is probably functionally replaced by numerous basolateral membrane infoldings. As in teleosts, abundant vesicles are present in the apical and the central regions of the cell. Numerous mitochondria are peripherally disposed in close relationships with the basolateral infoldings. Also present are a conspicuous Golgi apparatus and numerous endoplasmic reticulum cisternae.

Chloride cells and other gill epithelium components have been extensively studied in cyclostomes (Morris, 1957, 1972; Pickering and Morris, 1973, 1976; Morris and Pickering, 1975; Nakao and Uchinomiya, 1974; Freeman, 1974; Youson and Freeman, 1976; Nakao, 1977; Peek and Youson, 1979a,b) in different species at different stages of their life. It is well known that, for instance, the adult of *Petromyzon marinus,* when living in seawater (parasitical form), experiences osmoregulatory problems quite different from the larva (ammocoetes), which lives on a sedentary filter-feeding basis in freshwater streams. This larval stage lasts $5 \pm 2$ years (Thomas, 1962). After that stage, larvae finally metamorphose and, if not land-locked, start their catadromous migration. Thus, the epithelial chloride cells, as far as they are involved in cyclostome osmoregulation in addition to gut skin and kidney, have to fit saltwater hypo-osmoregulation problems during adult life and freshwater hyperosmoregulation problems during the larval stage.

In ammocoetes (Youson and Freeman, 1976), three types of cells constitute the interlamellar epithelium (also called osmoregulatory epithelium; Nakao and Uchinomiya, 1978): superficial, intermediate, and basal, according to their situation. Superficial cells are of three types. The first type, present in the early larval stages (stages 1 and 2), has large numbers of mitochondria. The endoplasmic reticulum is made up of short single segments, and the Golgi complex is small. The apical cytoplasm contains varying numbers of vesicles that do not seem to be involved in the secretion process. These cells are cuboidal or columnar in shape. The second type of superficial cell, still present in the premetamorphosis stage (stage 5), is rich in mitochondria but less than in the first type. The apical cytoplasm is very rich in vesicles that do not contain any visible matrix. Some other vesicles are apparently secretory, containing a more or less dense matrix. Golgi and endoplasmic reticulum are also present but of moderate importance. The third type of superficial cell is present throughout the entire development cycle. This type of cell is characterized by numerous mucous droplets that appear to secrete their content outside. A moderate number of mitochondria

**Fig. 38.** Chloride cell-like system in an elasmobranch (*Raja erinacea*). Note the convex (a) and the concave (b) apical surface. These two figures suggest either two different types or two different states of the same type. Note the richness in mitochondria and the laterobasal infoldings of the membrane in both cases. Bars = 1 μm.

are present. Intermediate cells possess a large central nucleus and numerous mitochondria dispersed within the cytoplasm. Rough endoplasmic reticulum and Golgi are of moderate importance. They progressively decrease in number and are completely absent in the adult stage. Basal cells separate the epithelium from the blood spaces. They do not have special characteristics and could be considered as nondifferentiated cells.

Experimental incorporation of [$^3$H]thymidine (autoradiography) causes an intense labeling in the preadult stage and very little during the early larval stage and the adult stage. In the interlamellar epithelium, most labeled cells belong to the group of basal cells. In contrast, a cell's degenerative process is particularly present during the metamorphosis. These data suggest that an active process of new cell formation occurs in preadult stages, as well as cell degeneration. Instances of cell division are also observed during all stages of metamorphosis (stages 4 and 5) (Peek and Youson, 1979a).

Typical chloride cells are already present in young freshwater adults. In the adult lamprey, chloride cells are characterized by an important tubular system (Nakao, 1974; Youson and Freeman, 1976). Adaptation to seawater leads to further proliferations of tubules and mitochondria (Peek and Youson, 1979a,b). Chloride cells differentiate from intermediate cells. They are columnar and elongated at stage 5. The tubules develop within the peripheral cytoplasm of intermediate cells; contacts with the basolateral membrane are rare, if they occur. Tubules proliferate toward the interior of the cell and by the end of stage 7 occupy most parts of the cell. Concomitantly, tubules become more and more branched and oriented in parallel direction with the sero–mucosal axis. Mitochondria develop in association with the tubules and become very numerous. As in teleost chloride cells, extracellular tracers enter the tubular system but not the apical vesicles, suggesting a complete anatomic separation of the two systems. Interestingly, most of the adult chloride cells develop within the intermediate epithelial zone, overlapped by superficial cells; later on they come into contact by their apex with the external medium. Cells are joined by junctional complexes, and there is no mention of a particular type of leaky junction or of the presence of accessory cells as in teleosts. Newly differentiating chloride cells are seen just beneath the functional chloride cells. These incompletely differentiated cells contain a large central nucleus, numerous free ribosomes, and a rich population of mitochondria. Some peripherally disposed tubules are also present. Presumably these intermediate chloride cells stem from nondifferentiated basal cells, as has also been suggested in teleosts (Laurent and Dunel, 1980). The origin of the tubules in adult lamprey chloride cells is still unclear. The endoplasmic reticulum as well as the Golgi complex should be involved. Although a significant increase of Golgian saccules and microvesicles were observed (Peek and Youson, 1979b), it has been concluded that the importance of the Golgi apparatus might hardly account for so considerable a tubular proliferation. Similar questions were raised in order to elucidate the tubules' differentiation of chloride and pseudobranch cells in the trout fry (Dunel and Laurent, 1973; Dunel, 1975). The conclusion of these studies was that the tubular system nevertheless develops from vesiculization of the Golgi apparatus and that coalescence with the plasma membrane occurs later.

Degeneration of chloride cells, a constant process in adults, is witnessed by images of vacuolization and by the presence of electron-opaque bodies. Disorganized tubules and mitochondria also characterize those degenerating cells. Degeneration of chloride cells in freshwater stages suggests that these cells are actively functioning rather than representing quiescent, nonworking cells. It also suggests that the saltwater chloride cells are new developing cells rather than transformed freshwater cells. This point throws light on the dynamics of epithelium development in teleosts. In other words, the concept of the de novo formation of the chloride–accessory cells complex deserves some attention and needs experimental confirmation.

The gill epithelium of Dipnoi once again contrasts with that of other groups of fish. The poorly developed arborescent gills have only a single type of epithelium that is more adapted for hydromineral than gaseous exchanges (Oduleye, 1977). *Lepidosiren* (Wright, 1973; Laurent *et al.*, 1978), as well as *Protopterus* (Laurent *et al.*, 1978), have a thick epithelium made up of several layers of cells lying on a thick basement membrane. Large intercellular channels are formed by the infoldings of basolateral membranes of epithelial cells. Most epithelial cells contain numerous small round mitochondria surrounded by an electron-opaque cytoplasmic matrix. The outermost cells have microvilli at their apical surface and are linked to each other by long zonulae occludentes, suggesting a tight junction according to the current concept (Claude and Goodenough, 1973). They also have numerous mitochondria and many vesicles. This structural background suggests a transport function. It might be also noticed that dipnoan epithelial cells, another type of mitochondria-rich cell, look like the elasmobranch type (Laurent and Dunel, 1980).

In conclusion, the filament epithelium shows a marked similarity throughout the different groups of fish. If we set aside the neuroepithelial cells present in all the species studied so far including lower group representatives, the main feature of this epithelium is the presence of ion-transporting cells in freshwater as well as in saltwater species. However, they display some specific differences, for example the structure of their basolateral membranes.

Neuroepithelial cells have been described in several species of teleosts (Section III,A,5). Their presence in lower groups is still unknown except in elasmobranchs. In the dogfish (*Scyliorhinus canicula*), they have organization similar to teleosts, except that the dense-cored vesicles are larger (> 1000 nm, as usual in many nervous structures from this group). A large population of mitochondria is packed within the basal part of the cell, where most of the dense-cored vesicles are also concentrated. Large amounts of glycogen granules are also seen in this area. Golgi cisternae and rough endoplasmic reticulum are particularly well developed in this elasmobranch. In contrast, neuroepithelial cells in *Scyliorhinus* are often grouped in clusters of four to six cells, all of them lying on the basal lamina. Functionally important might be the fact that the cluster in its entirety is surrounded by large systems of intercellular spaces with interdigitating finger-like processes from the neighboring cells (Fig. 39).

## B. The Lamellar Epithelium

### 1. ORGANIZATION AND CELLULAR COMPONENTS IN TELEOSTS

This epithelium lines the terminal branching of the gills. We have already described the structure of the vascular space, that is, the so-called pillar capillary including the basal lamina on which pillar cells are anchored. On its vascular side, the basal lamina is normally underlaid by a broad zone of interstitial connective tissue containing fibrils of collagen. On its other side, the basal lamina lines the secondary (lamellae) epithelium. This epithelium is often called respiratory epithelium in contrast with the non-respiratory or osmoregulatory one that surrounds the core of the filament. There is no doubt that most of the gill respiratory exchanges occur across this epithelium (Hughes, 1964). The resistance to gas transfer depends on its structural characteristics, including its mean thickness. That explains the early interest in comparative studies bearing on the form and size of epithelial cells (Bevelander, 1935). Modern studies involving morphometric measurements developed an attempt to correlate structural characteristics of this epithelium with habitat and behavioral patterns of fish.

From a developmental standpoint, it is well established that this epithelium stems from nondifferentiated cells of the filament epithelium (Morgan, 1974b; Dunel, 1975). This means that some cells keep the potentiality of their development in cell types similar to some of the filament epithelium (Laurent and Dunel, 1980; Boyd *et al.*, 1980). The question has been debated whether the basal lamina is produced by pillar cells (Hughes and Grimestone, 1965) or by epithelial cells (Leeson and Leeson, 1966). It has been suggested that because of the intimacy of contact between the basal

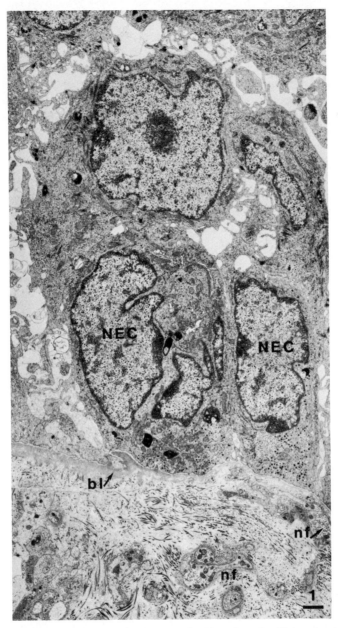

**Fig. 39.** A cluster of neuroepithelial cells (NEC) within the filament epithelium of *Scyliorhinus canicula*. This structure is very close to some neuroepithelial bodies of the superior airways of higher vertebrates. Note that the cluster lies on the basal lamina but does not reach the epithelial surface. Note also nerve profiles (nf) within the connective tissue underneath the basal lamina (bl). The cell group is surrounded by large extracellular spaces. Bar = 1 μm.

lamina and the surrounding connective tissue, the former takes its origin from the latter rather than from the epithelium (Newstead, 1967). The basal lamina has already been described with the pillar cells in Section II,B,2.

The first studies with the electron microscope of the lamella and its epithelium concern *Haplochromis multicolor* (Schulz, 1962), *Lebistes reticulatus* and *Callionymus lyra* (Hughes and Shelton, 1962), and *Pollachius virens* (Rhodin, 1964). They all emphasize the existence of two distinct epithelial layers—an inner one (basal or serosal) and an outer one (apical or mucosal)—and between them, an intercellular space that communicates with the central venous sinus through the intercellular spaces of the filament epithelium (Rhodin, 1964). Cells of both layers are differently characterized, and it has been pointed out from electron-microscopic studies that the mucosal side cells are particularly rich in organelles, indicative of a great activity (Hughes and Grimstone, 1965; Newstead, 1967; Laurent and Dunel, 1980). Most of the mucosal layer consists of pavement cells. They possess a well-developed Golgi apparatus, an abundant rough endoplasmic reticulum, and numerous vesicles of different sizes that often open through the basal or the apical plasma membranes (caveolae). The functional meaning of this obviously intensive activity is unknown. Among possibilities, it could be maintained that one of the major properties of such an epithelium is to be as impermeable as possible to ions and water, thus avoiding an exaggerated gain (in seawater) or loss (in fresh water) of ions across a surface that is actually about twice as large as the external surface area of the fish body (Hughes and Morgan, 1973). How this impermeability could be obtained depends on the membrane structure, the arrangement of intercellular junctions, and the presence of a cell coat. The membrane structure has been studied by freeze-fracture (Sardet, 1977). These studies revealed that the membrane of the mucosal layer has very smooth fracture faces and contains particles often aggregated in patches on the P face with corresponding pits on the E face. Those particles are of homogenous size (85 Å) and are arranged in hexagonal arrays. Vesicles attached to the membrane by a peduncular process (pinocytotic vesicles) also have aggregated particles on their P face. It remains uncertain whether this structural feature is the support of a particular function related to the aquatic environment.

Superficial cells are linked by tight junctions and desmosomes, but no gap junctions are observed. Examination by freeze-fracture of the tight junctions in *Mugil capito* adapted to fresh water or seawater revealed that these junctions are long and consist of five to nine anastomosed strands. This pattern of arrangement, which does not vary with the milieu where the fish lives, suggests that the lamellar epithelium is "tight," according to the usual classification (Claude and Goodenough, 1973), and therefore probably impermeable to electrolytes. These epithelial junctions do not let lanthanum penetrate. Another peculiarity is that the apical cell surface is underlaid with microfilaments and displays a continuous coat of fuzzy material. That is a feature common to several other epithelia. This coat contains acidic groups of polysaccharides. Several studies deal with the scanning electron micrography of the lamellar surface (Olson and Fromm, 1973; Lewis and Potter, 1976b; Rajbanshi, 1977; Kendall and Dale, 1979; Hossler et al., 1979a,b; Hughes, 1979; Lewis, 1979b; Dunel and Laurent, 1980). They show the presence of microridges that are thought to anchor the mucous coat. This question is studied in detail elsewhere in this volume (see Chapter 1).

The innermost or serosal layer consists of non- or poorly differentiated cells lying directly on the basal lamina. The nucleus : cytoplasm volume ratio is high, and the presence of numerous free ribosomes and well-developed rough endoplasmic reticulum suggests that these cells are starting to differentiate (Laurent and Dunel, 1980). Presumably they serve as replacements for worn-out superficial layer cells as do the basal cells in the filament epithelium. They also might occasionally differentiate into chloride cells (Laurent and Dunel, 1980).

Between the superficial and the basal layers are extracellular spaces (sometimes erroneously called lymphatic spaces). This extracellular compartment shows large variations in size that are difficult to relate to known parameters. Some extracellular lacunae are large enough to contain lymphocytes. From serial histological section examinations it is clear that these lacunae communicate with primary epithelium extracellular spaces. This arrangement suggests that as a result of a probably negative hydrostatic venous pressure gradient, the extracellular fluid drains off from the lamellar epithelium into the central venous sinus.

It has been reported that the physiological functions of respiration, excretion, and the maintenance of acid–base balance might occur through the lamellar epithelium (Girard and Payan, 1980). For instance, the so-called respiratory cells, the constituent of the serosal layer, are the site for $Na^+/NH_4^+$ and $Cl^-/HCO_3^-$ exchanges (De Renzis, 1975; Payan, 1978). Measurements of mannitol permeability give a value about 400 times higher for filament than for lamellar epithelium in seawater-adapted trout. This fact might be related with the occurence of leaky junctions between chloride and accessory cells (Sardet et al., 1979). In seawater, $Na^+$ and $Cl^-$ excretion occurs across the filament epithelium. A net influx of these two ions is supposed to occur in fresh water through the lamellar epithelium (see Girard and Payan, 1980).

## 2. The Lamellar Epithelium and the Milieu

Attempts were made to correlate structural changes in the lamellar epithelium with the external milieu characteristics. Transfer from fresh water to seawater or careful examination of the lamellar epithelium from freshwater and marine fishes do not support the conclusion of seawater-linked structural characteristics (Laurent and Dunel, 1980). Neither the organization of intra-membranous elements nor the morphological characteristics of the junction undergoes the effect of salinity changes (Sardet et al., 1979). Changes in water characteristics could affect the composition and the distribution of mucus over the lamellae. Mucous cells are predominantly located within the epithelium of the filaments on their leading edge and their trailing edge as well (see Section III,A,3). A few additional cells are scattered in the interlamellar areas, but localization within the lamellar epithelium seems to be quite rare (Morgan and Tovell, 1973), although it has been observed in some coldwater teleosts (Boyd et al., 1980) and in the icefish Chaenocephalus aceratus (Hughes and Byczkowska-Smyk, 1974). In addition, it has been observed that mucous cells are present on the lamellae of the white sucker and the brown bullhead acclimated for a short period of time in acidic water (pH 4) (P. Laurent and H. Hōbe, unpublished).

Mucus, secreted mainly from filament epithelium mucous cells, flows along the gill surface and finally covers the lamellae. It is generally accepted that the lamellar superficial structures play an important role by anchoring the mucous film (Hughes, 1979). Possible interplay has often been considered between mucus and the amplitude of gas exchanges across the lamellar epithelium. In relation to this, it has been mentioned that due to the presence of a mucous layer, the development of microridges does not functionally increase the exchange surface area. Mucus, which has an intermediate value of diffusion constant ($2.6 \times 10^{-5}$ cm$^2$ min$^{-1}$ atm$^{-1}$) between water and the tissue barrier, probably affects gill gas exchanges by addition of a nonconnective layer to the lamellae (Ultsch and Gros, 1979). Such a mucous barrier, which is affected by changes in salinity (see Section III,A,3), might also affect ionic permeability by physicochemical "trapping" or any other chemical actions.

Modifications of the lamellar epithelium in relation to the oxygen partial pressure in the external medium have been investigated by morphometric studies. From measurements of the harmonic mean epithelial thickness (water to blood distance), it is possible to evaluate the diffusing capacity of the gills (Hughes, 1972). Changes in thickness can be roughly correlated with the degree of hypoxia and the temperature. Experiments results show that the rainbow trout acclimated at 10°C does not undergo significant changes in epithelial cell size after 5 days of hypoxia, whereas at 18°C a significant decrease of epithelial cell volume was observed (Soivio and Tuurala, 1981). This modification is interpreted as the result of lamellar vasculature distension, which in its turn increases the functional gill surface, decreases the water to blood diffusion distance, and adjusts the lamellar orientation in the respiratory water flow. A longer adaptation to hypoxia (4 weeks), especially in a species that has a rather thick lamellar epithelium in normoxia (i.e., Ictalurus melas), at 20°C leads to a significant (50%) decrease in thickness (P. Laurent and G. Bombarde, unpublished results). The existence of "tertiary" lamellae or infoldings of the lamellae reported in *Barbus sophor* might have some implication in term of gas exchange (Hughes and Mittal, 1980).

Although it has already been mentioned in this chapter that chloride cells normally develop within the filament epithelium or alongside the base of the filaments (interlamellar region), there are experimental or natural circumstances where chloride cells develop up to on the lamellae. Then, chloride cells become an important component of an atypical lamellar epithelium to such an extent that gas exchanges are partially impeded. Keeping a fish in deionized water for 4 weeks leads to a strong stimulation of chloride cell proliferation (Matteij and Stroband, 1971). As shown by Laurent and Dunel (1980), chloride cells develop from the serosal cell layers. In the trout as well as in the eel held in deionized water for 2 weeks, chloride cells display an ovoid shape, a clear cytoplasmic matrix, numerous small and round mitochondria, and a loosely branched tubular system. Those cells have an extremely large contact area of their apical membrane with the external milieu. Proliferation of chloride cells is a reversible phenomenon. Animals transferred from deionized water into tap water or 10% salt water display a rapid covering of the chloride cell apical membrane by pavement cells. Thereafter, many chloride cells, now locked within the epithelium, form large lysosomes and degenerate. Pathological conditions such as skin wounds, fungal diseases, and confinement in a crowded hatchery also produce aberrant proliferation of chloride cells. In the former case, proliferation might be induced by an exaggerated and uncontrolled ion loss. In the second condition, accumulation of waste or any kind of acid–base imbalance might be implicated.

Observations from the wild are also consistent with experimental results. In gill removed from trout indigenous to natural hard water ($[Ca^{2+}] = 4$–5 mEq liter$^{-1}$), chloride cells are restricted to the primary lamellae. In gill removed from trout raised in soft water ($[Ca^{2+}] = 0.15$–0.20 mEq liter$^{-1}$) or even acclimated for 15 days to natural soft water, chloride cells covered almost the entire surface of both filaments and lamellae (Fig. 29). On return to tap water, there was a rapid covering of chloride cells characterized by a very large convex apical surface and a less dense tubular network, features suggestive of cell swelling. These results suggest that proliferation of this cell type would therefore represent an adaptive response to dilute external media (Laurent *et al.*, 1984b).

The reason that chloride cells then invade the lamellar epithelium is still unclear. Two possibilities might be considered: (1) a rapid migration from the filament epithelium where no more room is vacant or (2) a differentiation in situ of some quiescent cells of the innermost layer of the lamellar epithelium.

Thus, it could be concluded from those histophysiological data that a transfer of fish from fresh water into seawater induces a chloride cell proliferation within the filament epithelium. This transfer is known to stimulate an outward ion flux. However, transfer from fresh into deionized water or any other circumstances favoring an increased ion loss lead to the development of chloride cells even and perhaps mainly within the lamellar epithelium. Two reasons might be put forward: in deionized water, the absence of $Ca^{2+}$ increases the membrane permeability (Ogawa, 1974, 1975; Pic and Maetz, 1975, 1979). In contrast, there is now evidence for a $Ca^{2+}$ uptake from the external milieu by chloride cells, which should have a function in fresh water more related to $Ca^{2+}$ than to $Cl^-$ or $Na^+$ uptake (Payan et al., 1981). In low-$Ca^{2+}$ medium, chloride cells have their apical membrane largely open to the outside, and this is particularly relevant to this mechanism.

Other variations in the lamellar epithelium arrangement have been observed in teleosts. These variations are supposedly correlated with ecological or behavioral characteristics. For instance, the water to blood distance in a fast-swimming, $O_2$ high-consuming fish like *Scomber scombrus* or tuna is small because of a particularly thin lamellar epithelium (Hughes, 1972), in contrast with sluggish, $O_2$ low consumers like the toadfish (Hughes and Gray, 1972). Another example of variation in lamellar epithelium structure has been reported in the icefish, *Chaenocephalus aceratus,* an Arctic fish in which numerous mucous cells occupy a large part of the lamellar surface. These mucous cells reduce the area available for gas exchanges in addition to an already thick epithelial layer (Hughes and Byczkowska-Smyk, 1974).

Finally, and in addition to what has already been said about chloride cell proliferation in soft or acidic water, artificial adverse conditions lead to drastic modifications of the lamellar epithelium. Pollution by heavy metal or organic compounds is responsible for lethal structural disorders (Davis, 1972; Hughes and Perry, 1976; Hughes et al., 1979; Skidmore, 1970; Tuurala and Soivio, 1982). It is interesting to note that in the fingerlings of rainbow trout, chloride cells are abundant on the lamellae (Kimura and Kudo, 1979). This situation, which contrasts with adults, needs further studies in larval or nonmature forms.

## 3. Variations of the Lamellar Epithelium in Lower Groups

In elasmobranchs, the lamellar epithelium structure is not very different from the description given for teleosts. Lamellae are covered by a double layer of epithelial cells lying on a basal lamina. Between these two layers are intercellular spaces containing leukocytes and macrophages. The cells of the outer layer display a prominent terminal web close to their apical surface, often intermingled with dark vesicles. External cells join to each other by tight junctions in association with the terminal web. Desmosomes are present at this point precisely. Another type of cell has been described, characterized by cored vesicles or multivesicular bodies; its apical side projects outward (Hughes and Wright, 1970; Wright, 1973).

In Chondrostei, the outermost layer cells display large vesicles filled with a structured material and located just beneath the apical membrane. These microbodies, in *Acipenser*, are reminiscent of peroxisomes. They are in such close contact with the apical plasmalemma that they look if as they are open outward. The plasma membrane is, on its external side, covered by a filamentous material (Laurent and Dunel, 1980).

Among crossopterygians, the lamellar epithelium in *Latimeria* and in *Neoceratodus* displays the same type of organization: a bilayered epithelium whose outer surface is covered with microvilli (Hughes, 1980). It is noteworthy that in this fish also there are large populations of dense vesicles in close association with the apical membrane. Thus, apparently those vesicles are common characteristics in lower groups of fish and differentiate them from teleosts.

In Holostei, the lamellae display several structural characteristics supposedly related to a bimodal respiration (Dunel and Laurent, 1980; Olson, 1981; Daxboeck *et al.*, 1981). Epithelial cells of the outer layer also display numerous vesicles of different sizes and opacities. In addition, osmiophilic lamellar bodies of 0.4 to 0.6 μm in diameter are often seen isolated or in clusters. They are similar to the pneumocyte lamellated bodies also present in the lung of *Amia* and dipnoan lungfish. Interestingly, the gills in *Amia* are supposed to carry out gas exchanges during air exposure.

In cyclostomes, lamellae are covered by two layers of epithelium as in higher groups (Peek and Youson, 1979a). The space between these two layers is homologous in teleosts and cyclostomes. Since most investigators agree that there is no lymphatic tissue in lampreys (Fänge, 1972), the often used denomination of lymph space is irrelevant, at least in cyclostomes. Interestingly, the size of these spaces declines after metamorphosis in *Lampetra* (Lewis, 1976) and is particularly large in the larval lamprey *Gestria australis* (Lewis and Potter, 1982). In addition, differences have been found in the gill morphometrics of larval and adult stages that might relate to different behaviors of these two forms (Lewis and Potter, 1976a).

## IV. CONCLUDING REMARKS

This chapter has been long enough to outline the characteristics of the gill internal morphology in some detail.

From what has been said, a classification of the gill apparatus can be made on the basis of anatomy, giving three main groups: (1) the septal gills, which are encountered in cyclostomes and elasmobranchs, (2) the branchial arch gill, present in Holostei, Chondrostei, and Teleostei, and (3) the dipnoan type, which approaches the arrangement seen in the external amphibian gills.

In spite of these obvious external differences, the internal gill morphology is remarkably constant, for example, an association of an arterio-arterial and an arteriovenous circulation. The constancy of this organization might be explained on the bases of the gill functions that deal in all groups of fish with respiration and with acid–base and ionic regulation. In all cases, the structures involved in ionic regulation—chloride cells or other types of mitochondria-rich cells—are facing the venous compartment.

## ACKNOWLEDGMENTS

The author is greatly indebted to Dr. S. Dunel-Erb, who composed most of the illustration plates, and to the following departmental staff members: Mr. J. C. Barthe, our photographer, Mrs. C. Chevalier and Mr. G. Bombarde, our technicians, Mrs. G. Biellmann, who completed the references, and Mrs. M. Heinrich, who typed the manuscript.

## REFERENCES

Acrivo, C. (1935a). Sur l'organisation et la structure du corps caverneux chez *Scyllium canicula* Cuv. *Biol. Histol. Appl. Physiol. Pathol.* **12**, 362–372.

Acrivo, C. (1935b). Uber die Neubildung von Kiemenlamellen bei *Scyllium canicula* Cuv. *Zool. Anz.* **109**, 173–176.

Acrivo, C. (1938). Beobachtungen über die Morphologie und Struktur der Kiemenlamellen der Ganoiden. *Bul. Soc. Sci. Cluj. Biol.* **9**, 9–31.

Ahuja, S. K. (1970). Chloride cell and mucous cell response to chloride and sulphate enriched media in the gills of *Cambusia affines* and *Catla catla*. *J. Exp. Biol.* **173**, 231–250.

Allen, W. F. (1907). Distribution of the subcutaneous vessels in the head region of the ganoids, *Polyodon* and *Lepisosteus*. *Proc. Wash. Acad. Sci.* **9**, 79–158.

Allis, E. P., Jr. (1912). The pseudobranchial and carotid arteries in *Esox, Salmo* and *Gadus*, together with a description of the arteries in the adult Amia. *Anat. Anz.* **41**, 113–142.

Allis, E. P., Jr. (1920). The branches of the branchial nerves of fishes, with special reference to *Polyodon spathula*. *J. Comp. Neurol.* **32**, 137–153.

Bailly, Y. (1983). Recherches sur quelques innervations branchiales des Poissons. Thesis, Cytology. Université P. et M. Curie, Paris.

Ball, J. N. (1969). Fish prolactin and growth hormone. *In* "Fish Physiology" (W. S. Hoar and D. J. Randall, eds.), Vol. 1, pp. 207–240. Academic Press, New York.

Bern, H. A. (1975). Prolactin and osmoregulation. *Am. Zool.* **15**, 937–948.

Bettex-Galland, M., and Hughes, G. M. (1973). Contractile filmentous material in the pillar cells of fish gills. *J. Cell Sci.* **13**, 359–370.

Bevelander, G. (1934). The gills of *Amia calva* specialized for respiration in an oxygen deficient habitat. *Copeia* pp. 123–127.

Bevelander, G. (1935). A comparative study of the branchial epithelium in fishes, with special reference to extra-renal excretion. *J. Morphol.* **57**, 335–352.

Bietrix, E. (1895). Etude de quelques faits relatifs à la morphologie du système circulatoire à propos du réseau branchial des Poissons. Thèse Fac. Med., Univ. Paris.

Bijtel, J. H. (1943). Het bewegingsapparaat der kiewfilamenten bij de Teleostei. *Versl. Ned. Akad. Wet., Afd. Natuurkd.* 52.

Bijtel, J. H. (1949). The structure and mechanism of movement of the gill-filaments in Teleostei. *Arch. Neerl. Zool.* **8**, 267–288.

Bird, D. J., and Eble, A. F. (1979). Cytology and polysaccharide cytochemistry of the gill of the american eel, *Anguilla rostrata. Biol. Bull. (Woods Hole, Mass.)* **157**, 104–111.

Boland, E. J., and Olson, K. R. (1979). Vascular organization of the catfish gill filament. *Cell Tissue Res.* **198**, 487–500.

Booth, J. H. (1979). The effects of oxygen supply, epinephrine, and acetylcholine on the distribution of blood flow in trout gills. *J. Exp. Biol.* **83**, 31–39.

Boyd, R. B., De Vries, A. L., Eastman, J. T., and Pietra, G. G. (1980). The secondary lamellae of the gills of cold water (high latitude) Teleosts. A comparative light and electron microscopic study. *Cell Tissue Res.* **213**, 361–367.

Bradley, T. J. (1981). Improved visualization of apical vesicles in chloride cells of fish gills using an osmium quick-fix technique. *J. Exp. Zool.* **217**, 185–198.

Burden, C. E. (1956). The failure of hypophysectomized *Fundulus heteroclitus* to survive in fresh water. *Biol. Bull. (Woods Hole, Mass.)* **110**, 8–28.

Burggren, W., Dunn, J., and Barnard, K. (1979). Branchial circulation and gill morphometrics in the sturgeon *Acipenser transmontanus*, an ancient Chondrosteian fish. *Can. J. Zool.* **57**, 2160–2170.

Burne, R. H. (1927). A contribution to the anatomy of the ductless glands and lymphatic system of the angler fish, *Lophius piscatorius. Philos. Trans. R. Soc. London, Ser. B* **215**, 1–57.

Burne, R. H. (1929). A system of fine vessels associated with the lymphatics in the cod (*Gadus morrhua*). *Philos. Trans. R. Soc. London, Ser. B* **217**, 335–366.

Chilmonczyk, S., and Monge, D. (1980). Rainbow trout gill pillar cells: Demonstration of inert particle phagocytosis and involvement in viral infection. *J. Reticuloendothel. Soc.* **28**, 327–332.

Chusovitina, L. S. (1963). Cellules à chlorures chez les Acipenséridés. *Dokl. Akad. Nauk SSSR* **151**, 441.

Claude, P., and Goodenough, D. A. (1973). Fracture faces of zonulae occludentes from "tight" and "leaky" epithelia. *J. Cell Biol.* **58**, 390–400.

Cole, F. J. (1923). The vascular system of Myxine. *Br. Assoc. Adv. Sci., Rep.* **91**, 450–451.

Conte, F. P. (1969). Salt secretion. *In* "Fish Physiology" (W. S. Hoar and D. J. Randall, eds.), Vol. 1, pp. 241–283. Academic Press, New York.

Conte, F. P., and Lin, D. (1967). Kinetics of cellular morphogenesis in gill epithelium during sea water adaptation of *Oncorhynchus* (Walb.). *Comp. Biochem. Physiol.* **23**, 945–957.

Cooke, I. R. C. (1980). Functional aspects of the morphology and vascular anatomy of the gills of the Endeavour dogfish, *Centrophorus scalpratus* (Mc Culloch) (Elasmobranchii: Squalidae). *Zoomorphologie* **94**, 167–183.

Cooke, I. R. C., and Campbell, G. (1980). The vascular anatomy of the gills of the smooth toadfish, *Torquigiener glaber* (Teleosteii:Tetraodontidae). *Zoomorphologie* **94**, 151–166.

Copeland, D. E. (1948). The cytological basis of chloride transfer in the gills of *Fundulus heteroclitus. J. Morphol.* **82**, 201–228.

Das, S., and Srivasta, G. J. (1978). Response of gill to various changes in salinity in fresh water ⸱eleost (*Colisa fasciatus*). *Z. Mikrosk.-Anat. Forsch.* **92**, 770–780.

Datta Munshi, J. S. (1968). The accessory respiratory organs of *Anabas testudineus* (Bloch) (Anabantidae, Pisces). *Proc. Linn. Soc. London* **170**, 107–126.

Datta Munshi, J. S., and Singh, B. N. (1968). A study of gill epithelium of certain freshwater teleostean fishes with special references to the air-breathing fishes. *Indian J. Zool.* **9**, 91–107.

Davis, J. C. (1972). An infrared photographic technique useful for studying vascularization of fish gills. *J. Fish. Res. Board Can.* **29**, 109–111.

Daxboeck, C., Barnard, D. K., and Randall, D. J. (1981). Functional morphology of the gills of the bowfin, *Amia calva* L., with special reference to their significance during air exposure. *Respir. Physiol.* **43**, 349–364.

De Kock, L. L. (1963). A histological study of the head region of two salmonids with special references to pressure and chemoreceptors. *Acta Anat.* **55**, 39–50.

De Renzis, G. (1975). The branchial chloride pump in the goldfish *Carassius auratus*: Relationship between $Cl^-/HCO_3^-$ and $Cl^-/Cl^-$ exchanges and the effect of thyocyanate. *J. Exp. Biol.* **63**, 587–602.

Dornesco, G. T., and Miscalenco, D. (1963). Contribution à l'étude des branchies de la carpe (*Cyprinus carpio* L.). *Morphol. Jahrb.* **105**, 553–570.

Dornesco, G. T., and Miscalenco, D. (1967). Etude comparative des branchies de plusieurs espèces de l'ordre des perciformes. *Anat. Anz.* **121**, 182–208.

Dornesco, G. T., and Miscalenco, D. (1968a). Nouvelle contribution à l'étude comparative des branchies de quelques espèces de l'ordre des Perciformes. *Zool. Jahrb., Abt. Anat. Ontog. Tiere* **85**, 228–244.

Dornesco, G. T., and Miscalenco, D. (1968b). Etude comparative des branchies de quelques espèces de l'ordre des Clupeiformes. *Morphol. Jahrb.* **112**, 261–276.

Dornesco, G. T., and Miscalenco, D. (1968c). La structure des branchies de quelques Cyprins. *Ann. Sci. Nat., Zool. Biol. Anim.* [12] **10**, 291–300.

Doyle, W. L. (1960). The principal cells of the salt-gland of marine birds. *Exp. Cell Res.* **21**, 386–393.

Doyle, W. L. (1977). Cytological changes in chloride cells following altered ionic media. *J. Exp. Zool.* **199**, 427–434.

Doyle, W. L., and Gorecki, D. (1961). The so-called chloride cell of the fish gill. *Physiol. Zool.* **34**, 81–85.

Dröscher, W. (1882). Beiträge zur Kenntnisse der histologischen Struktur der Kiemen der Plagiostomen. *Arch. Naturgesch.* **48**, 120–177.

Dunel, S. (1975). Contribution à l'étude structurale et ultrastructurale de la pseudobranchie et de son innervation chez les Téléostéens. Ph.D. Thesis, University of Strasbourg.

Dunel, S., and Laurent, P. (1973). Ultrastructure comparée de la pseudobranchie chez les Téléostéens marins et d'eau douce. I. L'épithélium pseudobranchial. *J. Microsc. (Paris)* **16**, 53–74.

Dunel, S., and Laurent, P. (1977). La vascularisation branchiale chez l'anguille: Action de l'acétylcholine et de l'adrénaline sur la répartition d'une résine polymérisable dans les différents compartiments vasculaires. *C.R. Hebd. Seances Acad. Sci.* **284**, 2011–2014.

Dunel, S., and Laurent, P. (1980). Functional organization of the gill vasculature in different classes of fish. *In* "Epithelial Transport in the Lower Vertebrates" (B. Lahlou, ed.), pp. 37–58. Cambridge Univ. Press, London and New York.

Dunel-Erb, S., and Laurent, P. (1980). Ultrastructure of marine teleost gill epithelia: SEM and TEM study of the chloride cell apical membrane. *J. Morphol.* **165**, 175–186.

Dunel-Erb, S., Bailly, Y., and Laurent, P. (1982a). Neuroepithelial cells in fish gill primary lamellae. *J. Appl. Physiol.; Respir., Environ. Exercise Physiol.* **53**(6), 1342–1353.

Dunel-Erb, S., Massabuau, J. C., and Laurent, P. (1982b). Organisation fonctionnelle de la branchie d'écrevisse. *C.R. Seances Soc. Biol. Ses Fil.* **176**, 248–258.

Duverney (1699). *Hist. Acad. R. Sci. Paris* p. 300.

Duvernoy, M. (1839). Du mécanisme de la respiration dans les poissons. *Ann. Sci. Nat., Zool. Biol. Anim.* **12.**

Ernst, S. A., Dodson, W. C., and Karnaky, K. J., Jr. (1980). Structural diversity of occluding junctions in the low-resistance chloride-secreting opercular epithelium of seawater-adapted killifish (*Fundulus heteroclitus*). *J. Cell Biol.* **87,** 488–497.

Falck, B., Hillarp, N. A., Thieme, G., and Torp, A. (1962). Fluorescence of catecholamines and related compounds condensed with formaldehyde. *J. Histochem.* **10,** 348–354.

Fänge, R. (1972). The circulatory system. *In* "The Biology of Lampreys" (M. V. Hardisty and I. Potter, eds.), Vol. 2, pp. 241–259. Academic Press, New York

Farrell, A. P. (1979). Gill blood flow in teleosts. Ph.D. Thesis, University of British Columbia.

Farrell, A. P. (1980a). Vascular pathways in the gill of ling cod, *Ophiodon elongatus. Can. J. Zool.* **58,** 796–806.

Farrell, A. P. (1980b). Gill morphometrics, vessel dimensions, and vascular resistance in ling cod, *Ophiodon elongatus. Can. J. Zool.* **58,** 807–818.

Farrell, A. P., Sobin, S. S., Randall, D. J., and Crosby, S. (1980). Interlamellar blood flow patterns in fish gills. *Am. J. Physiol.* **239,** R428–R436.

Florkowski, W. (1930). Die Verteilung der Lymphgefässe im Kopfe des Aales, (*Anguilla anguilla*). *Bull. Int. Acad. Pol. Sci. Lett. Cl. Sci. Math. Nat., Ser. B 2* **2,** 848–857.

Foskett, J. K., Logsdon, C. D., Turner, T., Machen, T. E., and Bern, H. A. (1981). Differentiation of the chloride extrusion mechanism during seawater adaptation of a teleost fish, the cichlid *Sarotherodon mossambicus. J. Exp. Biol.* **93,** 209–224.

Freeman, P. A. (1974). The morphology of the gills at various stages of the life cycle of the Great Lakes lamprey, *Petromyzon marinus* (L.). M.Sc.Thesis, University of Toronto, Toronto, Ontario, Canada.

Fromm, P. O. (1974). Circulation in trout gills: Presence of 'blebs' in afferent filamental vessels. *J. Fish. Res. Board Can.* **31,** 1793–1796.

Gannon, B. J. (1978). Vascular casting. *In* "Principles and Techniques of Scanning Electron Microscopy" (M. A. Hayat, ed.), Vol. 6, pp. 170–193. Van Nostrand Reinhold Co., New York.

Gannon, B. J., Campbell, G., and Randall, D. J. (1973). Scanning electronmicroscopy of vascular casts for the study of vessel connections in a complex vascular bed. The Trout gill. *Proc.—Annu. Meet., Electron Microsc. Soc. Am.* **31,** 442–443.

Getman, H. C. (1950). Adaptation changes in the chloride cells of *Anguilla rostrata. Biol. Bull. (Woods Hole, Mass.)* **99,** 439–445.

Girard, J. P., and Payan, P. (1977a). Kinetic analysis and partitioning of sodium and chloride influxes across the gills of seawater adapted trout. *J. Physiol. (London)* **267,** 519–536.

Girard, J. P., and Payan, P. (1977b). Kinetic analysis of sodium and chloride influxes across the gills of the trout in fresh water. *J. Physiol. (London)* **273,** 195–209.

Girard, J. P., and Payan, P. (1980). Ion exchanges through respiratory and chloride cells in freshwater- and seawater-adapted teleosteans. *Am. J. Physiol.* **238,** R260–R268.

Goette, A. (1901). Uber die Kiemen der Fische. *Z. Wiss. Zool.* **69,** 533–577.

Goodrich, E. S. (1909). Vertebrata Craniata. I. Cyclostomes and fishes. *In* "A Treatise on Zoology" (E. R. Lankester, ed.), Vol. IX. A. & C. Black, London.

Goodrich, E. S. (1930). "Studies of the Structure and Development of Vertebrates," Vol. II (reprinted by Dover, New-York, 1958).

Grodzinski, Z. (1926). Uber das Blutgefässsystem von *Myxine glutinosa* L. *Bull. Int. Acad. Pol. Sci. Lett., Cl. Sci. Math. Nat., Ser. B* **38,** 123–157.

Grodzinski, Z. (1932). Bemerkungen über das Lymphgefässsystem der Myxime glutinosa. *Bull. Int. Acad. Pol. Sci. Lett., Cl. Sci. Math. Nat., Ser. B* **2,** 44, 123–157.

Harb, J. H., and Copeland, D. E. (1969). Fine structure of the pseudobranch of the flounder, *Paralichthys lethostigma. Z. Zellforsch. Mikrosk. Anat.* **101**, 167–174.

Harder, W. (1964). Anatomie der Fische. *In* "Handbuch der Binnenfischerei Mitteleuropas" (Demoll, Maier, and Wundsch, eds.). E. Schweitzerbart'sche Verlagsbuchhandlung, Stuttgart.

Holbert, P. W., Boland, E. J., and Olson, K. R. (1979). The effect of epinephrine and acetylcholine on the distribution of red cells within the gills of the channel catfish (*Ictalurus punctatus*). *J. Exp. Biol.* **79**, 135–146.

Hootman, S. R., and Philpott, C. W. (1980). Accessory cells in teleost branchial epithelium. *Am. J. Physiol.* **238**, R199–R206.

Hossler, F. E. (1980). The gill arch of the mullet, *Mugil cephalus*. III. Rate of response to the salinity change. *Am. J. Physiol.* **238**, R160–R164.

Hossler, F. E., Ruby, J. R., and McIlwain, T. D. (1979a). The gill arch of the mullet, *Mugil cephalus*. I. Surface ultrastructure. *J. Exp. Zool.* **208**, 379–398.

Hossler, F. E., Ruby, J. R., and McIlwain, T. D. (1979b). The gill arch of the mullet, *Mugil cephalus*. II. Modification in surface ultrastructure and Na, K-ATPase content during adaptation to various salinities. *J. Exp. Zool.* **208**, 399–406.

Hughes, G. M. (1964). Fish respiratory homeostasis. *Symp. Soc. Exp. Biol.* **18**, 81–107.

Hughes, G. M. (1972). Morphometrics of fish gills. *Respir. Physiol.* **14**, 1–25.

Hughes, G. M. (1979). Scanning electron microscopy of the respiratory surfaces of trout gills. *J. Zool.* **188**, 443–453.

Hughes, G. M. (1980). Ultrastructure and morphometry of the gills of *Latimeria chalumnae*, and a comparison with the gills of associated fishes. *Proc. R. Soc. London Ser. B* **208**, 309–328.

Hughes, G. M., and Byczkowska-Smyk, W. (1974). Ultrastructure of the secondary gill lamella of the icefish, *Chaenocephalus aceratus. J. Zool.* **174**, 79–87.

Hughes, G. M., and Datta Munshi, J. S. (1978). Scanning electron microscopy of the respiratory surfaces of *Saccobranchus* (=Heteropneustes) *fossilis* (Bloch). *Cell Tissue Res.* **195**, 99–109.

Hughes, G. M., and Gray, I. E. (1972). Dimensions and ultrastructure of toadfish gills. *Biol. Bull. (Woods Hole, Mass.)* **143**, 150–161.

Hughes, G. M., and Grimstone, A. V. (1965). The fine structure of the secondary lamellae of the gills of *Gadus pollachius. Q. J. Microsc. Sci.* **106**, 343–353.

Hughes, G. M., and Mittal, A. K. (1980). Structure of the gills of Barbus sophor (Ham), a cyprinid with tertiary lamellae. *J. Fish Biol.* **16**, 461–467.

Hughes, G. M., and Morgan, M. (1973). The structure of fish gills in relation to their respiratory function. *Biol. Rev. Cambridge Philos. Soc.* **48**, 419–475.

Hughes, G. M., and Perry, S. F. (1976). Morphometic study of trout gills: A light-microscopic method suitable for the evaluation of pollutant action. *J. Exp. Biol.* **64**, 447–460.

Hughes, G. M., and Shelton, G. (1962). Respiratory mechanisms and their nervous control in fish. *Adv. Comp. Physiol. Biochem.* **1**, 275–364.

Hughes, G. M., and Weibel, E. R. (1972). Similarity of supporting tissue in fish gills and the mammalian reticuloendothelium. *J. Ultrastruct. Res.* **39**, 106–114.

Hughes, G. M., and Wright, D. E. (1970). A comparative study of the ultrastructure of the water–blood pathways in the secondary lamellae of teleost and elasmobranch fishes. Benthic forms. *Z. Zellforsch. Mikrosk. Anat.* **104**, 478–493.

Hughes, G. M., Perry, S. F., and Brown, V. M. (1979). A morphometic study of effects of nickel, chromium and cadmium on the secondary lamellae of rainbow trout gills. *Water Res.* **13**, 665–676.

Hulbert, W. C., Moon, T. W., and Hockachka, P. W. (1978). The osteoglossid gill: Correlations of structure, function and metabolism with transition to air breathing. Can. J. Zool. **56,** 801–808.

Ishimatsu, A., Itazawa, Y., and Takeda, T. (1979). On the circulatory systems of the snakeheads *Channa maculata* and *C. argus* with reference to bimodal breathing. *Jpn. J. Ichthyol.* **26**(2), 167–180.

Jackson, C. M. (1901). An investigation of the vascular system of *Bdellostoma dombeyi*. *J. Cincinn. Soc. Nat. Hist.* **20,** 13.

Jossifov, S. M. (1906). Sur les voies principales et les organes de propulsion de la lymphe chez certains poissons. *Arch. Anat. Microsc.* **8,** 398–424.

Karnaky, K. J., Jr., and Kinter, W. B. (1977). Killifish opercular skin: A flat epithelium with a high density of chloride cells. *J. Exp. Zool.* **199,** 355–364.

Karnaky, K. J., Jr., Ernst, S. A., and Philpott, C. W. (1976a). Teleost chloride cell. I. Response of pupfish *Cyprinodon variegatus* gill Na-K-ATPase and chloride cell fine structure to various salinity environments. *J. Cell Biol.* **70,** 144–156.

Karnaky, K. J., Jr., Kinter, L. B., Kinter, W. B., and Sterling, C. E. (1976b). Teleost chloride cell. II. Autoradiographic localization of gill Na-K-ATPase in killifish, *Fundulus heteroclitus* adapted to low and high salinity environments. *J. Cell Biol.* **70,** 157–177.

Kempton, R. T. (1969). Morphological feature of functional significance in the gills of the spiny dogfish, *Squalus acanthias*. *Biol. Bull. (Woods Hole, Mass.)* **136,** 226–240.

Kendall, M. W., and Dale, J. E. (1979). Scanning and transmission electron microscopic observations of rainbow trout (*Salmo gairdneri*) gill. *J. Fish. Res. Board Can.* **36,** 1072–1079.

Kessel, R. G., and Beams, H. W. (1962). Electron microscope studies on the gill filaments of *Fundulus heteroclitus* from sea water and fresh water with special reference to the ultrastructural organization of the "chloride cell." *J. Ultrastruct. Res.* **6,** 77–87.

Keys, A. B. (1931). Chloride and water secretion and absorption by the gills of the eel. *Z. Vergl. Physiol.* **15,** 364–388.

Keys, A. B., and Wilmer, E. N. (1932). "Chloride secreting cells" in the gills of fishes with special reference to the common eel. *J. Physiol. (London)* **76,** 368–378.

Kimura, N., and Kudo, S. (1979). The fine structure of gill filaments in the fingerlings of rainbow trout *Salmo gairdneri*. *Jpn. J. Ichthyol.* **26**(3), 289–301.

Krischner, L. B. (1969). Ventral aortic pressure and sodium fluxes in perfused eel gills. *Am. J. Physiol.* **217,** 596–604.

Korte, G. E. (1979). Unusual association of chloride cells with an other cell type in the skin of the glass catfish, *Kryptopterus bicirrhis*. *Tissue Cell* **11,** 63–68.

Kryzanovsky, S. G. (1934). Die Pseudobranchie. Morphologie und biologische Bedeutung. *Zool. Jahrb., Abt. Allg. Zool. Physiol. Tiere* **78,** 171–238.

Lasker, R., and Threadgold, L. T. (1968). Chloride cells in the skin of the larval sardine. *Exp. Cell Res.* **52,** 582–590.

Laurent, P. (1981). Circulatory adaptation to diving in amphibious fish. *In* "Advances in Animal and Comparative Physiology " (G. Pethes and V. L. Frenys, eds.), Vol. 20, pp. 305–306. Pergamon Press and Akademiai Kiado.

Laurent, P. (1982). Structure of vertebrate gills. *In* "Gills" (D. F. Houlihan, J. C. Rankin, and T. J. Shuttleworth, eds.), pp. 25–43. Cambridge Univ. Press, London and New York.

Laurent, P., and Dunel, S. (1976). Functional organization of the teleost gill I. Blood pathways. *Acta Zool. (Stockholm)* **57,** 189–209.

Laurent, P., and Dunel, S. (1978). Relations anatomiques des ionocytes (cellules à chlorure) avec le compartiment veineux branchial: Définition de deux types d'épithélium de la branchie des poissons. *C.R. Hebd. Seances Acad. Sci., Ser. D* **286,** 1447–1450.

Laurent, P., and Dunel, S. (1980). Morphology of gill epithelia in fish. *Am. J. Physiol.* **238**, R147–R159.

Laurent, P., DeLaney, R. G., and Fishman, A. P. (1978). The vasculature of the gills in the aquatic and aestivating lungfish (*Protopterus aethiopicus*). *J. Morphol.* **156**, 173–208.

Laurent, P., Delaney, R. G., and Fishman, A. P. (1984a). A morphofunctional study of the circulatory adaptation to bimodal respiration in the Dipnoan lungfish. (Submitted.)

Laurent, P., Dunel-Erb, S., and Höbe, H. (1984b). The role of environmental calcium relative to sodium chloride in determining gill morphology of soft water trout and catfish. (Submitted.)

Lauweryns, J. M., Cokelaere, M., and Theunynck, P. (1972). Neuroepithelial bodies in the respiratory mucosa of various mammals. A light optical, histochemical and ultrastructural investigation. *Z. Zellforsch. Mikrosk. Anat.* **135**, 569–592.

Lauweryns, J. M., Cokelaere, M., and Theunynck, P. (1973). Serotonin producing neuroepithelial bodies in rabbit respiratory mucosa. *Science* **180**, 410–413.

Leatherland, J. E., and Lam, J. J. (1968). Effect of prolactin on the density of the marine form (*trachurus*) of the three spined stickleback, *Gasterosteus aculeatus. Can. J. Zool.* **46**, 1095–1097.

Leeson, C. R., and Leeson, T. S. (1966). "Histology." Saunders, Philadelphia.

Legris, G. J., Will, P. C., and Hopfer, U. (1981). Effects of serotonin on ion transport in intestinal and respiratory epithelium. *Ann. N.Y. Acad. Sci.* **372**, 345–346.

Lewis, S. V. (1976). Respiration and gill morphology of the paired species of Lampreys *Lampetra fluviatilis* (L.) and *Lampetra planeri* (Bloch). Ph.D. Thesis, University of Bath.

Lewis, S. V. (1979a). The morphology of the accessory air-breathing organs of the catfish, *Clarias batrachus*: A SEM study. *J. Fish Biol.* **14**, 187–191.

Lewis, S. V. (1979b). A scanning electron microscope study of the gills of the air-breathing catfsih, *Clarias batrachus* L. *J. Fish Biol.* **15**, 381–384.

Lewis, S. V., and Potter, I. C. (1976a). Gill morphometrics of the lampreys, *Lampetra fluviatilis* (L.) and *Lampetra planeri* (Bloch). *Acta Zool. (Stockholm)* **57**, 103–112.

Lewis, S. V., and Potter, I. C. (1976b). A scanning electron microscope study of gills of the lamprey, *Lampetra fluviatilis* (L.). *Micron* **7**, 205–211.

Lewis, S. V., and Potter, I. C. (1982). A light and electron microscope study of the gills of larval lampreys (*Geotria australis*) with particular reference to the water-blood pathway. *J. Zool.* **198**, 157–176.

Liem, K. F. (1961). Tetrapod parallelism and other features in the functional morphology of the blood vascular system of *Fluta alba. J. Morphol.* **108**, 131–143.

Liu, C. K. (1942). Osmotic regulation and chloride secreting cells in the paradise fish, *Macropodus opercularis. Sinensia* **13**, 15–20.

MacKinnon, M., and Enesco, H. (1980). Cell renewal in the gills of the fish *Barbus conchonius. Can. J. Zool.* **58**, 650–653.

Marinelli, W., and Strenger, A. (1956). "Vergleichende Anatomie und Morphologie der Wirbeltiere. 2. Myxine glutinosa L." Deuticke, Wien.

Masoni, A., and Garcia Romeu, F. (1972). Accumulation et excrétion de substances organiques par les cellules à chlorure de la branchie d'*Anguilla anguilla* L. adaptée à l'eau de mer. *Z. Zellforsch. Mikrosk. Anat.* **133**, 389–398.

Mattheij, J. A. M., and Sprangers, J. A. P. (1969). The site of prolactin secretion in the adenohypophysis of the stenohaline teleost, *Anoptichthys jordani,* and the effects of this hormone on mucous cells. *Z. Zellforsch. Mikrosk. Anat.* **99**, 411–419.

Mattheij, J. A. M., and Stroband, H. W. J. (1971). The effects of osmotic experiments and prolactin on the mucous cells in the skin and the ionocytes in the gills of the teleost *Cichlasoma biocellatum. Z. Zellforsch. Mikrosk. Anat.* **121**, 93–101.

Monro, A. (1785). "The Structure and Physiology of Fishes Explained and Compared with those of Man and other Animals." C. Elliot, Edinburgh.

Morgan, M. (1971). Gill development, growth and respiration in the trout, *Salmo gairdneri* (Richardson). Ph.D. Dissertation, University of Bristol.

Morgan, M. (1974a). The development of gill arches and gill blood vessels of the rainbow trout, *Salmo gairdneri*. *J. Morphol.* **142**, 351–363.

Morgan, M. (1974b). Development of secondary lamellae of the gills of the trout, *Salmo gairdneri* (Richardson). *Cell Tissue Res.* **151**, 509–523.

Morgan, M., and Tovell, P. W. A. (1973). The structure of the gill of the trout, *Salmo gairdneri* (Richardson). *Z. Zellforsch. Mikrosk. Anat.* **142**, 147–162.

Moroff, T. (1904). Über die Entwicklung der Kiemen bei Fischen. *Arch. Mikrosk. Anat.* **64**, 189–213.

Morris, R. (1957). Some aspects of the structure and the cytology of the gills of *Lampetra fluviatilis*. *Q. J. Microsc. Sci.* **98**, 473–485.

Morris, R. (1972). Osmoregulation. *In* "The Biology of Lampreys" (M. W. Hardisty and I. C. Potter, eds.), Vol. 2, pp. 193–239. Academic Press, New York.

Morris, R., and Pickering, A. D. (1975). Ultrastructure of the presumed ion-transporting cells in the gill of ammocoete lampreys, *Lampetra fluviatilis* (L) and *Lampetra planeri* (Bloch). *Cell Tissue Res.* **163**, 327–341.

Muir, B. S. (1970). Contributions to the study of blood pathways in teleost gills. *Copeia* **1**, 19–28.

Muir, B. S., and Kendall, J. L. (1968). Structural modifications in the gills of tunas and some other oceanic fishes. *Copeia* **2**, 388–398.

Müller, J. (1839). Vergleichende Anatomie der Myxinoiden. III. Über das Gefässsystem. *Abh. Akad. Wiss. Berlin* pp. 175–303.

Murakami, T. (1971). Application of the scanning electron microscope to the study of the fine distribution of the blood vessels. *Arch. Histol. Jpn.* **32**, 445–454.

Nakao, T. (1974). Fine structure of the agranular cytoplasmic tubules in the lamprey chloride cells. *Anat. Rec.* **178**, 49–62.

Nakao, T. (1977). Electron microscopic studies of coated membranes in two types of gill epithelial cells of lampreys. *Cell Tissue Res.* **178**, 385–396.

Nakao, T. (1978). An electron microscopic study of the cavernous bodies in the lamprey gill filaments. *Am. J. Anat.* **151**, 319–336.

Nakao, T., and Uchinomiya, K. (1974). Intracisternal tubules of lamprey chloride cells. *J. Electron Microsc.* **23**, 51–55.

Nakao, T., and Uchinomiya, K. (1978). A study on the blood vascular system of the lamprey gill filament. *Am. J. Anat.* **151**, 239–264.

Newstead, J. D. (1967). Fine structure of the respiratory lamellae of teleostean gills. *Z. Zellforsch. Mikrosk. Anat.* **79**, 396–428.

Nilsson, S. (1983). "Autonomic Nerve Function in the Vertebrates." Springer-Verlag, Berlin and New York.

Nilsson, S., and Pettersson, K. (1981). Sympathetic nervous control of blood flow in the gill of the atlantic cod, *Gadus morhua*. *J. Comp. Physiol.* **144**, 157–163.

Norris, H. V. (1925). Observations upon the peripheral distribution of the cranial nerves of Ganoid fishes. *J. Comp. Neurol.* **39**, 345–432.

O'Donoghue, C. H. (1914). Notes on the circulatory system of Elasmobranchs. I. The venous system of the dogfish, *Scyllium canicula*. *Proc. Zool. Soc. London*, pp. 435–455.

Oduleye, S. O. (1977). Unidirectional water and sodium fluxes and respiratory metabolism in the african lungfish, *Protopterus annectens*. *J. Comp. Physiol.* **119**, 127–139.

Ogawa, M. (1974). The effects of bovine prolactin, sea water and environmental calcium on water influx in isolated gills of the euryhaline teleosts, *Anguilla japonica* and *Salmo gairdneri. Comp. Biochem. Physiol. A* **49A**, 545–554.

Ogawa, M. (1975). The effects of prolactin, cortisol and calcium-free environment on water influx in isolated gills of Japanese eel, *Anguilla Japonica. Comp. Biochem. Physiol. A* **52A**, 539–543.

Olivereau, M. (1970). Réaction des cellules à chlorures de la branchie après passage en eau de mer chez l'Anguille européenne. *C.R. Seances Soc. Biol. et Fil.* **164**, 1951–1955.

Olson, K. R. (1980). Application of corrosion casting procedures in identification of perfusion distribution in a complex microvasculature. *Scanning Electron Microsc.* **3**, 357–364.

Olson, K. R. (1981). Morphology and vascular anatomy of the gills of a primitive air-breathing fish, the bowfin (*Amia calva*). *Cell Tissue Res.* **218**, 499–517.

Olson, K. R., and Fromm, P. O. (1973). A scanning electron microscopic study of secondary lamellae and chloride cells of rainbow trout (*Salmo gairdneri*). *Z. Zellforsch. Mikrosk. Anat.* **143**, 439–449.

Olson, K. R., and Kent, B. (1980). The microvasculature of the Elasmobranch gill. *Cell Tissue Res.* **209**, 49–63.

Payan, P. (1978). A study of the $Na^+/NH_4^+$ exchange across the gill of the perfused head of the trout (*Salmo gairdneri*). *J. Comp. Physiol.* **124**, 181–188.

Payan, P., Mayer-Gostan, N., and Pang, P. K. T. (1981). Site of calcium uptake in the fresh water trout gill. *J. Exp. Zool.* **216**, 345–347.

Peek, W. D., and Youson, J. H. (1979a). Transformation of the interlamellar epithelium of the gills of anadromous sea lamprey, *Petromyzon marinus* L. during metamorphosis. *Can. J. Zool.* **57**, 1318–1332.

Peek, W. D., and Youson, J. H. (1979b). Ultrastructure of chloride cells in young adults of the anadromous sea lamprey, *Petromyzon marinus* L., in fresh water and during adaptation to sea water. *J. Morphol.* **160**, 143–164.

Pettengill, P., and Copeland, D. (1948). Alkaline phosphatase activity in the chloride cell of *Fundulus heteroclitus* and its relation to osmotic work. *J. Exp. Zool.* **108**, 235–241.

Pettersson, K., and Nilsson, S. (1979). Nervous control of the branchial vascular resistance of the Atlantic Cod, *Gadus morhua. J. Comp. Physiol.* **129**, 179–183.

Philpott, C. W. (1962). The comparative morphology of the chloride secreting cells of three species of *Fundulus* as revealed by the electron microscope. *Anat. Rec.* **142**, 267–268.

Philpott, C. W. (1966). The use of horseradish peroxidase to demonstrate functional continuity between the plasmalemma and the unique tubular system of the chloride cell. *J. Cell Biol.* **31**, 88A (abstr.).

Philpott, C. W. (1980). Tubular system membranes of teleost chloride cells: Osmotic response and transport sites. *Am. J. Physiol.* **238**, R171–R184.

Philpott, C. W., and Copeland, D. E. (1963). Fine structure of chloride cells from three species of *Fundulus. J. Cell Biol.* **18**, 389–404.

Pic, P., and Maetz, J. (1975). Différences de potentiel trans-branchial et flux ioniques chez *Mugil capito* adapté à l'eau de mer. Importance de l'ion Ca. *C.R. Hebd. Seances Acad. Sci., Ser. D* **280**, 983–986.

Pic, P., and Maetz, J. (1979). Etude de la spécificité du contrôle des mécanismes d'excrétion branchiale de $Na^+$ par $Ca^{++}$ et divers cations multivalents chez *Mugil capito* adapté à l'eau de mer. *C.R. Hebd. Seances Acad. Sci., Ser. D* **289**, 319–322.

Pickering, A. D., and Morris, R. (1973). Localization of ion-transport in the intestine of the migrating river lamprey, *Lampetra fluviatilis* L. *J. Exp. Biol.* **58**, 165–176.

Pickering, A. D., and Morris, R. (1976). Fine structure of the interplatelet area in the gills of the macrophtalmia stage of the river lamprey, *Lampetra fluviatilis* L. *Cell Tissue Res.* **168**, 433–443.

Pisam, M. (1981). Membranous systems in the "chloride cell" of teleostean fish gill; their modifications in response to the salinity of the environment. *Anat. Rec.* **200**, 401–414.

Pisam, M., Chretien, M., and Clermont, Y. (1983). Two anatomical pathways for the renewal of surface glycoproteins in chloride cells of fish gills. *Anat. Rec.* **207**, 385–397.

Pisam, M., Sardet, C., and Maetz, J. (1980). Polysaccharidic material in chloride cell of teleostean gill: Modifications according to salinity. *Am. J. Physiol.* **238**, R213–R218.

Pohla, H., Lametschwandtner, A., and Adam, H. (1977). Die Vaskularisation der Kiemen von *Myxine glutinosa* L. (Cyclostomata). *Zool. Scr.* **6**, 331–341.

Potts, W. T. W. (1977). Fish gills. *In* "Transport of Ions and Water in Epithelia" (B. L. Gupta, R. B. Moreton, J. L. Ochsman, and B. J. Wall, eds.), pp. 453–480. Academic Press, New York.

Purser, G. L. (1926). On *Calamoichthys calabaricus*. I. The alimentary and respiratory systems. *Trans. R. Soc. Edinburgh* **54**, 767–784.

Rajbanshi, V. K. (1977). The architecture of the gill surface of the catfish, *Heteropneustes fossilis* (Bloch): Sem study. *J. Fish Biol.* **10**, 325–329.

Randall, D. J., Cameron, J. N., Daxboeck, C., and Smatresk, N. (1981). Aspects of bimodal gas exchange in the bowfin, *Amia calva* L. (Actinopterygii: Amiiformes). *Respir. Physiol.* **43**, 339–348.

Rankin, J. C., and Maetz, J. (1971). A perfused teleostean gill preparation: Vascular actions of neurohypophyseal hormones and catecholamines. *J. Endocrinol.* **51**, 621–635.

Rauther, M. (1937). Kiemen der Anamnier-Kiemendarmderivate der Cyclostomen und Fische. *In* "Handbuch der vergleichenden Anatomie der Wirbeltiere," Vol. III. Urban & Schwarzenberg, Berlin.

Rhodin, J. A. G. (1964). Structure of the gills of the marine pollack (*Pollachius virens*). *Anat. Rec.* **148**, 420.

Rhodin, J. A. G. (1968). Ultrastructure of mammalian venous capillaries, venules, and small collecting veins. *J. Ultrastruct. Res.* **25**, 452–500.

Rhodin, J. A. G., and Silversmith, C. (1972). Fine structure of elasmobranch arteries, capillaries and veins in the spiny dogfish, *Squalus acanthias. Comp. Biochem. Physiol. A* **42A**, 59–64.

Richards, B. D., and Fromm, P. O. (1969). Patterns of blood flow through filaments and lamellae of isolated-perfused rainbow trout (*Salmo gairdneri*) gills. *Comp. Biochem. Physiol.* **29**, 1063–1070.

Richards, B. D., and Fromm, P. O. (1970). Sodium uptake by isolated gills of rainbow trout (*Salmo gairdneri*). *Comp. Biochem. Physiol.* **33**, 303–310.

Riess, J. A. (1881). Der Bau der Kiemenblätter bei den Knochenfischen. *Arch. Naturgesch.* **47**, 518–550.

Ristori, M. T., and Laurent, P. (1977). Action de l'hypoxie sur le système vasculaire branchial de la tête perfusée de Truite. *C.R. Seances Soc. Biol. et Fil.* **171**, 809–813.

Ritch, R., and Philpott, C. W. (1969). Repeating particles associated with an electrolyte-transport membrane. *Exp. Cell Res.* **55**, 17–24.

Romer, A. S. (1959). "The Vertebrate Story." University of Chicago Press, Chicago.

Romer, A. S., and Parsons, T. S. (1977). "The Vertebrate Body." Saunders, Philadelphia, Pennsylvania.

Rowing, C. G. M. (1981). Interrelationships between arteries, veins and lymphatics in the head region of the eel, *Anguilla anguilla* L. *Acta Zool. (Stockholm)* **62**, 159–170.

Sardet, C. (1977). Ordered arrays of intramembrane particles on the surface of fish gills. *Cell Biol. Int. Rep.* **1**, 409–418.

Sardet, C. (1980). Freeze fracture of the gill epithelium of euryhaline teleost fish. *Am. J. Physiol.* **238**, R207–R212.

Sardet, C., Pisam, M., and Maetz, J. (1979). The surface epithelium of teleostean fish gills.

Cellular and junctional adaptations of the chloride cell in relation to salt adaptation. *J. Cell Biol.* **80**, 96–117.

Satchell, G. H., and Jones, M. P. (1967). The function of the conus arteriosus in the Port Jackson shark, *Heterodontus portusjacksoni*. *J. Exp. Biol.* **46**, 373–382.

Schulz, H. (1962). Some remarks on the sub-microscopic anatomy and pathology of the blood-air pathway in the lung. *In* "Pulmonary Structure and Function, Ciba Foundation Symposium" (A.V.S. de Reuck and M. O'Connor, eds.), pp. 205–210. Churchill, London.

Schwerdtfeger, W. K., and Bereiter-Hahn, J. (1978). Transient occurence of chloride cells in the abdominal epidermis of the guppy, *Poecilia reticulata,* Peters, adapted to sea water. *Cell Tissue Res.* **191**, 463–471.

Sewertzoff, A. N. (1911). Die Kiemenbogennerven der Fische. *Anat. Anz.* **38**, 487–494.

Sheldon, W. F., Sheldon, W., and Sheldon, L. (1962). The pulse wave in *Squalus acanthias*. *Bull. Mt. Desert Isl. Biol. Lab.* **4**, 79 (cited from Kempton, 1969).

Shirai, N., and Utida, S. (1970). Development and degeneration of the chloride cell during sea water and fresh water adaptation of the Japanese eel, *Anguilla japonica*. *Z. Zellforsch. Mikrosk. Anat.* **103**, 247–264.

Singh, B. R., Yadav, A. N., Ojha, J., and Datta Munshi, J. S. (1981). Gross structure and dimensions of the gills of an intestinal air-breathing fish (*Lepidocephalichthys guntea*). *Copeia* **1**, 224–229.

Skidmore, J. F. (1970). Respiration and osmoregulation in rainbow trout with gills damaged by zinc sulphate. *J. Exp. Biol.* **52**, 484–494.

Smith, D. G. (1977). Sites of cholinergic vasoconstriction in trout gills. *Am. J. Physiol.* **233**, R222–R229.

Smith, D. G., and Chamley-Campbell, J. (1981). Localization of smooth-muscle myosin in branchial pillar cells of snapper (*Chrysophys auratus*) by immunofluorescence histochemistry. *J. Exp. Zool.* **215**, 121–124.

Smith, D. G., and Johnson, D. W. (1977). Oxygen exchange in a simulated trout gill secondary lamella. *Am. J. Physiol.* **233**, R145–R161.

Smith, H. W. (1929). The excretion of ammonia and urea by the gills of the fish. *J. Biol. Chem.* **81**, 727–742.

Smith, H. W. (1930). The absorption and excretion of water and salts by marine teleosts. *Am. J. Physiol.* **93**, 485–505.

Soivio, A., and Tuurala, H. (1981). Structural and circulatory responses to hypoxia in the secondary lamellae of *Salmo gairdneri* gills at two temperatures. *J. Comp. Physiol.* **145**, 37–43.

Steen, J. B., and Kruysse, A. (1964). The respiratory function of teleostean gills. *Comp. Biochem. Physiol.* **12**, 127–142.

Strauss, L. P. (1963). A study of the fine structure of the so-called chloride cell in the gill of the guppy *Lebistes reticulatus* P. *Physiol. Zool.* **36**, 183–198.

Sutterlin, A. M., and Saunders, R. L. (1969). Proprioceptors in the gills of teleosts. *Can. J. Zool.* **47**, 1209–1212.

Taxi, J. (1971). Ultrastructural data on the cytology and cytochemistry of the autonomic nervous system. *Philos. Trans. R. Soc. London, Ser. B* **261**, 311–312.

Thiery, G., and Rambourg, A. (1976). A new staining technique for studying thick sections in the electron microscope. *J. Microsc. Biol. Cell.* **26**, 103–106.

Thomas, M. L. H. (1962). Observations on the ecology of *Petromyzon marinus* L. and *Entosphenus lamottei* (Le Seur) in the Great Lakes Watershed. M.S.A. Thesis, University of Toronto, Toronto, Ontario, Canada.

Threadgold, L. T., and Houston, A. H. (1964). An electronmicroscopic study of the "chloride cell" of *Salmo salar* L. *Exp. Cell Res.* **34**, 1–23.

Tomonaga, S., Sakai, K., Tashiro, J., and Awaya, K. (1975). High-walled endothelium in the gills of the hagfish. *Zool. Mag.* **84**, 151–155.

Tranzer, J. P., and Richards, J. G. (1976). Ultrastructural cytochemistry of biogenic amines in nervous tissue: Methodologic improvements. *J. Histochem. Cytochem.* **24**, 1178–1193.

Tuurala, H., and Soivio, A. (1982). Structural and circulatory changes in the secondary lamellae of *Salmo gairdneri* gills after sublethal exposures to dehydroabietic acid and zinc. *Aquat. Toxicol.* **2**, 21–29.

Ultsch, G. R., and Gros, G. (1979). Mucus as a diffusion barrier to oxygen: Possible role in $O_2$ uptake at low pH in carp (*Cyprinus carpio*) gills. *Comp. Biochem. Physiol.* A **62A**, 685–689.

Virabhadrachari, V. (1961). Structural changes in the gills, intestine and kidney of *Etroplus maculatus* (Teleostei) adapted to different salinities. *Q. J. Microsc. Sci.* **102**, 361–369.

Vogel, W. (1978a). The origin of Fromm's arteries in trout gills. *Z. Mikrosk.-Anat. Forsch.* **92**, 565–570.

Vogel, W. (1978b). Arteriovenous anastomoses in the afferent region of trout gill filaments (*Salmo gairdneri* Richardson, Teleostei). *Zoomorphologie* **90**, 205–212.

Vogel, W. O. P. (1981). Struktur und organisationsprinzip im Gefässsystem der Knochenfische. *Gegenbaurs Morph. Jahrb.*, Leipzig **127**, 772–784.

Vogel, W., and Claviez, M. (1981). Vascular specialization in fish, but no evidence for lymphatics. *2. Naturforsch., C:Biosci.* **36C**, 490–492.

Vogel, W., and Kock, K. H. (1981). Morphology of gill vessels in icefish. *Arch. Fischerei wiss.* **31**, 139–150.

Vogel, W., Vogel, V., and Kremers, H. (1973). New aspects of the intrafilamental vascular system in gills of a euryhaline teleost, *Tilapia mossambica*. *Z. Zellforsch. Mikrosk. Anat.* **144**, 573–583.

Vogel, W., Vogel, V., and Schlote, W. (1974). Ultrastructural study of arterio-venous anastomses in gill filaments of *Tilapia mossambica*. *Cell Tissue Res.* **155**, 491–512.

Vogel, W., Vogel, V., and Pfautsch, M. (1976). Arterio-venous anastomoses in rainbow trout gill filaments. *Cell Tissue Res.* **167**, 373–385.

Wendelaar-Bonga, S. E., Greven, J. A. A. and Veenhuis, M. (1976). The relationship between the ionic composition of the environment and the secretory activity of the endocrine cell types of Stannius corpuscles in the teleost *Gasterosteus aculeatus*. *Cell Tissue Res.* **175**, 297–312.

Wilder, B. G. (1877). On the respiration of *Amia*. *Proc. Am. Assoc. Adv. Sci.* **26**, 306–313.

Wright, D. E. (1973). The structure of the gills of the elasmobranch, *Scyliorhinus canicula* L. *Z. Zellforsch. Mikrosk. Anat.* **144**, 489–509.

Wu, H. W., and Chang, H. W. (1947). On the arterial system of the gills and the suprabranchial cavities in *Ophiocephalus argus*, with special reference to the correlation with the bionomics of the fish. *Sinensia* **17**, 1–16.

Youson, J. H., and Freemann, P. A. (1976). Morphology of the gills of larval and parasitic adult sea lamprey, *Petromyzon marinus* L. *J. Morphol.* **149**, 73–104.

Zaccone, G. (1981). Effect of osmotic stress on the chloride and mucous cells in the gill epithelium of the fresh-water teleost, *Barbus filamentosus* (Cypriniformes, Pisces). A structural and histochemical study. *Acta Histochem.* **68**, 147–159.

# 3

# INNERVATION AND PHARMACOLOGY OF THE GILLS

*STEFAN NILSSON*

Department of Zoophysiology
University of Göteborg
Göteborg, Sweden

## I. GENERAL INTRODUCTION

The skeletal, vascular, and nervous anatomy of the gill region in fish has been the source of great enthusiasm among comparative anatomists for the past 100 years, and from their studies much has been learned about the phylogenetic relationship and development of the structures in the head and neck region of the vertebrates. Only recently, however, has the anatomic arrangement of gill vasculature been worked out in detail, and although the gross anatomy of the branchial nerves has long been known, the knowledge of the *function* of the branchial innervation in fish is still fragmentary.

185

Although perfused teleost gills have been the target of numerous studies regarding the effects of catecholamines and other drugs on the transfer of water, gases, ions, and other compounds, the relative importance of vasomotor effects of the drugs compared to direct effects on the permeability of the branchial epithelia requires much further attention. A better understanding of the hemodynamic effects of the drugs is therefore crucial, and further research to elucidate the sites of action of the vasomotor nerves and circulating vasoactive substances and their role under different conditions in vivo should be given high priority.

The aim of this chapter is first to outline broadly the basic anatomy of the branchial innervation, in order then to be able to discuss in more detail some of the functions of the different sensory and motor components of the branchial nerves. The autonomic innervation of the gill vasculature will receive special attention together with the effects of some drugs on the branchial blood flow. Detailed descriptions of the vascular anatomy of fish gills and the possible role of the pseudobranch is given elsewhere in this volume (Chapter 2, this volume and Chapter 9, Volume XB, this series), and there are also detailed discussions on the function of oxygen receptors and the dynamics of oxygen transfer in the gills (Chapter 5, this volume). An outline of the direct effects of catecholamines on ion transfer in chloride cells is given by Zadunaisky (Chapter 5, Volume XB, this series).

## II. ORGANIZATION OF THE BRANCHIAL NERVES

Eleven pairs of cranial nerves are present in fish: terminal (0), olfactory (I), optic (II), oculomotor (III), trochlear (IV), trigeminal (V), abducens (VI), facial (VII), acoustic (VIII), glossopharyngeal (IX), and vagus (X). There are no accessory vagus (XI) or hypoglossal (XII) nerves in fish; instead the nerves leaving the central nervous system (CNS) behind the vagus in the "occipital region" are referred to as spinal nerves (Stannius, 1849, Romer, 1962). Detailed accounts of the cranial nerve anatomy in fish can be found, for example, in Stannius (1849), Allis (1920), Norris (1925), Goodrich (1930), Young (1931, 1933), and Romer (1962). A useful and painstakingly detailed anatomic description of the branchial innervation in the percoid fish, *Polycentrus schomburgkii*, is given by Freihofer (1978).

The cranial nerves 0, I, II, III, IV, VI, and VIII are of limited interest in the discussion of the innervation of the gill region, and the trigeminal nerve (V) is primarily of interest as a motor nerve to the anterior jaw muscles. The branchial nerves proper are branches of the facial (VII), glossopharyngeal (IX), and vagus (X) nerves, and in gnathostome fish the glossopharygeal and

## Table I

### A Summary of the Different Nerve Components of the Cranial Nerves in Fish[a]

| | Presence in cranial nerve number | | | | | | | | | | | Examples of innervation |
|---|---|---|---|---|---|---|---|---|---|---|---|---|
| | 0 | I | II | III | IV | V | VI | VII | VIII | IX | X | |
| **Somatic sensory system** | | | | | | | | | | | | |
| General cutaneous | x | | | | | x | | x | | x | x | Cutaneous sensors (free nerve endings in fish) |
| | | | x | | | | | | | | | Eyes |
| Special cutaneous | | | | | | | | x | | x | x | Lateral line system |
| | | | | | | | | | x | | | Ear |
| | | | | | | x | | x | | x | x | Proprioceptors of striated muscle |
| **Somatic motor system** | | | | x | x | | x | | | | | Eye muscles |
| **Visceral sensory system** | | | | | | | | | | | | |
| General visceral | | x | | | | | | | | | | Olfactory organ |
| | | | | | | | | | | x | x | Baro- and chemoreceptors |
| Special visceral | | | | | | | | x | | x | x | Taste buds and gustatory fibers |
| **Visceral motor system** | | | | | | | | | | | | |
| Special visceral | | | | | | x | | x | | x | x | Branchial muscles of jaw and gills |
| Autonomic | | | | x | | | | E[b] | | E | x | Glands, smooth muscle of iris, and vasculature[c] |

[a] Based primarily on Goodrich (1930) and Romer (1962).

[b] E, elasmobranchs.

[c] Spinal autonomic ("sympathetic") fibers from the cephalic sympathetic chain join the cranial nerves in ganoid and teleost fish.

vagus form large trunks that enter the dorsal part of the gill arches. The branchial nerves are composed of a variety of fibers, which are both sensory (afferent) and motor (efferent), but anatomic studies indicate that most fibers are sensory (De Kock, 1963).

The different components of the vertebrate cranial nerves can be conveniently summarized into four main groups: somatic sensory, somatic motor, visceral sensory, and visceral motor nerve fibers (Table I). From the table it can be seen that the cranial nerves innervating the gill region (VII, IX, and X) are particularly complex. It should be noted that in contrast to the arrangement of the spinal nerves, which have a ventral motor root and a dorsal sensory root connecting to the CNS, both the sensory and special visceral motor systems in the cranial nerves VII, IX, and X (Table I) are of dorsal root origin. Therefore, the nerves to the striated muscle of the jaws and gill region are classified as part of the *special visceral* motor system, and only the

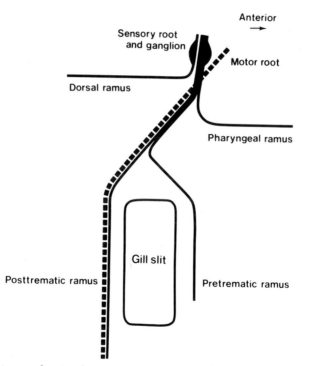

**Fig. 1.** Diagram showing the components of a "typical" branchial nerve, such as the glossopharyngeal of a fish. Three nerve components are indicated: (1) the somatic sensory component of the dorsal ramus, (2) the visceral sensory component of the pharyngeal ramus and the pre- and posttrematic rami of the gill slit, and (3) the visceral motor component of the posttrematic ramus. (Based primarily on Goodrich, 1930 and Romer, 1962.)

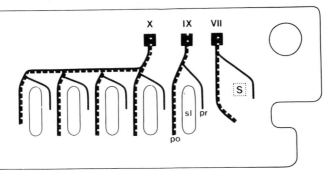

**Fig. 2.** Simplified diagram showing the basic arrangement of the facial (VII), glosso-pharyngeal (IX), and vagus (X) branchial nerves in a gnathostome fish. The figure shows visceral sensory (solid lines) and motor (broken lines) elements only; the pharyngeal and dorsal rami are not shown. po, Posttrematic ramus; pr, pretrematic ramus; S, spiracle; sl, gill slit.

"true" ventral root cranial motor nerves to the eye muscles (III, IV, and VI) belong to the somatic motor system (Goodrich, 1930; Romer, 1962). Some of the components in a "typical" cranial nerve (the glossopharyngeal of a fish) are summarized in Fig. 1.

The *facial nerve* (VII) innervates the region of the spiracular gill slit (Fig. 2) and sends motor fibers to the most anterior gill sac (cyclostomes; Lindström, 1949) or jaw muscles (hyomandibular and opercular muscles; gnathostomes; Goodrich, 1930). Sensory components in the facial nerve include the lateral line system of the head, innervation of taste buds (which in fish are present on the surface of the mouth, lips, and gill arches as well as over large parts of the body surface), and the sensory innervation of the pseudobranch in some forms (Laurent and Dunel, 1966; Dunel, 1975; Freihofer, 1978; see also Chapter 9, Volume XB, this series).

The *glossopharyngeal nerve* (IX) is the main sensory nerve of the pseudobranch in many teleost fish, and the posttrematic ramus provides the main sensory and motor innervation of the first gill arch (Figs. 1–3). Cranial autonomic ("parasympathetic") fibers may be present in the glossopharyngeal of elasmobranch fish, although there is no conclusive evidence for an innervation of the branchial vasculature in this group (see later). There is no evidence for cranial autonomic (parasympathetic) fibers in the glossopharyngeal nerve of teleosts (Burnstock, 1969; K. Pettersson and S. Nilsson, unpublished).

The second and the more posterior gill arches are innervated by the *vagus* (X), which carries both sensory and motor fibers. Cranial autonomic (parasympathetic) vagal fibers to the gill vasculature have been demonstrated in teleosts (Pettersson and Nilsson, 1979)

In the actinopterygian fish the sympathetic chains continue into the

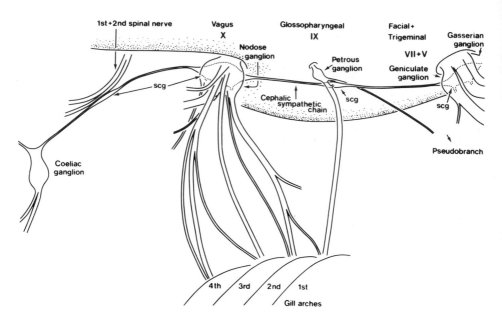

**Fig. 3.** Basic anatomy of the branchial nerves on the right side of the cod, *Gadus morhua*. Note that the branchial nerves are divided into both pre- and posttrematic rami (see Fig. 2). The cranial nerve ganglia on the trigeminal (Gasserian ganglion), facial (geniculate ganglion), glossopharyngeal (petrous ganglion), and vagus (nodose ganglion) are indicated. Note also the presence of a cephalic part of the sympathetic chain, with ganglia (scg) sending gray rami communicantes into the vagus (X), glossopharyngeal (IX), and trigeminal–facial (V + VII) complex. The size of the ganglia in the figure has been exaggerated for clarity. (Modified from Pettersson and Nilsson, 1979.)

head, and spinal autonomic ("sympathetic") fibers enter all or some of the cranial nerves III, V, VII, IX, and X via gray rami communicantes (Fig. 3). No such fibers enter the heads of elasmobranch fish. In the present descriptions the terms cranial autonomic and spinal autonomic are used instead of parasympathetic and sympathetic, respectively, since the original anatomic classification of autonomic nerves made in mammals (Langley, 1921) is not always valid in the nonmammalian vertebrates (Lutz, 1931; Young, 1936; Campbell, 1970; Nilsson, 1983).

## III. SENSORY PATHWAYS

Several types of receptors sensitive to different mechanical and a wide variety of chemical stimuli have been demonstrated in the branchial region of fish. Although it is practical to classify these receptors as proprioceptors, nociceptors, baroreceptors, and chemoreceptors in the present descriptions, it should immediately be emphasized that the demarcation line between the different groups is sometimes fluid. Thus, there may be little difference between some of the receptors classified as proprioceptors and the nociceptors sensitive to light mechanical stimuli such as particles in the inspired water. It may also be difficult to distinguish between specific chemoreceptors, such as $O_2$ and $CO_2$ receptors, and chemoreceptive nociceptors, reacting to noxious substances in the water.

## A. Proprioceptors

In the elasmobranch *Squalus acanthias* (= *S. lebruni*), experimental inflation of the pharynx has been shown to elicit an inhibition of the rate and amplitude of the respiratory movements (Lutz, 1930; Satchell, 1959). The reflex, which shows many similarities with the classical Hering–Breuer reflex of mammals (Hering and Breuer, 1868), can be mimicked by electrical stimulation of afferent fibers in the prespiracular branch of the facial (VII) nerve, or the branchial (IX and X) nerves (Satchell, 1959). Curarization of the animal leads to a reduction of the respiratory amplitude due to paralysis of the respiratory muscles, and parallel to the development of this reduction in breathing amplitude there is an increase in the frequency of respiration (Satchell, 1959). The pause between each respiratory cycle present during normal respiration in *Squalus* is abolished by sectioning of the branchial nerves, confirming the presence in the branchial branches of the glossopharyngeal (IX) and vagus (X) of afferent inhibitory fibers involved in the reflexogenic control of the breathing rhythm (Satchell, 1959). No such inhibitory reflex was, however, seen in *Scyliorhinus* (Butler *et al.*, 1977).

The localization of the receptors responsible for the described inhibition of respiration in *Squalus* was investigated by Satchell and Way (1962). These authors suggested that the slowly adapting proprioceptors located at the base of the branchial processes within the pharyngeal cavity are involved in the reflex. By carefully separating and recording the frequency of action potentials in individual nerve fibers in the branchial nerves, Satchell and Way (1962) demonstrated nerve discharges in response to deflection of the branchial processes, with a discharge rate proportional to the logarithm of the amplitude of the mechanical stimulus. It was concluded that the role of the branchial process receptors is to control the respiratory rate in relation to the amplitude of each respiratory cycle (Satchell and Way, 1962).

Proprioceptors affected by pharyngeal dilation, possibly related to the type previously described, are also important in the afferent limb of the reflex arc involved in the linking of the heartbeat to the respiratory frequency in elasmobranchs. In this reflex, the breathing cycle affects the vagal inhibitory tonus on the heart to produce synchrony in the cardiac and respiratory cycles (Satchell, 1960; Johansen and Burggren, 1980).

There is evidence for the presence of branchial proprioceptors also in teleosts. Two sets of proprioceptors with afferent nerve fibers in branchial nerves IX and X were demonstrated in the gill arches of the sea raven (*Hemitripterus americanus*) and Atlantic salmon (*Salmo salar*): the first group is activated by displacement of the gill filaments, whereas the second group is associated with the gill rakers (Sutterlin and Saunders, 1969). The receptive field of the gill filament receptor unit extended over 5 to 10 gill filaments and included also the corresponding filaments of the opposite hemibranch of the same gill arch. It is possible that the filament proprioceptors are involved in reflex control of the adductor (and in some cases the abductor) muscles of the filaments. These muscles adjust the angle of the two rows of filaments on each gill arch to maintain a continuous gill curtain despite the gill movements during the respiratory cycle (Pasztor and Kleerekoper, 1962).

The receptors of the gill rakers (or gill raker pads in *Hemitripterus*) responded to mechanical stimulation, and the receptive field was restricted to one raker or raker pad. A role of these proprioceptors in the control of respiration during swallowing of large prey was suggested (Sutterlin and Saunders, 1969).

## B. Nociceptors

There exists in the branchial region of fish a group of receptors (nociceptors) that are activated by potentially noxious or damaging stimuli, mechanical or chemical. Some types of mechanical nociceptors could well be identical with the proprioceptor types described previously. Gentle mechanical stimulation of the gill filaments, pharyngeal wall, or respiratory openings will result in a cough or expulsion reflex during which the respiratory surfaces of the gills are "backflushed" to remove the irritant (Ballintijn, 1969; Young, 1972; Satchell and Maddalena, 1972). In the Port Jackson shark, *Heterodontus portusjacksoni*, two types of coughs could be discerned: one orobranchial cough triggered by chemical stimuli and one parabranchial cough triggered by mechanical stimuli such as the presence of particles in the inspired water. It was suggested that the orobranchial cough serves to prevent the access of noxious substances to the gill surfaces, whereas the parabranchial cough clears the respiratory channels of debris and parasites (Satchell and Maddalena, 1972).

In mammals, a type of juxtapulmonary capillary receptor (type J receptor) has been described by Paintal (1969, 1970). Activation of these receptors causes bradycardia, hypotension, and inhibition of respiration, and it has been argued that the major function of the receptor system in mammals is the protection of the alveoli against edema. Type J receptors can be directly activated by phenyldiguanide (PDG) and 5-hydroxytryptamine (5-HT) and edema induced by agents such as alloxan also produces the discharge of type J receptor fibers in mammals (Paintal, 1969, 1970).

A special type of nociceptor, showing similarities with the type J receptor of the mammalian lung, has been demonstrated in the dogfish, *Squalus acanthias,* by Satchell and co-workers (Satchell, 1978a,b; Poole and Satchell, 1979). As in mammals, these receptors are stimulated by PDG and 5-hydroxytryptamine, and edema induced by injection of alloxan produces discharge of the receptors. The receptive fields in the dogfish gill ranged from parts of a single gill filament up to 15 filaments, and PDG injected into the duct of Cuvier produced bradycardia, apnea, hypotension, and inhibited swimming. It was concluded that the dogfish gill type J nociceptors could be of importance to protect the secondary lamellae from interstitial edema (Satchell, 1978a,b).

## C. Baroreceptors

In mammals, receptors sensitive to the arterial blood pressure (baroreceptors) are located in the aortic arch and carotid arteries, and there are also baroreceptors within the heart itself. The baroreceptors are free nerve endings that are sensitive to stretching of the elastic arterial wall and thus to the intraarterial blood pressure (Folkow and Neil, 1971). In reptiles and amphibians the baroreceptors appear to be located primarily in the truncus arteriosus and pulmocutaneous artery, respectively (Ishii and Ishii, 1978; Berger *et al.,* 1980; Smith *et al.,* 1981). The efferent limb of the tetrapod baroreceptor reflex arc consists of vagal inhibitory nerve fibers to the heart, and an elevated arterial blood pressure thus produces a reflex bradycardia. In mammals, a reflexogenic inhibition of the adrenergic (sympathetic) tonus on the heart and vasculature has also been demonstrated (see Folkow and Neil, 1971).

Baroreceptors have been demonstrated in the branchial vasculature of gnathostome fish. In the elasmobranchs, *Squalus acanthias* and *Mustelus canis,* the branchial nerves (IX and X) carry baroreceptor fibers. In the intact fish, bursts of impulses synchronous with the heartbeat could be recorded in the branchial nerves, and during perfusion of the gill apparatus with an elevated pressure a continuous high impulse rate ensued (Irving *et al.,* 1935). In *Squalus,* bradycardia or cardiac arrest was produced by stimulation of the cranial ends of the cut branchial nerves, or elevation of the blood pressure in the gill vasculature. As in tetrapods, inhibitory vagal fibers to the heart constituted the efferent limb of the reflex arc (Lutz and Wyman, 1932).

The first clear evidence for the presence of baroreceptors in the gills of teleosts comes from the work of Mott (1951), who demonstrated an inhibition of the heart of the eel (*Anguilla anguilla*) during stimulation of the cranial ends of the cut branchial nerves. This effect was abolished by vagotomy or administration of atropine, again showing the involvement of the vagal cholinergic cardiac innervation as the efferent limb of the reflex. A bradycardia in response to elevation of the pressure in the first or second pair of gill arches during branchial perfusion from a Mariotte bottle was also demonstrated (Mott, 1951). This type of experiment was repeated with similar results by Ristori (1970), working with the carp (*Cyprinus carpio*). All four pairs of gill arches contain baroreceptors, but the sensitivity declines in a gradient from the first to the fourth pair in the carp (Ristori and Dessaux, 1970).

In addition to the baroreceptors of the branchial arches, there is also evidence in favor of a baroreceptive function of the pseudobranch (see Chapter 9, Volume XB, this series).

Evidence for a baroreceptor reflex bradycardia due to elevated blood pressure induced by injection of adrenaline in vivo has been obtained in several teleosts (Fig. 4), such as *Cyprinus carpio* (Laffont and Labat, 1966), *Salmo gairdneri* (Randall and Stevens, 1967; Wood and Shelton, 1980a), *Ophiodon elongatus* (Stevens *et al.*, 1972), *Gadus morhua* (Helgason and Nilsson, 1973; Pettersson and Nilsson, 1980), and *Anguilla japonica* (Chan

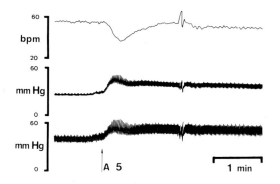

**Fig. 4.** Bradycardia in response to elevated blood pressure induced by injection of adrenaline (A) 5 μg/kg in Atlantic cod, *Gadus morhua*. Injection of hyoscine (4 mg/kg) or bilateral vagotomy before injection of the adrenaline abolished the bradycardia, while the blood pressure response remained unchanged or even increased slightly. Upper channel, heart rate (beats per minute); middle channel, dorsal aortic (celiac artery) blood pressure, lower channel, ventral aortic blood pressure. (From Helgason and Nilsson, 1973.)

and Chow, 1976; see also Jones and Randall, 1978). The location of the baroreceptors is not demonstrated by these studies, but the bradycardia induced is blocked by vagotomy or atropine (hyoscine), demonstrating the involvement of the vagal innervation of the heart. Adrenaline penetrates the blood–brain barrier in fish (Peyraud-Waitzenegger *et al.*, 1979), and the possibility of a direct effect of adrenaline on the CNS should not be neglected.

There is no clear evidence for an involvement of the adrenergic system (adrenergic nerve fibers or circulating catecholamines) in the baroreceptor reflex in fish. In fact no adrenergic effect on the heart was seen during induced hypotension (hemorrhage or α-adrenoceptor blockade) in the rainbow trout (*Salmo gairdneri*), although it is known that adrenergic stimulation can elevate the heart rate in this species above resting levels (Wood *et al.*, 1979; Wood and Shelton, 1980b).

The exact location of baroreceptors in fish is not known. A peculiar arrangement of small blood vessels in the walls of the afferent branchial arteries (and a few other arteries) in salmonid fish was described by De Kock (1963), and these *rete-mirabile jackets* receive an ample supply of nerve fibers. Whether these nerve fibers are vasomotor or sensory is not clear.

It seems that the most dense innervation of the branchial vasculature is found at the junction of the efferent filamental arteries and the branchial artery in both elasmobranch and bony fish (Boyd, 1936; Dunel and Laurent, 1980). The nerve terminals in the efferent artery sphincters of bony fish contain small clear vesicles ("c-type" nerve terminals) and are interpreted as cholinergic motor fibers involved in the autonomic vasomotor control of the gill vasculature (Dunel and Laurent, 1977, 1980).

It may be worth noting, however, that the ultrastructural distinction between the c-type nerve terminals and a second group, the "p-type" nerve terminals, can be difficult, and it has been demonstrated that these two types of nerve profiles sometimes represent extremes of the same group of nerve fibers (Gibbins, 1982). The p-type nerve terminals, which are normally characterized by their content of large granular vesicles (see Gibbins, 1982), are thought to store different materials (e.g., neuropeptides). Primary sensory neurons in mammals have been shown to contain substance P (a neuropeptide) (Hökfelt *et al.*, 1975, 1980), and a further investigation of the possibility of substance P-like immunoreactivity within the nerve terminals of the efferent filamental artery sphincters (and elsewhere in the gills) should be of great interest. It must be emphasized, however, that the bulk of evidence at the moment favors the view that the neurons innervating the efferent filamental artery sphincters are cholinergic autonomic neurons constricting the vasculature (Dunel and Laurent, 1977, 1980; see later).

Evidence for baroreceptors in the psuedobranch has also been presented (Laurent, 1967; see also Chapter 9, Volume XB, this series).

## D. Chemoreceptors

The concept of chemoreceptors includes, strictly speaking, all types of receptors sensitive to chemical stimuli—that is, also including olfactory and gustatory receptors. In fish, gustatory receptors (taste buds or end buds and free nerve endings) and probably also other types of chemoreceptors with afferent fibers in the facial (VII), glossopharyngeal (IX), or vagus (X) nerves can be found spread out over the pharynx, lips, gills, and also the skin covering the head and body (De Kock, 1963; Freihofer, 1978; Vasilevskaya and Polyakova, 1979, 1981; Walker et al., 1981). However, in the literature the term chemoreceptor has in some cases come to mean a receptor sensitive to changes in the tension of respiratory gases, especially oxygen. The present account will focus on the function and localization of chemoreceptors sensitive to changes in oxygen tension (oxygen receptors) (see also Chapter 5, this volume).

There are oxygen receptors that are stimulated by a reduction (and in some cases by an elevation) in the oxygen tension, and this initiates reflexes that affect either (a) the ventilation (breathing rate and/or amplitude) or (b) the cardiovascular system (heart rate and vascular resistance). A direct effect of hypoxia on the branchial vasculature has also been concluded (Satchell, 1962; Ristori and Laurent, 1977). The cardiac effect of hypoxia is a bradycardia in practically all cases studied, and it has been shown both in elasmobranch and teleost fish that, as in the baroreceptor reflex, the efferent limb of the reflex consists of vagal cholinergic inhibitory fibers to the heart. A central nervous coordination of chemo- and baroreceptor functions was concluded by Wood and Shelton (1980b) from experiments with the rainbow trout, *Salmo gairdneri*. A summary of the ventilatory and cardiovascular effects of hypoxia in a few species is given in Table II.

There are few reports on the presence and functions of oxygen receptors in cyclostomes. One observation by Wikgren (1953) in *Lampetra fluviatilis* indicates an increase in the number of gill sacs engaged in breathing movements during progressive hypoxia. This observation suggests the presence of oxygen receptors in *Lampetra*, but their localization is unknown.

There is good evidence for the presence of chemoreceptor reflexes in elasmobranch fish. Thus, hypoxia (or anoxia) is accompanied by a small increase in breathing rate or amplitude (Table II; Satchell, 1961; Piiper et al., 1970; Butler and Taylor, 1971). Hypoxia or anoxia also induces a marked bradycardia, and there is a tendency for the ventral and dorsal aortic blood pressure to fall (Fig. 5; Table II). A moderate bradycardia in response to hypoxia in *Scyliorhinus canicula* is compensated for by an increase in stroke volume, keeping the cardiac output largely constant (Short et al., 1977). It is known that prolonged hypoxia releases catecholamines into the blood, in-

**Table II**

Some Examples of Ventilatory and Cardiovascular Effects of Rapidly Induced Hypoxia in Some Fish Species[a,b]

| Species | Hypoxic $P_{O_2}$ (mm Hg) | Ventilatory effects | | Cardiovascular effects[c] | | | Reference |
|---|---|---|---|---|---|---|---|
| | | Rate | Amplitude | Heart rate | VAP | DAP | |
| Elasmobranchs | | | | | | | |
| Squalus acanthias | Anoxia | + | n.d.[d] | − | n.d. | − | Satchell (1961) |
| Scyliorhinus stellaris | 54 | 0(−) | + | − | − | − | Piiper et al. (1970) |
| Scyliorhinus canicula | <70 | + | n.d. | − | 0(−) | 0(−) | Butler and Taylor (1971) |
| Teleosts | | | | | | | |
| Tinca tinca | 45 | + | + | − | n.d. | n.d. | Randall and Shelton (1963) |
| Salmo gairdneri | <80 | + | + | − | + | + | Randall and Smith (1967); Holeton and Randall (1967) |
| Hemitripterus americanus | 30 | + | + | 0(−) | + | + | Saunders and Sutterlin (1971) |

[a] Chemoreceptor reflexes involving oxygen receptors in the gills (and/or elsewhere) are responsible for at least the ventilatory and cardiac adjustments. The possibility of direct effects (e.g., on the branchial vasculature) by reduced arterial or ambient oxygen tension should not be neglected (cf. Satchell, 1962; Ristori and Laurent, 1977). The ventilatory effects are generally more pronounced in teleosts than in elasmobranchs.

[b] +, Increase; −, decrease; 0(−) no effect or a small decrease.

[c] DAP, dorsal aortic blood pressure; VAP, ventral aortic blood pressure.

[d] n.d., no data available.

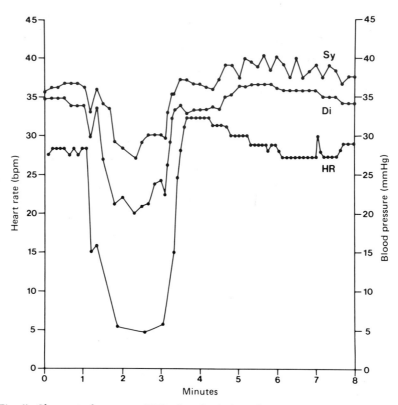

**Fig. 5.** Change in heart rate (HR), diastolic (Di), and systolic (Sy) dorsal aortic blood pressure caused by a 2-min exposure to deoxygenated water of the spiny dogfish, *Squalus acanthias*. (From Satchell, 1961.)

creasing the plasma concentrations of adrenaline and noradrenaline more than 10-fold (Butler *et al.*, 1978), but there seems to be no adrenergic stimulation of the heart involved in the enhanced stroke volume. Instead the increase in stroke volume during bradycardia may be explained by Starling's law of the heart (Short *et al.*, 1977).

The bradycardia produced by rapidly induced hypoxia or anoxia can be only partially blocked by sectioning the branchial nerves IX and X in *Squalus acanthias,* and it was concluded that additional receptors outside the gills, presumably in the CNS, are responsible for the remaining reflexogenic response (Satchell, 1961). In a careful study of the branchial innervation in *Scyliorhinus canicula*, Butler and co-workers demonstrated the presence of oxygen receptors also in areas innervated by trigeminal (V) and facial (VII) branches. In these studies, the oxygen tension of the inspired water was rapidly (within 1 min) lowered to about 30 mm Hg. The hypoxia produced an initial large transient bradycardia (heart rate 32% of initial), and the heart rate was then stabilized at about 65% of the initial value. Bilateral sectioning of the branchial branches of cranial nerves IX or X had no effect on the response, but when *both* IX and X were sectioned, the initial large transient decrease in heart rate was abolished. Bilateral sectioning of cranial nerves V, VII, IX, and X removed all response to hypoxia, and completely released the vagal inhibitory tonus on the heart (Butler *et al.*, 1977). The results show the lack of centrally located oxygen receptors involved in the cardiac reflex in this species.

In teleosts, Powers and Clark (1942) concluded a respiratory control involving receptors located in areas innervated by IX (*Salvelinus fontinalis*, *Salmo gairdneri*, and *Lepomis macrochirus*). Randall and Smith (1967), working with *Salmo gairdneri*, demonstrated a marked increase in the breathing rate and amplitude in response to hypoxia, while the heart rate decreased (Table II). It was clear from these experiments that a cardiorespiratory synchrony develops as a response to hypoxia at the low heart rate (cf. Section III,A). The bradycardia and cardiorespiratory synchrony were abolished by atropine, which suggests that the response is dependent on the vagal cholinergic inhibitory innervation of the heart.

Water-breathing fish may well, in contrast to the air-breathing vertebrates, encounter a medium with low oxygen tension. A rapid detection of the hypoxia would seem advantageous to the animal, which could then move away from the hypoxic region. In air breathers, a change in the arterial $P_{O_2}$ is likely to be due to endogenous events, and the oxygen receptors are located in the arterial bloodstream (see Johansen, 1970, 1971). The location of the oxygen receptors in the water-breathing fish has been studied by several workers, and the bulk of evidence favors at least two different locations of such receptors (cf. Holeton, 1977).

The first set of oxygen receptors detects the arterial $P_{O_2}$, and these receptors are probably located in the efferent arterial bloodstream, possibly within the central nervous system. The main effect mediated by these "arterial blood oxygen receptors" appears to be an increase in ventilation during hypoxia and a ventilatory decrease during hyperoxia (Davis, 1971; Dejours, 1973; Bamford, 1974; Holeton, 1977; Wilkes *et al.*, 1981).

The second set of oxygen receptors are located superficially in contact with the inspired water, at least in *Salmo gairdneri*, most likely on the anterodorsal surface of the anterior pair of gill arches. These receptors are involved in the bradycardia induced by hypoxia (Randall and Smith, 1967; Holeton, 1977; Daxboeck and Holeton, 1978; Smith and Jones, 1978). It is possible that some types of the numerous "end buds" found on the surface of the buccal cavity, pharynx, and gill arches (De Kock, 1963) represent the receptors sensitive to the oxygen tension in the inspired water (see also Chapter 5, this volume).

There is in addition good evidence for oxygen receptors in the pseudo-branch of some species (Laurent, 1967, 1969; Laurent and Rouzeau, 1972), but extirpation or denervation of the pseudobranch failed to block the effects of hypoxia on either heart rate or respiration, and flushing the pseudobranch only with hypoxic water did not elicit the reflex response (Randall and Jones, 1973; Bamford, 1974; Daxboeck and Holeton, 1978; Smith and Jones, 1978; see also Chapter 9, Volume XB, this series).

In lungfish, a marked redistribution of the blood from the branchial to the pulmonary circuit takes place after a breath as a result of reflex induced changes in the branchial compared to the pulmonary vascular resistance (Johansen et al., 1968; Johansen, 1971). The ventilatory response to hypoxia during water breathing is much more pronounced in Neoceratodus, which is a facultative air breather, than in Protopterus, which is an obligatory air breather (Johansen et al., 1968; Johansen, 1971). Hypoxia produces a decrease in the time between air breaths at the surface in Protopterus, suggesting the presence of oxygen receptors (Lahiri et al., 1970). Injection of hypoxic blood into the anterior (second) or posterior (fifth) afferent gill arteries shows that the major oxygen-sensitive area of the gill apparatus is associated with the anterior part (Lahiri et al., 1970).

## IV. PHARMACOLOGY OF THE BRANCHIAL VASCULATURE

Ever since the early studies of the pike (Esox lucius) by Krawkow (1913), it has been clear that the vascular resistance of fish gills can be altered by various drugs. Later studies have expanded our knowledge of pharmacological effects on the gill vasculature, and attempts have been made to demonstrate nervous and hormonal control of the branchial vascular resistance in vivo and the physiological significance of such control. In these studies, pharmacological agents (chemical tools) other than the naturally occurring neurotransmitters and hormones have frequently been used, but a quantitative pharmacological analysis of the action of these drugs is in many cases wanting. It should therefore be emphasized, before going into the details of drug effects on fish gill vasculature, that the selectivity of the concentrations of many of the drugs commonly used is often poorly established, and a reinterpretation of some observed effects may be necessary in the future. For an outline of the quantitative pharmacological approach to the use of chemical tools in the study of lower vertebrate autonomic neurotransmission, see Nilsson (1983).

Nervous and hormonal mechanisms that could possibly elicit a control of the branchial vasculature have been demonstrated, but the knowledge of the physiological significance of these mechanisms in vivo is unclear. For instance, the possibility of a control of the cod (*Gadus morhua*) gill vasculature via both circulating catecholamines and adrenergic nerve fibers has been demonstrated (see later), but the relative importance of the two systems under different physiological conditions remains unknown.

Most of the earlier studies dealing with the branchial vascular resistance in teleosts are concerned with an arterio-arterial pathway only, but later work has also focused interest on the presence of an arteriovenous pathway and mechanisms regulating a shunting of blood between the two pathways. Thus, the blood entering the gill filaments via the afferent filamental arteries passes the secondary lamellae and leaves *either* via the efferent filamental arteries and efferent branchial arteries to the dorsal aorta and systemic circulation (arterio-arterial pathway), *or* from the efferent filamental arteries via a nutritive vasculature of the gill filaments and arches or arteriovenous anastomoses that connect to the filamental venous compartment (arteriovenous pathway). A simplified summary of the main vascular connections is offered in Fig. 10 (see also Chapter 2, this volume).

A number of different preparations have been used in the study of the nervous and hormonal control of gill vasculature. In most cases gill arch or whole-head preparations perfused at constant flow or constant pressure have been studied to determine changes in branchial vascular resistance, and in vivo preparations have also been used. For convenience, the estimations of branchial vascular resistance obtained in preparations that do not distinguish between arterial and venous outflow will be referred to as *overall branchial vascular resistance*, as opposed to *arterio-arterial vascular resistance* or *arteriovenous vascular resistance* from preparations where the efferent arterial flow and the venous flow have been measured separately.

## A. Cholinergic and Adrenergic Drugs

### 1. CYCLOSTOMES

Very little information is available about the control of the branchial vasculature of cyclostomes. In the study by Reite (1969), the effects of acetylcholine and catecholamines on the perfused gill apparatus of *Myxine glutinosa* are described. Acetylcholine was found to increase the overall branchial vascular resistance, and immediately after administration of the drug a transient increase of the resistance due to a tubocurarine-sensitive stimulation of striated gill sac muscle was observed. Adrenaline and noradrenaline produced biphasic responses: an overall vasodilation was most evident during the early part of each experiment, and this effect could be blocked by the β-adrenoceptor antagonist propranolol. Later during the same experiment, the major response to the catecholamines was instead an overall vasoconstriction, and this effect could be blocked by the α-adrenoceptor antagonists phentolamine and dihydroergotamine. The presence of both an α-adrenoceptor-mediated vasoconstrictor and a β-adrenoceptor-mediated vasodilator component in the response to catecholamines has also been demonstrated in gnathostome fish (see later).

## 2. ELASMOBRANCHS

In an early study of the responses to drugs of the perfused gill apparatus of *Squalus acanthias*, Östlund and Fänge (1962) failed to detect any vasomotor effects of adrenaline. The reason for this could be the use of diluted seawater as the perfusion medium, and later studies have demonstrated marked overall vasodilatory effects of adrenaline on the perfused gills of both *Squalus* (Capra and Satchell, 1977; D. H. Evans and J. B. Claiborne, unpublished) and *Scyliorhinus canicula* (Davies and Rankin, 1973). The dilator response to the catecholamines is probably mediated primarily by β-adrenoceptors, as judged from the antagonistic effect of propranolol (Davies and Rankin, 1973; Capra and Satchell, 1977; D. H. Evans and J. B. Claiborne, unpublished). In some experiments a transient increase in the overall vascular resistance was demonstrated, and it seems that this effect is mediated by α-adrenoceptors, as judged by the mimicking effect of the α-adrenoceptor agonist methoxamine and the antagonistic properties of phentolamine (Capra and Satchell, 1977; D. H. Evans and J. B. Claiborne, unpublished).

It is of interest to note that the concentration of circulating catecholamines in plasma from "stressed" dogfish is high enough to produce a marked overall dilation of the gill vasculature (Davies and Rankin, 1973; see later).

## 3. DIPNOANS

The effects of acetylcholine and catecholamines on the vascular resistance in the gills of lungfish have been studied by Johansen and coworkers in *Protopterus aethiopicus* and *Neoceratodus forsteri* (Johansen and Reite, 1968; Johansen et al., 1968). In *Neoceratodus*, gill breathing is the dominant mode of respiration, and the vascular resistance of the gills in this species is higher than in *Protopterus*, which is an obligatory air breather (Johansen et al., 1968). In both species the overall branchial vascular resistance, measured as perfusion backpressure, is increased by acetylcholine and decreased by adrenaline (in *Protopterus*, also noradrenaline). High bolus doses of adrenaline (50–100 μg) sometimes caused constriction of the vasculature in *Protopterus* (Johansen and Reite, 1968). The dilatory responses to adrenaline and noradrenaline were insensitive to propranolol and phentolamine (2–5 μg/ml) present in the perfusion medium; however, no details of exposure time are given in the article, and the concentration of the competitive antagonists may not have been high enough to abolish the response to the doses of catecholamines administered (Johansen and Reite, 1968).

## 4. TELEOSTS

Acetylcholine increases the overall vascular resistance of teleost gills (Östlund and Fänge, 1962; Bergman *et al.*, 1974; Wood, 1975; Smith, 1977; Dunel and Laurent, 1977). In a study of the separate effects of this drug on the arterio-arterial and the arteriovenous branchial resistance, Payan and Girard (1977) demonstrated a shunting toward the arteriovenous pathway. This effect could be due simply to constriction of the arterio-arterial pathway downstream of the branching of the two pathways (Dunel and Laurent, 1977; Smith, 1977), and it is not clear if an active vasodilation of the arteriovenous vasculature takes place in response to acetylcholine (see Section VI,A). Cholinergic control of the gills *in vivo* is certainly due to acetylcholine released from cholinergic nerve endings, and there is no reason to expect the presence of "circulating acetylcholine" in the blood plasma.

Acetylcholine appears to act via stimulation of muscarinic cholinoceptors of the vascular smooth muscle, as judged from the antagonistic effects of atropine (Östlund and Fänge, 1962; Bergman *et al.*, 1974; Wood, 1975). In the holostean fish *Amia calva*, Johansen (1972) concluded a cholinergic vasoconstrictor tonus on the gills of the intact fish, since atropine lowered the branchial vascular resistance. Bergman *et al.* (1974) also describe antagonistic effects of hexamethonium and phenoxybenzamine, but this blockade is likely to be nonspecific. Atropine-like properties of hexamethonium have been demonstrated in fish (Edwards, 1972), and phenoxybenzamine is known to possess a number of nonspecific effects, such as anticholinergic, antihistaminergic, and antiserotoninergic (5-hydroxytryptamine) properties (see Carrier, 1972; Nickerson, 1970; Day, 1979). No antagonistic effects of another nicotinic cholinoceptor antagonist, tubocurarine, were seen by Wood (1975).

Numerous studies of the effects of adrenaline on the teleost branchial vascular resistance have been performed ever since the first demonstration by Krawkow (1913) of a vasodilatory effect of adrenaline in the pike (*Esox lucius*) gills. Thus, an overall vasodilatory effect of adrenaline and noradrenaline has been demonstrated in *Anguilla anguilla* (Keys and Bateman, 1932; Östlund and Fänge, 1962; Steen and Kruysse, 1964; Reite, 1969; Kirschner, 1969; Rankin and Maetz, 1971; Bolis and Rankin, 1973; Forster, 1976a,b; Dunel and Laurent, 1977), *Anguilla japonica* (Chan and Chow,

1976), *Salmo gairdneri* (Richards and Fromm, 1969; Randall *et al.*, 1972; Bergman *et al.*, 1974; Wood, 1974, 1975; Payan and Girard, 1977), *Cyprinus carpio* and *Conger conger* (Belaud *et al.*, 1971), *Gadus morhua* (Östlund and Fänge, 1962; Reite, 1969; Pettersson and Nilsson, 1980; Wahlqvist, 1980, 1981), *Zoarces viviparus* and *Labrus berggylta* (Östlund and Fänge, 1962), *Myoxocephalus octodecimspinosus* (Claiborne and Evans, 1980), and *Pseudopleuronectes americanus* (D'Amico Martel and Cech, 1978). Although the predominant overall effect of adrenaline in most studies of teleosts is a vasodilation of the branchial vasculature that is mediated by β-adrenoceptors, the presence of an α-adrenoceptor-mediated vasoconstriction has also been demonstrated in several studies (Reite, 1969; Belaud *et al.*, 1971; Bergman *et al.*, 1974; Wood, 1975; Wood and Shelton, 1975, 1980a; Dunel and Laurent, 1977; Payan and Girard, 1977; Colin and Leray, 1979; Wahlqvist, 1980, 1981; Claiborne and Evans, 1980). In some studies, the vasoconstrictor effect dominates, and there appears to be a seasonal variation in the relation between the constrictor and the dilator response (Pettersson, 1983; Pärt *et al.*, 1982).

According to the classification of adrenoceptors originally proposed by Ahlquist (1948) into an α and a β category, the order of potency for the adrenoceptor agonists should be adrenaline > noradrenaline > phenylephrine > isoprenaline at the α-adrenoceptors, and isoprenaline > adrenaline ≥ noradrenaline > phenlyephrine at the β-adrenoceptors (Furchgott, 1967). Studies of the potency relationship for these amines have been attempted in concentration–response and dose–response studies of rainbow trout branchial vasculature (Bergman *et al.*, 1974; Wood, 1974, 1975), but in most studies the assumed (but not necessarily real) specificity of adrenoceptor antagonists has been used to classify the adrenoceptors.

Some early studies suggested that the vasodilatory effect of adrenaline on the teleost branchial vasculature was due to activation of α-adrenoceptors. These conclusions were based on results obtained with the noncompetitive α-adrenoceptor antagonist phenoxybenzamine (Dibenzyline) in studies of blood pressure and blood flow in rainbow trout (Randall and Stevens, 1967) and bowfin (*Amia calva*) (Johansen, 1972). The observed results may be better explained by a blocking action of phenoxybenzamine on the systemic vasculature (Bergman *et al.*, 1974), and unspecific effects of the antagonist cannot be ruled out. Phenoxybenzamine is known to possess anticholinergic, antihistaminergic, and antiserotoninergic properties (see earlier), and the noncompetitive mode of action makes in vivo studies particularly difficult to assess. It is, however, possible that the shunting of blood between an arterio-arterial and an arteriovenous pathway in the gills is dependent on α-adrenoceptors mediating *vasoconstriction* in the arteriovenous pathway, and such an effect of adrenaline would tend to increase the portion of the blood leaving the gills via the arterio-arterial pathway. The evidence for α-adrenoceptors mediating *vasodilation* in any part of the gill vasculature is, however, weak.

The bulk of the literature now favors a β-adrenoceptor-mediated mechanism as responsible for the adrenergic vasodilation of the teleost branchial vasculature. Particularly the studies by Bergman *et al.* (1974) and Wood (1974, 1975), where the relative potencies of phenylephrine, noradrenaline, adrenaline, and isoprenaline have been assessed, are of importance in classifying the adrenoceptors. The least potent of these amines was phenylephrine, which is a partial adrenoceptor agonist on α-adrenoceptors. The vasodilatory action of this drug could in part be due to a release of nervously stored catecholamines, a phenomenon known from studies on mammals (Trendelenburg, 1972) that has also been demonstrated in fish (Fänge *et al.*, 1976).

Unknown differences between adrenaline and noradrenaline in the affinity for the neuronal uptake mechanism of the adrenergic nerve terminals (uptake$_1$; Iversen, 1974), and hence in the concentration of the amine available at the receptor sites, make it impossible to determine the relative potency of adrenaline and noradrenaline unless the neuronal uptake is blocked (e.g., by cocaine) or the adrenergic nerve terminals destroyed (e.g., by surgical or chemical "sympathectomy") (Trendelenburg, 1963; Iversen, 1967; Holmgren and Nilsson, 1982). The difference between adrenaline and noradrenaline in vasodilator potency in trout gills was used to classify the β-adrenoceptors of the branchial vasculature as β$_1$-adrenoceptors (Wood, 1975) according to the terminology introduced by Lands *et al.* (1967). In view of possible differences in the neuronal uptake of the two catecholamines, further studies involving specific β$_1$- and β$_2$-adrenoceptor agonists and antagonists would be welcome.

The most effective vasodilator substance of the adrenergic agonists is isoprenaline (Bergman *et al.*, 1974; Wood, 1974, 1975). This substance stimulates both β$_1$- and β$_2$-adrenoceptors, and the α-adrenoceptor *agonistic* properties are low (see Nilsson, 1983). An α-adrenoceptor *blocking* capacity of isoprenaline, possibly related to the D isomer of the racemate, has been demonstrated in fish (Holmgren and Nilsson, 1974; Wood, 1975).

## B. Other Vasoactive Agents

Relatively few studies deal with branchial vascular effects of vasoactive agents other than the adrenergic and cholinergic drugs. Reite (1969) concluded that the general effects of histamine and 5-hydroxytryptamine (5-HT) on the circulatory system of cyclostomes and elasmobranchs were due to a nonspecific stimulation of adrenoceptors, but in teleost fish a marked overall constriction of the branchial vasculature was produced by 5-HT (Östlund and Fänge, 1962; Reite, 1969). The effect of 5-HT in this case is probably direct at serotoninergic receptors of the vascular smooth muscle; at least it is not due to activation of cholinergic neurons, since the effect persists in the presence of atropine (Reite, 1969).

A marked overall branchial vasoconstriction in response to adenosine and related nucleotides has been demonstrated in the trout, *Salmo gairdneri*. The drugs constrict the arterio-arterial pathway, thus increasing the venous outflow from the gills, and it appears that the effect is mediated via specific purinoceptors (purinergic receptors) of the vasculature (Colin and Leray, 1979; Colin *et al.*, 1979).

Several "peptide hormones" are known to affect the branchial vascular resistance of teleost fish. Vasoactive intestinal polypeptide (VIP), known from experiments with mammals to be a potent vasodilator substance (Said and Mutt, 1970), also produces an overall dilation of the branchial vasculature of the eel (*Anguilla anguilla*) (L. Bolis and J. C. Rankin, unpublished). An increased sensitivity to noradrenaline was found after pretreatment of isolated perfused gills with β-endorphin, and this effect was abolished by naloxone (Bolis *et al.*, 1980). Marked overall constrictor effects of posterior lobe pituitary hormones have been demonstrated [Keys and Bateman, 1932 ("pitressin"); Rankin and Maetz, 1971 (isotocin, arginine-vasotocin)].

The physiological significance of the effects on the branchial vasculature of the substances just listed awaits further clarification.

## V. AUTONOMIC NERVOUS CONTROL OF THE BRANCHIAL VASCULATURE

### A. Cyclostomes and Elasmobranchs

The knowledge of the anatomy and physiology of the autonomic innervation of the branchial vasculature is fragmentary in teleost fish and almost nonexistent in cyclostome and elasmobranch fish. In an electron-microscopic study of the innervation of the gill filaments of the lamprey, *Lampetra japonica*, Nakao (1981) demonstrated two types of nerve profiles. The first type, containing mainly small clear vesciles (diameter 30–100 nm) and a few large granular vesicles (diameter 60–180 nm), innervates mainly the striated muscle of the gill sac. The second type, characterized by its content of mainly large granular vesicles, was demonstrated in the walls of the afferent and the efferent filamental arteries. The possible effects of this innervation on the branchial vasculature remain to be elucidated.

In the dogfish, *Scyliorhinus canicula*, an overall vasoconstriction has been observed on stimulation of the vagal supply of the gill arch (D. T. Davies and J. C. Rankin, personal communication). Later studies suggest that all or part of this response is due to contraction of the striated muscle of the gill arch (J. D. Metcalfe, personal communication), but further studies of the vasomotor control of the branchial vasculature of elasmobranchs are obviously needed. Particular attention should be paid to the neurons described by Boyd (1936) in *Mustelus* at the junction between the efferent filamental and arch arteries.

## B. Teleosts

A dense innervation of small blood vessels forming a *rete-mirabile jacket* around the afferent branchial arteries in salmonids was described by De Kock (1963), but neither the nature of these fibers (vasomotor or sensory) nor their origin (cranial or spinal) is clear (cf. Section III,C).

In the gill filaments, ultrastructural studies have revealed a dense innervation of the sphincters at the base of the efferent filamental arteries (Dunel and Laurent, 1980; see also Fig. 10). These fibers contain small clear vesicles, typical of cholinergic neurons, and could be responsible for a vagal constrictory control of these sphincters (see later).

In *Tilapia mossambica*, an innervation of the arteriovenous anastomoses on the efferent side was described by Vogel *et al.* (1974), and it is not unlikely that nerve fibers running along the efferent filamental artery reach the efferent lamellar arterioles. A direct innervation of the pillar cells of the lamellae has been claimed by Gilloteaux (1969), but this suggestion lacks confirmation by other workers (Laurent and Dunel, 1980). Clearly the microanatomy of the vascular innervation in fish gills requires further attention. A fluorescence histochemical study of the distribution of adrenergic nerve fibers, which are known from physiological and pharmacological experiments to be present within the gills, would be of particular interest.

### 1. CRANIAL AUTONOMIC ("PARASYMPATHETIC") CONTROL

An increase in the vascular resistance of an isolated perfused gill arch in response to branchial nerve stimulation was first described for the Atlantic cod, *Gadus morhua*, by Nilsson (1973), and later confirmed in a more detailed study by Pettersson and Nilsson (1979). Part of the excitatory response to the branchial nerve stimulation could be abolished by atropine, suggesting the presence of cholinergic vasoconstrictor fibers in the branchial nerve. Both constrictor and dilator effects of adrenergic nerve fibers of spinal autonomic ("sympathetic") origin (see later) could also be demonstrated during stimulation of the branchial nerve (Pettersson and Nilsson, 1979). In the presence of both cholinergic (atropine) and adrenergic (phenoxybenzamine and propranolol) antagonists, a small constrictor response to branchial nerve stimulation often persisted. The nature of this remaining response is obscure: an effect of nonadrenergic, noncholinergic (NANC) vasomotor fibers in the vagal innervation of the gill vasculature is possible, and, although the slow development of the response during nerve stimulation would seem to speak against an involvement of skeletal muscle, this possibility also needs investigation (cf. Section V,A).

## 2. SPINAL AUTONOMIC ("SYMPATHETIC") CONTROL

The spinal autonomic innervation of the branchial vasculature has been studied in the Atlantic cod, *Gadus morhua* (Pettersson and Nilsson, 1979; Nilsson and Pettersson, 1981). The spinal autonomic pathways leave the CNS in the trunk segments via the white rami communicantes of the spinal nerves (Fig. 3) and run forward in the sympathetic chains into the head (Nilsson, 1976). Postganglionic fibers enter the branchial nerves via gray rami communicantes. These fibers appear to be solely adrenergic, and the effects are mediated via both $\alpha$- and $\beta$-adrenoceptors within the gill vasculature.

In the right-side gill apparatus of the cod, perfused at constant flow from a peristaltic pump, stimulation of the right sympathetic chain produced an increase in the efferent arterial outflow and a decrease in the inferior jugular

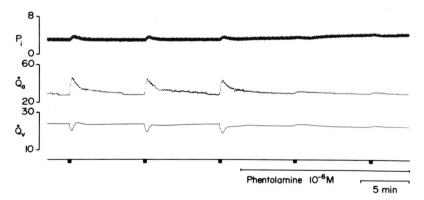

**Fig. 6.** Changes in inflow counterpressure ($P_i$), and arterial ($\dot{Q}_a$) and venous ($\dot{Q}_v$) flow during constant-flow perfusion of the right-side gill apparatus of the Atlantic cod, *Gadus morhua*, induced by electrical stimulation of the right sympathetic chain with 10 Hz, 1 msec pulse duration, and 8 V for 1 min every 8 min. Note the slight reversal of $P_i$ after administration of phentolamine, and the strong reduction in both arterial and venous flow responses. $P_i$ is expressed in kPa and the flow in drops per minute. (From Nilsson and Pettersson, 1981.)

vein outflow (Fig. 6). This is similar to the effects of adrenaline on the same type of preparation (*Salmo gairdneri*, Payan and Girard, 1977; *Myoxocephalus octodecimspinosus*, Claiborne and Evans, 1980; *Gadus morhua*, Nilsson and Pettersson, 1981). In contrast to the overall decrease of the branchial vascular resistance normally produced by exogenous adrenaline in cod gills (see, however, Pettersson, 1983), sympathetic chain stimulation produced an increase in the inflow pressure (Fig. 6). All the responses to sympathetic chain stimulation were reduced or reversed by the α-adrenoceptor antagonist phentolamine (Fig. 6), whereas the β-adrenoceptor antagonist propranolol had little effect (Nilsson and Pettersson, 1981). Although further quantitative pharmacological work to establish the selectivity of the concentrations of the antagonists used would be instructive, it seems that the responses to adrenergic nerve stimulation are chiefly mediated by α-adrenoceptor mechanisms.

## VI. CONTROL OF THE BRANCHIAL VASCULATURE BY CIRCULATING CATECHOLAMINES

In both cyclostomes and gnathostome fish, catecholamine-containing chromaffin tissue is strategically located in the large veins just outside the heart. In elasmobranchs the anterior chromaffin cell masses form the "axillary bodies," together with paravertebral ("sympathetic") ganglion cells within the posterior cardinal sinuses; whereas in teleosts and dipnoans the chromaffin cells line the posterior cardinal veins (for review, see Nilsson, 1983).

In elasmobranchs, circulating catecholamines released from the chromaffin tissue of the axillary bodies are considered to be of great importance in the control of the heart (Gannon *et al.*, 1972), branchial vasculature (Davies and Rankin, 1973), and systemic vasculature (Butler *et al.*, 1978). In the study by Davies and Rankin (1973), it could be directly shown that blood plasma from stressed dogfish (*Scyliorhinus canicula*) contains catecholamines in sufficient concentration to produce an overall dilation of the branchial vasculature.

Also in teleosts, it appears that the levels of circulating catecholamines reached during severe stress are high enough to affect the circulatory system. The chromaffin tissue of the cod, *Gadus morhua*, is innervated by preganglionic fibers of spinal autonomic origin (Nilsson, 1976), and stimulation of these fibers releases catecholamines into the posterior cardinal veins (Nilsson *et al.*, 1976). During perfusion of a cardinal vein–heart preparation of the cod in situ, Holmgren (1977) showed that stimulation of the nerve

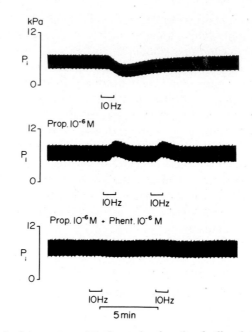

**Fig. 7.** Recording of inflow pressure ($P_i$) of an isolated perfused gill arch from the cod, *Gadus morhua*. The perfusion fluid was passed through the left cardinal vein before entering the peristaltic pump, and catecholamines were released from the chromaffin tissue of the cardinal vein by electrical stimulation of the nerve supply to this tissue with pulses of 8 V at 10 Hz as indicated. The predominant response to the humoral catecholamines is a decrease in branchial vascular resistance (top tracing), which is reversed to an overall vasoconstriction after propranolol ($10^{-6}$ $M$) (middle tracing) and abolished in the presence of both propranolol and phentolamine ($10^{-6}$ $M$) (bottom tracing). (From Wahlqvist, 1981.)

supply to the chromaffin tissue released catecholamines in sufficient quantities to affect heart rate significantly. The same type of preparation, but with the heart replaced by a peristaltic pump, was used by Wahlqvist (1981) to study the effects of catecholamines released from the chromaffin tissue on the branchial vascular resistance. In these experiments catecholamines were released by electrical stimulation of the nerve supply to the chromaffin tissue, and the perfusate leaving the cardinal vein was pumped into an isolated gill arch. The inflow pressure of the gill arch was monitored, and the release of catecholamines produced by nerve stimulation at different frequencies could thus be continuously bioassayed (Figs. 7 and 8). A frequency–response analysis showed a significant effect on the branchial vasculature at a stimulation frequency as low as 1 Hz, and a steep rise in the response between 1 and 4 Hz (Fig. 8). These results are consistent with the view that catecholamines could be released from the cardinal vein chromaffin tissue at

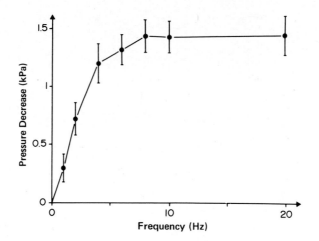

**Fig. 8.** Relationship between stimulation frequency and vasodilator response (indicated as decrease in $P_i$) of an isolated perfused gill arch in a preparation similar to that in Fig. 7. Stimulation of the nerve supply to the chromaffin tissue of the cardinal vein was made at frequencies between 1 and 20 Hz. Note the steep rise in the response between 1 and 4 Hz. Vertical bars, ± SD ($n = 12$). (From Wahlqvist, 1981.)

"physiological frequencies" in quantities high enough to affect the branchial vasculature.

Further evidence for an influence of circulating catecholamines on the branchial vasculature comes from comparisons between the concentration of adrenaline in blood plasma and the concentration–response curves for the dilatory effect on the gill vasculature (Fig. 9). Both in *Salmo gairdneri* (Wood, 1974; Nakano and Tomlinson, 1967) and in *Gadus morhua* (Wahlqvist, 1980; Wahlqvist and Nilsson, 1980), the concentrations of adrenaline in blood plasma during rest and stress lie within the response range of the perfused branchial vasculature (Fig. 9).

During severe stress induced by draining most of the water from the fish tank, the overall branchial vascular resistance increases as a result of a non-adrenergic effect (direct hypoxic vasoconstriction and/or influence of non-adrenergic nerves), and this effect is counteracted to some degree by an adrenergic dilation (Wahlqvist and Nilsson, 1980). The adrenergic influence was not affected by cutting the sympathetic chains and thus the adrenergic innervation of the branchial vasculature, but it was significantly reduced when the nerve supply to the chromaffin tissue of the cardinal veins was severed. These observations also speak in favor of an adrenergic control of the branchial vasculature of the cod via circulating catecholamines, particularly during severe stress (Wahlqvist and Nilsson, 1980).

In dipnoans, the sympathetic chains are very poorly developed (Jenkin,

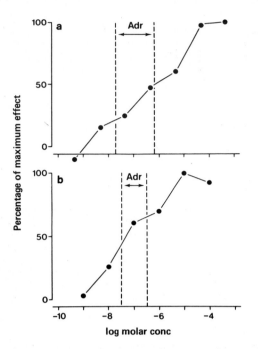

**Fig. 9.** Concentration–response curves for the vasodilator effect of adrenaline on the perfused branchial vasculature of *Salmo gairdneri* (a) and *Gadus morhua* (b), compared to the range of adrenaline concentrations in blood plasma of the same species in resting and stressed condition (vertical broken lines). Concentration–response curves redrawn from Wood, 1974 (*Salmo*) and Wahlqvist, 1980 (*Gadus*). Values of plasma adrenaline concentrations from Nakano and Tomlinson, 1967 (*Salmo*) and Wahlqvist and Nilsson, 1980 (*Gadus*).

1928; Abrahamsson *et al.*, 1979b), and in *Protopterus aethiopicus* general adrenergic control of the circulatory system via circulating catecholamines was concluded (Abrahamsson *et al.*, 1979a,b). In this species chromaffin tissue could be demonstrated in the atrium of the heart, in the left cardinal vein, and in the segmentally arranged intercostal arteries (Abrahamsson *et al.*, 1979a). The possible role of circulating amines in the control of the branchial vasculature is, however, unknown.

## VII. POSSIBLE SITES OF DRUG AND NERVE ACTION

A shunting of blood *between the different gill arches*, as well as changes in the relative resistances in the vascular beds of the gills and the accessory respiratory organ (air bladder, lung, etc.), is evident in bimodal breathers such as dipnoans and air-breathing actinopterygians (Johansen, 1970, 1971, 1972). In the Amazonian fish *Hoplerythrinus unitaeniatus,* there is a narrow connection between the dorsal aorta and the efferent branchial arteries of the third and fourth gill arches, and the air bladder artery branches off the efferent gill arteries proximal to this narrow connection. During air breathing there is a preferential perfusion of the third and fourth gill arches, which will thus increase the blood flow through the air bladder circuit (Farrell, 1978; Smith and Gannon, 1978). The control of the shunting may take place within the gills: acetylcholine directs the flow away from the first and second gill arches and into the third and fourth gill arches, while there is no effect of cholinergic or adrenergic agonists on the vascular resistance of the air bladder vasculature (Smith and Gannon, 1978).

A shunting between different gill arches was also observed in the skate, *Raja rhina,* by Satchell *et al.* (1970). In this study an increase in the proportion of blood entering the posterior gill arches was seen during swimming activity, and this effect was related to the increase in the respiratory water flow during swimming.

In teleosts, many studies have dealt with the changes in blood flow *within the individual gill filaments,* and it is now known that both the functional surface area of the gills and the shunting of blood between efferent arterial and branchial venous outflow from the gills can be affected by cholinergic and adrenergic agonists. The following sections attempt to summarize, from observations of several species of teleosts, some possible intrafilamental sites of action of nerves and circulating agents that affect branchial blood flow patterns.

## A. Control of Functional Surface Area

Both fresh- and saltwater teleosts live in media with an osmolarity that deviates from that of their blood plasma. Thus, loss of ions (freshwater teleosts) or water (saltwater teleosts) will occur at the gills, and the respiratory demand for a large functional surface area of the gills must be balanced by the disadvantage of osmotic losses (respiratory–osmoregulatory compromise) (Kirschner, 1969; Randall *et al.*, 1972; Wood and Randall, 1973). In the trout, *Salmo gairdneri,* Wood and Randall (1973) demonstrated increased loss of $Na^+$ during increased swimming activity. This effect was attributed to an augmented blood flow in the gills due to catecholamine release into the blood, since it had been shown that addition of noradrenaline or isoprenaline to the water in which the fish were swimming evoked an increase in the release of $^{22}Na$ from the fish (Randall *et al.*, 1972). It is also clear that direct effects on ion transfer can be produced by the catecholamines (see Chapter 5, Volume XB, this series).

To explain observed changes in the functional surface area of the gills, Steen and Kruysse (1964) suggested that blood could bypass the secondary lamellae from the afferent to the efferent filamental artery via a nonrespiratory shunt. In their model, the nonrespiratory shunt was believed to include the filamental venous sinus, which, in the eel, *Anguilla anguilla*, connects to both the afferent and the efferent filamental arteries by arteriovenous anastomoses. It is now known that the filamental venous sinus is a low-pressure compartment and that blood flow from this sinus into the efferent filamental artery therefore is impossible (see also Chapter 2, this volume).

Another possibility of a nonrespiratory shunt is that the blood passes the secondary lamellae through the basal channels, which are situated under the filamental surface and thus lack direct contact with the surrounding water (see Farrell *et al.*, 1980). It is not clear how changes in vessel diameters elicited by nerves or circulating agents could affect the intralamellar flow, or to what extent local effects of hypoxia or passive changes in branchial vascular resistance due to the compliance of the gill vasculature can affect the blood flow pattern in the lamellae (Farrell *et al.*, 1979, 1980).

Changes in the number of lamellae perfused at any given moment may also affect the functional surface area of the gills (lamellar recruitment model; Hughes, 1972; Booth, 1978, 1979a,b). Acetylcholine produces an overall vasoconstriction of the branchial vasculature and a reduction in the number of lamellae perfused (particularly the distally located lamellae) (*Salmo gairdneri:* Smith, 1977; Booth, 1979a,b; *Ictalurus punctatus:* Holbert *et al.*, 1979). It should be noted, however, that the increase in branchial vascular resistance does not necessarily relate directly to the de-recruitment of lamellae: in the lingcod, *Ophiodon elongatus*, Farrell *et al.* (1979) could show no simple relationship between the number of lamellae perfused and the branchial vascular resistance.

In addition to the reduction in the number of lamellae perfused, there is an increase in the lamellar blood space induced by acetylcholine (Booth, 1979b; Holbert *et al.*, 1979). There is a possibility that this effect is due to a constriction distal to the secondary lamellae, which would tend to increase the intralamellar pressure. In fact, the best available evidence for a cholinergic control of the branchial vasculature suggests that the main site of action is at the sphincters at the base of the efferent filamental arteries (Fig. 10). The reasoning is as follows:

1. These sphincters are innervated by c-type nerve terminals (Dunel and Laurent, 1980).
2. These sphincters constrict after addition of acetylcholine (Dunel and Laurent, 1977), altering the arteriovenous shunting pattern (see Section VII,B).

3. There is also evidence for a constrictory cholinergic nervous control of the arterio-arterial pathway (Pettersson and Nilsson, 1979).

In this context it should be stressed that there is no evidence for "circulating acetylcholine" in vertebrates (the blood plasma is rich in cholinesterase), and any cholinergic control in vivo must be due to a cholinergic *innervation* of the effector tissue. Demonstration of muscarinic receptors per se is not conclusive evidence for a cholinergic innervation, and exogenous acetyl-choline may also affect tissues that are not cholinergically controlled in vivo (see also Wood, 1975). It is thus essential to establish the presence of cholinergic (c-type) nerve terminals in the effector tissues thought to be cholinergically controlled.

The effects of adrenergic agonists on the functional surface area of teleost gills is well established, although, as pointed out in Section I, the relative importance of hemodynamic effects compared to possible direct cate-cholamine-induced permeability changes in the gill epithelium requires much further attention. An adrenergic control may, contrary to a cholinergic control, also be due to a humoral agent. This means that an adrenergic control by circulating catecholamines is possible also in the parts of the gill vasculature that are not innervated.

Adrenaline, as well as some other adrenoceptor agonists, produces an increase in the functional surface area of the gills measured as [$^{14}$C]urea fluxes in the rainbow trout (Bergman *et al.*, 1974), and it is also clear that oxygen uptake by the gills—both in vivo and in vitro—is enhanced by the administration of adrenaline (Steen and Kruysse, 1964; Peyraud-Waitzenegger, 1979; Pettersson and Johansen, 1982; Pettersson, 1983). At a superficial glance it would appear tempting to ascribe the increased functional surface area observed after adrenaline to the overall vasodilator β-adrenoceptor-mediated effect of this drug. Several studies agree, however, that the predominant cause of the increased oxygen uptake or [$^{14}$C]urea exchange is due to an α-adrenoceptor-mediated effect, although a smaller effect is also produced by the potent β-adrenoceptor agonist isoprenaline (Bergman *et al.*, 1974; Peyraud-Waitzenegger *et al.*, 1979; Pettersson, 1983). Hypoxia produces an overall branchial vasoconstriction (Holeton and Randall, 1967; Pettersson and Johansen, 1982) and an increased lamellar recruitment (Booth, 1979b). It has been concluded that the hypoxic as well as the α-adrenoceptor-mediated vasoconstriction of the branchial vasculature takes place distal to the secondary lamellae, since there is an increase in the lamellar recruitment after adrenaline, and since hypoxia in the respiratory water (which is not detected by the prelamellar vasculature) produces vasoconstriction. The effect of hypoxia may be proximal to the arteriovenous connections, possibly at the efferent lamellar arterioles, since there are no apparent changes in flow between the arterio-arterial and the arteriovenous pathways (Pettersson and Johansen, 1982; Pettersson, 1982). In addition, a β-adrenoceptor-mediated vasodilation of the afferent lamellar arterioles may help to increase the lamellar recruitment in response to adrenaline (Pettersson, 1982). In the lingcod, *Ophiodon elongatus*, Farrell (1980) concluded that the afferent lamellar arterioles are the main site of branchial vascular resistance, and if this is true also in other species changes of the vascular resistance of these vessels could significantly affect the lamellar blood flow.

The presence of contractile material in the pillar cells of the secondary lamellae was suggested by Bettex-Galland and Hughes (1973), and in a later immunohistochemical study, Smith and Chamley-Campbell (1981) demonstrated the presence of myosin in the pillar cells. There is thus a possibility that the pillar cells could actively modify the width of the secondary lamellae, but changes in lamellar blood space may also be explained by changes in the intralamellar pressure. This in turn is controlled by the relative vascular resistance of the afferent and efferent lamellar arterioles and the blood flow. Measurements of the intralamellar blood pressure with tip-transducer technique for comparison with afferent and efferent filamental artery pressures during different treatments (hypoxia, adrenaline, autonomic nerve stimulation) should be of great interest in the localization of the vascular effector units in the gill filaments and secondary lamellae.

## B. Control of Arteriovenous Shunting

In most teleosts, with the exception of the eel (Laurent and Dunel, 1976) and catfish (Boland and Olson, 1979), the blood passes through the secondary lamellae before being divided into an efferent arterial and a branchial venous outflow. In the eel and catfish there are also direct connections from the afferent filamental artery to the filamental venous "sinus," but nothing is known about the control of these afferent arteriovenous anastomoses (see also Chapter 2, this volume). In fish such as the eel where, under certain circumstances, cutaneous breathing is the dominant mode of respiration, a direct connection via the afferent arteriovenous anastomoses would avoid a deoxygenation in secondary lamellae exposed to hypoxia, of blood reaching the filamental venous sinus.

The *arterio-arterial pathway* supplies the systemic vascular beds with oxygenated blood via the efferent branchial arteries and dorsal aorta. The *arteriovenous pathway* consists basically of two parallel vascular units: the nutritional vasculature of the gill filament and arch (which is essentially a systemic vascular bed) and the efferent arteriovenous anastomoses (Fig. 10). The nutritional vasculature dervies arterial blood from the efferent filamental and branchial arteries, and drains into the filamental venous sinus and the branchial vein. Several functions have been ascribed to the arteriovenous pathway and the filamental venous system, including nutritional supply to the gill tissues, oxygen supply to the chloride cells of the filamental epithelium (Laurent and Dunel, 1980; Dunel and Laurent, 1980), storage of red blood cells (Girard and Payan, 1976; Boland and Olson, 1979), and a direct pathway for oxygenated blood to the heart, which may be important in teleosts, which lack a coronary supply (Girard and Payan, 1976). In fish that possess a coronary supply to the compact myocardium, such as *Salmo gairdneri*, coronary artery ablation does not affect the exercise performance (Daxboeck, 1982), showing the importance of the lacunar venous blood as an oxygen source for the myocardium.

A direct demonstration of an adrenergic control of the shunting between the dorsal aortic and venous outflow of perfusion fluid from the isolated perfused head of the rainbow trout was made by Payan and Girard (1977). In these experiments it was shown that the arterial outflow increased and the venous outflow decreased after adrenaline, and it was concluded that the effect was due to an α-adrenoceptor-mediated closure of the efferent arteriovenous anastomoses (Payan and Girard, 1977).

Similar conclusions were reached by Claiborne and Evans (1980) working with the isolated perfused head of the sculpin, *Myoxocephalus octodecimspinosus*. In a study of the cod (Nilsson and Pettersson, 1981), it was also shown that electrical stimulation of the sympathetic chain elicits an increase in arterial (suprabranchial artery) and a decrease in venous (inferior jugular vein) outflow accompanied by an overall increase in the branchial vascular resistance (Fig. 9). The effects were concluded to be due to an α-adrenoceptor-mediated constriction of the arteriovenous connections and the systemic vasculature. It was also concluded that the dorsal aortic pressure, which is regulated by circulating catecholamines and adrenergic nerve fibers, is probably of great importance in controlling the arteriovenous shunting in the cod (Nilsson and Pettersson, 1981). A direct (passive?) increase in the arteriovenous flow as a result of an increased pulse pressure has also been demonstrated in the isolated perfused trout head (Daxboeck and Davie, 1982).

The possible mode of innervation of the gill vasculature of the cod is summarized in Fig. 10. The main cholinergic innervation is probably at the sphincters at the base of the efferent filamental arteries, but so far there has been no elucidation of the importance of this innervation in the control of arteriovenous shunting. The predominant effect of the adrenergic innervation is an α-adrenoceptor-mediated constriction of the arteriovenous connections (nutritional vasculature and/or efferent arteriovenous anastomoses), which is recorded as an increase in the overall branchial vascular resistance

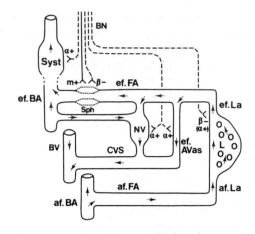

**Fig. 10.** Speculative summary (working model) of the arrangement of the branchial vasculature and possible sites of action of the autonomic fibers in the branchial (X) nerve innervating the gill vasculature of the cod, *Gadus morhua*. The blood enters the afferent filamental artery (af. FA) from the afferent branchial artery (af. BA) and leaves the filamental circulation either via the efferent branchial artery (ef. BA) (arterio-arterial pathway) or the branchial vein (BV) (arteriovenous pathway). The cranial autonomic fibers in the branchial nerve constrict the arterio-arterial pathway, possibly by contracting the sphincter (Sph) at the base of the efferent filamental artery (ef. FA). The adrenergic fibers originate from the cephalic sympathetic chain ganglia and act chiefly by constricting the efferent arteriovenous anastomoses (ef. AVas) and/or the filamental nutritional vasculature (NV) which drain into the central venous "sinus" (CVS). A β-adrenoceptor-mediated control by adrenergic fibers innervating the vasculature of the lamellae (L), probably the efferent lamellar arterioles (ef. La), may also be present. β-Adrenoceptors, which are affected by circulating catecholamines only (not shown in figure), may be present in the afferent lamellar arterioles (af. La) (cf. Pettersson, 1983). A general adrenergic control of the systemic blood pressure by circulating catecholamines and/or innervation of the systemic vasculature (Syst) is probably of importance in the control of arteriovenous shunting in the cod gills. It should be noted that circulating catecholamines may produce responses in effector tissues that are not innervated by adrenergic fibers. α+, β−, and m+ refer to α-adrenoceptors mediating vasoconstriction, β-adrenoceptors mediating vasodilation, and muscarinic cholinoceptors mediating vasoconstriction, respectively. (Slightly modified from Nilsson, 1983.)

(contrary to the main effect of exogenous adrenaline) and a shunting toward arterio-arterial flow (Figs. 9 and 10; Nilsson and Pettersson, 1981).

There is some evidence for a β-adrenoceptor-mediated decrease in the overall branchial vascular resistance during adrenergic nerve stimulation after α-adrenoceptor blockade with phentolamine (Nilsson and Pettersson, 1981). These β-adrenoceptors are probably located at the level of the lamellar arterioles, since there is in most cases no change in the relative arterial/venous outflow after isoprenaline (Nilsson and Pettersson, 1981; see also Claiborne and Evans, 1980). In some cases a slight increase in the arterial outflow could be detected without a change in the venous outflow, which would indicate an effect distal to the arteriovenous branching.

It is possible that although the major effect of the adrenergic (spinal autonomic) innervation of the gill vasculature is an α-adrenoceptor constrictor control of the arteriovenous pathway, the β-adrenoceptor-mediated vasodilator effect seen after the administration of catecholamines takes place in a part of the branchial vasculature that is not well innervated by adrenergic neurons—possibly at the level of the lamellar arterioles (Claiborne and Evans, 1980; Nilsson and Pettersson, 1981; Pettersson, 1983). Circulating catecholamines released from the chromaffin tissue of the posterior cardinal veins may be of particular importance in this control.

## ACKNOWLEDGMENTS

The critical reading of the manuscript by Drs. Susanne Holmgren, David J. Randall, and Chris M. Wood is gratefully acknowledged. Our own research concerning the autonomic innervation of the gill vasculature is currently supported by the Swedish Natural Science Research Council.

## REFERENCES

Abrahamsson, T., Holmgren, S., Nilsson, S., and Pettersson, K. (1979a). On the chromaffin system of the African lungfish, *Protopterus aethiopicus*. Acta Physiol. Scand. **107**, 135–139.

Abrahamsson, T., Holmgren, S., Nilsson, S., and Pettersson, K. (1979b). Adrenergic and cholinergic effects on the heart, the lung and the spleen of the African lungfish, *Protopterus aethiopicus*. Acta Physiol. Scand. **107**, 141–147.

Ahlquist, R. P. (1948). A study of the adrenotropic receptors. Am. J. Physiol. **153**, 586–600.

Allis, A. P. (1920). The branches of the branchial nerves of fishes, with special reference to *Polyodon spathula*. J. Comp. Neurol. **32**, 137–153.

Ballintijn, C. M. (1969). Muscle coordination of the respiratory pump of the carp (*Cyprinus carpio L.*). J. Exp. Biol. **50**, 569–591.

Bamford, O. S. (1974). Oxygen reception in rainbow trout (*Salmo gairdneri*). Comp. Biochem. Physiol. A **48A**, 69–76.

Belaud, A., Peyraud-Waitzenegger, M., and Peyraud, C. (1971). Etude comparée des réactions vasomotrices des branchies perfusées de deux téléostéens: La carpe et la congre. C. R. Seances Soc. Biol. Ses Fil. **165**, 1114–1118.

Berger, P. J., Evans, B. K., and Smith, D. G. (1980). Localization of baroreceptors and gain of the baroreceptor-heart rate reflex in the lizard *Trachydosaurus rugosus*. J. Exp. Biol. **86**, 197–209.

Bergman, H. L., Olson, K. R., and Fromm, P. O. (1974). The effects of vasoactive agents on the functional surface area of isolated-perfused gills of rainbow trout. J. Comp. Physiol. **94**, 267–286.

Bettex-Galland, M., and Hughes, G. M. (1973). Contractile filamentous material in the pillar cells of fish gills. J. Cell Sci. **13**, 359–370.

Boland, E. J., and Olson, K. R. (1979). Vascular organization of the catfish gill filament. Cell Tissue Res. **198**, 487–500.

Bolis, L., and Rankin, J. C. (1975). Adrenergic control of blood flow through fish gills: Environmental implications. *In* "Comparative Physiology II" (L. Bolis, K. Schmidt-Nielsen, and S. H. P. Maddrell, eds.), pp. 223–233. North-Holland Publ., Amsterdam.

Bolis, L., Rankin, J. C., and Trischitta, F. (1980). β-endorphin and control of blood flow in perfused brown trout gills. *In* "Proceedings of the International Meeting on Fish Gills, Giarre (Sicily)," p. 27. Institute of General Physiology, University of Messina.

Booth, J. H. (1978). The distribution of blood in the gills of fish: Application of a new technique to rainbow trout (*Salmo gairdneri*). *J. Exp. Biol.* **73,** 119–129.

Booth, J. H. (1979a). The effects of oxygen supply, epinephrine, and acetylcholine on the distribution of blood flow in trout gills. *J. Exp. Biol.* **83,** 31–39.

Booth, J. H. (1979b). Circulation in trout gills: The relationship between branchial perfusion and the width of the lamellar blood space. *Can. J. Zool.* **57,** 2183–2185.

Boyd, J. D. (1936). Nerve supply to the branchial arch arteries of vertebrates. *J. Anat.* **71,** 157–158.

Burnstock, G. (1969). Evolution of the autonomic innervation of visceral and cardio-vascular systems in vertebrates. *Pharmacol. Rev.* **21,** 247–324.

Butler, P. J., and Taylor, E. W. (1971). Response of the dogfish (*Scyliorhinus canicula* L.) to slowly induced and rapidly induced hypoxia. *Comp. Biochem. Physiol. A* **39A,** 307–323.

Butler, P. J., Taylor, E. W., and Short, S. (1977). The effect of sectioning cranial nerves V, VII, IX and X on the cardiac response of the dogfish *Scyliorhinus canicula* to environmental hypoxia. *J. Exp. Biol.* **69,** 233–245.

Butler, P. J., Taylor, E. W., Capra, M. F., and Davison, W. (1978). The effect of hypoxia on the levels of circulating catecholamines in the dogfish *Scyliorhinus canicula*. *J. Comp. Physiol.* **127,** 325–330.

Campbell, G. (1970). Autonomic nervous systems. *In* "Fish Physiology" (W. S. Hoar and D. J. Randall, eds.), Vol. 4, pp. 109–132. Academic Press, New York.

Capra, M. F., and Satchell, G. H. (1977). The adrenergic responses of isolated saline-perfused prebranchial arteries and gills of the elasmobranch *Squalus acanthias*. *Gen. Pharmacol.* **8,** 67–71.

Carrier, O. (1972). "Pharmacology of the Peripheral Autonomic Nervous System." Year Book Med. Publ., Chicago, Illinois.

Chan, D. K. O., and Chow, P. H. (1976). The effects of acetylcholine, biogenic amines and other vasoactive agents on the cardiovascular functions of the eel, *Anguilla japonica*. *J. Exp. Zool.* **196,** 13–23.

Claiborne, J. B., and Evans, D. H. (1980). The isolated, perfused head of the marine teleost fish, *Myoxocephalus octodecimspinosus:* Hemodynamic effects of epinephrine. *J. Comp. Physiol.* **138,** 79–85.

Colin, D. A., and Leray, C. (1979). Interaction of adenosine and its phosphorylated derivatives with putative purinergic receptors in the gill vascular bed of rainbow trout. *Pfluegers Arch.* **383,** 35–40.

Colin, D. A., Kirsch, R., and Leray, C. (1979). Haemodynamic effects of adenosine on gills of the trout (*Salmo gairdneri*). *J. Comp. Physiol.* **130,** 325–330.

D'Amico Martel, A. L., and Cech, J. J., Jr. (1978). Peripheral vascular resistance in the gills of the winter flounder *Pseudopleuronectes americanus*. *Comp. Biochem. Physiol. A* **59A,** 419–423.

Davies, D. T., and Rankin, J. C. (1973). Adrenergic receptors and vascular responses to catecholamines of perfused dogfish gills. *Comp. Gen. Pharmacol.* **4,** 139–147.

Davis, J. C. (1971). Circulatory and ventilatory responses of rainbow trout (*Salmo gairdneri*) to artificial manipulation of gill surface area. *J. Fish. Res. Board Can.* **28,** 1609–1614.

Daxboeck, C. (1982). Effect of coronary artery ablation on exercise performance in *Salmo gairdneri*. *Can. J. Zool.* **60,** 375–381.

Daxboeck, C., and Davie, P. S. (1982). Effects of pulsatile perfusion on flow distribution within an isolated saline-perfused trout head preparation. *Can. J. Zool.* **60,** 994–999.

Daxboeck, C., and Holeton, G. F. (1978). Oxygen receptors in the rainbow trout *Salmo gairdneri. Can. J. Zool.* **56,** 1254–1259.

Day, M. D. (1979). "Autonomic Pharmacology (Experimental and Clinical Aspects)." Churchill-Livingstone, Edinburgh and London.

Dejours, P. (1973). Problems of control of breathing in fishes. *In* "Comparative Physiology" (L. Bolis, S. H. P. Maddrell, and K. Schmidt-Nielsen, eds.), pp. 117–133. North-Holland/Am. Elsevier, Amsterdam and New York.

De Kock, L. L. (1963). A histological study of the head region of two salmonids with special reference to pressor- and chemo-receptors. *Acta Anat.* **55,** 39–50.

Dunel, S. (1975). Contribution à l'étude structurale et ultrastructurale de la pseudobranchie et de son innervation chez les téléostéens. Thèse, Université Lois Pasteur de Strasbourg, Strasbourg.

Dunel, S., and Laurent, P. (1977). Physiologie comparée: La vascularisation branchiale chez l'Anguille: Action de l'acetylcholine et de l'adrénaline sur la répartition d'une résine polymérisable dans les différents compartiments vasculaires. *C. R. Hebd. Seances Acad. Sci.* **284,** 2011–2014.

Dunel, S., and Laurent, P. (1980). Functional organization of the gill vasculature in different classes of fish. *In* "Jean Maetz Symposium: Epithelial Transport in the Lower Vertebrates" (B. Lahlou, ed.), pp. 37–58. Cambridge Univ. Press, London and New York.

Edwards, D. J.(1972). Reactions of the isolated plaice stomach to applied drugs. *Comp. Gen. Pharmacol.* **3,** 235–242.

Fänge, R., Holmgren, S., and Nilsson, S. (1976). Autonomic nerve control of the swimbladder of the goldsinny wrasse, *Ctenolabrus rupestris. Acta Physiol. Scand.* **97,** 292–303.

Farrell, A. P. (1978). Cardiovascular events associated with air breathing in two teleosts, *Hoplerythrinus unitaeniatus* and *Arapaima gigas. Can. J. Zool.* **56,** 953–958.

Farrell, A. P. (1980). Gill morphometrics, vessel dimensions, and vascular resistance in ling cod, *Ophiodon elongatus. Can. J. Zool.* **58,** 807–818.

Farrell, A. P., Daxboeck, C., and Randall, D. J. (1979). The effect of input pressure and flow on the pattern and resistance to flow in the isolated perfused gill of a teleost fish. *J. Comp. Physiol.* **133,** 233–240.

Farrell, A. P., Sobin, S. S., Randall, D. J., and Crosby, S. (1980). Intralamellar blood flow patterns in fish gills. *Am. J. Physiol.* **239,** R428–R436.

Folkow, B., and Neil, E. (1971). "Circulation." Oxford Univ. Press, London and New York.

Forster, M. E. (1976a). Effects of catecholamines on the heart and on branchial and peripheral resistances of the eel, *Anguilla anguilla (L.). Comp. Biochem. Physiol. C* **55C,** 27–32.

Forster, M. E. (1976b). Effects of adrenergic blocking drugs on the cardiovascular system of the eel, *Anguilla anguilla (L.). Comp. Biochem. Physiol. C* **55C,** 33–36.

Freihofer, W. C. (1978). Cranial nerves of a percoid fish, *Polycentrus schomburgkii* (family Nandidae), a contribution to the morphology and classification of the order perciformes. *Occas. Pap. Calif. Acad. Sci.* **128,** 1–78.

Furchgott, R. F. (1967). The pharmacological differentiation of adrenergic receptors. *Ann. N.Y. Acad. Sci.* **139,** 553–570.

Gannon, B. J., Campbell, G. D., and Satchell, G. H. (1972). Monoamine storage in relation to cardiac regulation in the Port Jackson shark, *Heterodontus portusjacksoni. Z. Zellforsch. Mikrosk. Anat.* **131,** 437–450.

Gibbins, I. L. (1982). Lack of correlation between ultrastructural and pharmacological types of non-adrenergic autonomic nerves. *Cell. Tissue Res.* **221,** 551–582.

Gilloteaux, J. (1969). Note sur l'innervation des branchies chez *Anguilla anguilla* L. *Experientia* **25**, 270.

Girard, J. P., and Payan, P. (1976). Effect of epinephrine on vascular space of gills and head of rainbow trout. *Am. J. Physiol.* **230**, 1555–1560.

Goodrich, E. S. (1930). "Studies on the Structure and Development of Vertebrates." Macmillan, New York.

Helgason, S. S., and Nilsson, S. (1973). Drug effects on pre- and post-branchial blood pressure and heart rate in a free-swimming marine teleost, *Gadus morhua. Acta Physiol. Scand.* **88**, 533–540.

Hering, E., and Breuer, J. (1868). Die Selbststeuerung der Athmung durch die Nervus Vagus. *Sitzungsber. Kais. Akad. Wiss. Wien, Math.-Naturwiss. Kl., Abt. 1* **57**, 672–704.

Hökfelt, T., Kellert, J.-O., Nilsson, G., and Pernow, B. (1975). Substance P: Localization in the central nervous system and in some primary sensory neurons. *Science* **190**, 889–890.

Hökfelt, T., Johansson, O., Ljungdahl, A., Lundberg, J. M., and Schultzberg, M. (1980). Peptidergic neurons. *Nature (London)* **284**, 515–521.

Holbert, P. W., Boland, E. J., and Olson, K. R. (1979). The effect of epinephrine and acetylcholine on the distribution of red cells within the gills of the channel catfish (*Ictalurus punctatus). J. Exp. Biol.* **79**, 135–146.

Holeton, G. F. (1977). Constancy of arterial blood pH during CO-induced hypoxia in the rainbow trout. *Can. J. Zool.* **55**, 1010–1013.

Holeton, G. F., and Randall, D. J. (1967). Changes in blood pressure in the rainbow trout during hypoxia. *J. Exp. Biol.* **46**, 297–305.

Holmgren, S. (1977). Regulation of the heart of a teleost, *Gadus morhua*, by autonomic nerves and circulating catecholamines. *Acta Physiol. Scand.* **99**, 62–74.

Holmgren, S., and Nilsson, S. (1974). Drug effects on isolated artery strips from two teleosts, *Gadus morhua* and *Salmo gairdneri. Acta Physiol. Scand.* **90**, 431–437.

Holmgren, S., and Nilsson, S. (1982). Neuropharmacology of adrenergic neurons in teleost fish. *Comp. Biochem. Physiol. C* **72C**, 289–302.

Hughes, G. M. (1972). Morphometrics of fish gills. *Respir. Physiol.* **14**, 1–25.

Irving, L., Solandt, D. Y., and Solandt, O. M. (1935). Nerve impulses from branchial pressure receptors in the dogfish. *J. Physiol. (London)* **84**, 187–190.

Ishii, K., and Ishii, K. (1978). A reflexogenic area for controlling the blood pressure in toad (*Bufo vulgaris formosa). Jpn. J. Physiol.* **28**, 423–431.

Iversen, L. L. (1967). "The Uptake and Storage of Noradrenaline in Sympathetic Nerves." Cambridge Univ. Press, London and New York.

Iversen, L. L. (1974). Uptake mechanisms for neurotransmitter amines. *Biochem. Pharmacol.* **23**, 1927–1935.

Jenkin, P. M. (1928). Note on the nervous system of *Lepidosiren paradoxa. Proc. R. Soc. Edinburgh* **48**, 55–69.

Johansen, K. (1970). Air breathing in fishes. *In* "Fish Physiology" (W. S. Hoar and D. J. Randall, eds.), Vol. 4, pp. 361–411. Academic Press, New York.

Johansen, K. (1971). Comparative physiology: Gas exchange and circulation in fishes. *Annu. Rev. Physiol.* **33**, 569–612.

Johansen, K. (1972). Heart and circulation in gill, skin and lung breathing. *Respir. Physiol.* **14**, 193–210.

Johansen, K., and Burggren, W. (1980). Cardiovascular function in the lower vertebrates. *In* "Hearts and Heart-like Organs" (G. H. Bourne, ed.), Vol. 1, pp. 61–117. Academic Press, New York.

Johansen, K., and Reite, O. B. (1968). Influence of acetylcholine and biogenic amines on branchial, pulmonary and systemic vascular resistance in the African lungfish, *Protopterus aethiopicus. Acta Physiol. Scand.* **74**, 465–471.

Johansen, K., Lenfant, C., and Hanson, D. (1968). Cardiovascular dynamics in the lungfishes. Z. Vergl. Physiol. 59, 157–186.

Jones, D. R., and Randall, D. J. (1978). The respiratory and circulatory systems during exercise. In "Fish Physiology" (W. S. Hoar and D. J. Randall, eds.), Vol. 7, pp. 425–501. Academic Press, New York.

Keys, A., and Bateman, J. B. (1932). Branchial responses to adrenaline and to pitressin in the eel. Biol. Bull. (Woods Hole, Mass.) 63, 327–336.

Kirschner, L. B. (1969). Ventral aortic pressure and sodium fluxes in perfused eel gills. Am. J. Physiol. 217, 596–604.

Krawkow, N. P. (1913). Über die Wirkung von Giften aur die Gefässe isolierter Fischkiemen. Pfluegers Arch. Gesamte Physiol. Menschen Tiere 151, 583–603.

Laffont, J., and Labat, R. (1966). Action de l'adrénaline sur la fréquence cardiaque de le carpe commune. Effect de la température du milieu sur l'intensité de la réaction. J. Physiol. (Paris) 58, 351–355.

Lahiri, S., Szidon, J. P., and Fishman, A. P. (1970). Potential respiratory and circulatory adjustments to hypoxia in the African lungfish. Fed. Proc., Fed. Am. Soc. Exp. Biol. 29, 1141–1148.

Lands, A. M., Arnold, A., McAuliff, J., Luduena, F. P., and Brown, T. G. (1967). Differentiation of receptor systems activated by sympathomimetic amines. Nature (London) 214, 597–598.

Langley, J. N. (1921). "The Autonomic Nervous System," Part 1. Heffer, Cambridge.

Laurent, P. (1967). Neurophysiologie—La pseudobranchie des téléostéens: Preuves électrophysiologiques de ses fonctions chémoréceptrice et baroréceptrice. C.R. Hebd. Seances Acad. Sci. 264, 1879–1882.

Laurent, P. (1969). Action du pH et de la $pO_2$ sur le potentiel de membrane des cellules de l'épithelium récepteur dans la pseudobranchie d'un poisson téléostéen. Rev. Can. Biol. 28, 149–155.

Laurent, P., and Dunel, S. (1966). Recherches sur l'innervation de la pseudobranchie des téléostéens. Arch. Anat. Microsc. Morphol. Exp. 55, 633–656.

Laurent, P., and Dunel, S. (1976). Functional organization of the teleost gill. I. Blood pathways. Acta Zool. (Stockholm) 57, 189–209.

Laurent, P., and Dunel, S. (1980). Morphology of gill epithelia in fish. Am. J. Physiol. 238, R147–R149.

Laurent, P., and Rouzeau, J.-D. (1972). Afferent neural activity from pseudobranch of teleosts. Effets of $P_{O_2}$, pH, osmotic pressure and $Na^+$ ions. Respir. Physiol. 14, 307–331.

Lindström, T. (1949). On the cranial nerves of the cyclostomes with special reference to n. trigeminus. Acta Zool. (Stockholm) 30, 315–458.

Lutz, B. R. (1930). Reflex cardiac and respiratory inhibition in the elasmobranch, Scyllium canicula. Biol. Bull. (Woods Hole, Mass.) 59, 170–178.

Lutz, B. R. (1931). The innervation of the stomach and rectum and the action of adrenaline in elasmobranch fishes. Biol. Bull. (Woods Hole, Mass.) 61, 93–100.

Lutz, B. R., and Wyman, L. C. (1932). Reflex cardiac inhibition of branchio-vascular origin in the elasmobranch, Squalus acanthias. Biol. Bull. (Woods Hole, Mass.) 62, 10–16.

Mott, J. C. (1951). Some factors affecting the blood circulation in the common eel (Anguilla anguilla). J. Physiol (London) 114, 387–398.

Nakano, T., and Tomlinson, N. (1967). Catecholamines and carbohydrate concentrations in rainbow trout (Salmo gairdneri) in relation to physical disturbance. J. Fish. Res. Board Can. 24, 1701–1715.

Nakao, T. (1981). An electron microscopic study on the innervation of the gill filaments of a lamprey, Lampetra japonica. J. Morphol. 169, 325– 336.

Nickerson, M. (1970). Drugs inhibiting adrenergic nerves and structures innervated by them.

*In* "The Pharmacological Basis of Therapeutics" (L. S. Goodman and A. Gilman, eds.), 4th ed., pp. 549–584 Macmillan, New York.

Nilsson, S. (1973). On the autonomic nervous control of organs in teleostean fishes. *In* "Comparative Physiology" (L. Bolis, K. Schmidt-Nielsen, and S. H. P. Maddrell, eds.), pp. 325–331. North-Holland Publ., Amsterdam.

Nilsson, S. (1976). Fluorescent histochemistry and cholinesterase staining in sympathetic ganglia of a teleost, *Gadus morhua. Acta Zool. (Stockholm)* **57**, 69–77.

Nilsson, S. (1983). "Autonomic Nerve Function in the Vertebrates. " Vol. 13. Springer-Verlag, Berlin and New York.

Nilsson, S., and Pettersson, K. (1981). Sympathetic nervous control of blood flow in the gills of the Atlantic cod, *Gadus morhua. J. Comp. Physiol.* **144**, 157–163.

Nilsson, S., Abrahamsson, T., and Grove, D. J. (1976). Sympathetic nervous control of adrenaline release from the head kidney of the cod, *Gadus morhua. Comp. Biochem. Physiol. C* **55C**, 123–127.

Norris, H. W. (1925). Observations upon the peripheral distribution of the cranial nerves of certain ganoid fishes (*Amia, Lepidosteus, Polyodon, Scaphirhyncus* and *Acipenser*). *J. Comp. Neurol.* **39**, 345–432.

Östlund, E., and Fänge, R. (1962). Vasodilation by adrenaline and noradrenaline, and the effects of some other substances on perfused fish gills. *Comp. Biochem. Physiol.* **5**, 307–309.

Paintal, A. S. (1969). Mechanism of stimulation of type J pulmonary receptors. *J. Physiol. (London)* **203**, 511–532.

Paintal, A. S. (1970). The mechanism of excitation of type J receptors and the J reflex. *In* "Breathing; Hering Breuer Centenary Symposium" (R. Porter, ed.), pp. 59–76. Churchill, London.

Pärt, P., Kiessling, A., and Ring, O. (1982). Adrenaline increases vascular resistance in perfused rainbow trout (*Salmo gairdnerii* Rich.) gills. *Comp. Biochem. Physiol. C* **72C**, 107–108.

Pasztor, V. M., and Kleerekoper, H. (1962). The role of the gill filament musculature in teleosts. *Can. J. Zool.* **40**, 785–802.

Payan, P., and Girard, J.-P. (1977). Adrenergic receptors regulating patterns of blood flow through the gills of trout. *Am. J. Physiol.* **232**, H18–H23.

Pettersson, K. (1983). Adrenergic control of oxygen transfer in perfused gills of the cod, *Gadus morhua. J. Exp. Biol.* **102**, 327–335.

Pettersson, K., and Johansen, K. (1982). Hypoxic vasoconstriction and the effects of adrenaline on gas exchange efficiency in fish gills. *J. Exp. Biol.* **97**, 263–272.

Pettersson, K., and Nilsson, S. (1979). Nervous control of the branchial vascular resistance of the Atlantic cod, *Gadus morhua. J. Comp. Physiol.* **129**, 179–183.

Pettersson, K., and Nilsson, S. (1980). Drug induced changes in cardio-vascular parametres in the Atlantic cod, *Gadus morhua. J. Comp. Physiol.* **137**, 131–138.

Peyraud-Waitzenegger, M. (1979). Simultaneous modifications of ventilation and arterial $P_{O_2}$ by catecholamines in the eel, *Anguilla anguilla* L.: Participation of $\alpha$ and $\beta$ effects. *J. Comp. Physiol.* **129**, 343–354.

Peyraud-Waitzenegger, M., Savina, A., Laparra, J., and Morfin, R. (1979). Blood-brain barrier for epinephrine in the eel (*Anguilla anguilla* L.). *Comp. Biochem. Physiol. C* **63C**, 35–38.

Piiper, J., Baumgarten, D., and Meyer, M. (1970). Effects of hypoxia upon respiration and circulation in the dogfish *Scyliorhinus stellaris. Comp. Biochem. Physiol.* **36**, 513–520.

Poole, C. A., and Satchell, G. H. (1979). Nociceptors in the gills of the dogfish *Squalus acanthias. J. Comp. Physiol.* **130**, 1–7.

Powers, E. B., and Clark, R. T. (1942). Control of normal breathing in fishes by receptors located in the regions of the gills and innervated by the IXth and Xth cranial nerves. *Am. J. Physiol.* **138**, 104–107.

Randall, D. J., and Jones, D. R. (1973). The effect of deafferentiation of the pseudobranch on the respiratory response to hypoxia and hyperoxia in the trout (*Salmo gairdneri*). *Respir. Physiol.* **17**, 291–301.

Randall, D. J., and Shelton, G. (1963). The effects of changes in environmental gas concentrations on the breathing and heart rate of a teleost fish. *Comp. Biochem. Physiol.* **9**, 229–239.

Randall, D. J., and Smith, J. C. (1967). The regulation of cardiac activity in fish in a hypoxic environment. *Physiol. Zool.* **40**, 104–113.

Randall, D. J., and Stevens, E. D. (1967). The role of adrenergic receptors in cardiovascular changes associated with exercise in salmon. *Comp. Biochem. Physiol.* **21**, 415–424.

Randall, D. J., Baumgarten, D., and Malyusz, M. (1972). The relationship between gas and ion transfer across the gills of fishes. *Comp. Biochem. Physiol. A* **41A**, 629–637.

Rankin, J. C., and Maetz, J. (1971). A perfused teleostean gill preparation: Vascular actions of neurohypophysal hormones and catecholamines. *J. Endocrinol.* **51**, 621–635.

Reite, O. B. (1969). The evolution of vascular smooth muscle responses to histamine and 5-hydroxytryptamine. *Acta Physiol. Scand.* **75**, 221–239.

Richards, B. D., and Fromm, P. O. (1969). Patterns of blood flow through filaments and lamellae of isolated-perfused rainbow trout (*Salmo gairdneri*) gills. *Comp. Biochem. Physiol.* **29**, 1063–1070.

Ristori, M. T. (1970). Réflexe de barosensibilité chez un poisson téléostéen (*Cyprinus carpio* L.). *C.R. Seances Soc. Biol. Ses Fil.* **164**, 1512–1516.

Ristori, M. T., and Dessaux, G. (1970). Sur l'existence d'un gradient de sensibilité dans les récepteurs branchiaux de *Cyprinus carpio*. *C.R. Seances Soc. Biol. Ses Fil.* **164**, 1517–1519.

Ristori, M. T., and Laurent, P. (1977). Action de l'hypoxie sure le système vasculaire branchial de la tête perfusée de truite. *C.R. Seances Soc. Biol. Ses Fil.* **171**, 809–813.

Romer, A. S. (1962). "The Vertebrate Body," 3rd ed. Saunders, Philadelphia, Pennsylvania.

Said, S. I., and Mutt, V. (1970). Polypeptide with broad biological activity. Isolation from small intestine. *Science* **169**, 1217–1218.

Satchell, G. H. (1959). Respiratory reflexes in the dogfish. *J. Exp. Biol.* **36**, 62–71.

Satchell, G. H. (1960). The reflex co-ordination of the heart beat with respiration in the dogfish. *J. Exp. Biol.* **37**, 719–731.

Satchell, G. H. (1961). The response of the dogfish to anoxia. *J. Exp. Biol.* **38**, 531–543.

Satchell, G. H. (1962). Intrinsic vasomotion in the dogfish gill. *J. Exp. Biol.* **39**, 503–512.

Satchell, G. H. (1978a). Type J receptors in the gills of fish. *In* "Studies in Neurophysiology" (C. Porter, ed.), pp. 131–142. Cambridge Univ. Press, London and New York.

Satchell, G. (1978b). The J reflex in fish. *In* "Respiratory Adaptations, Capillary Exchange and Reflex Mechanisms" (A. S. Paintal and P. Gill-Kumar, eds.), pp. 432–441. Vallabhbhai Patel Chest Institute, University of Delhi, Delhi.

Satchell, G. H., and Maddalena, D. J. (1972). The cough or expulsion reflex in the Port Jackson shark, *Heterodontus portusjacksoni*. *Comp. Biochem. Physiol. A* **41A**, 49–62.

Satchell, G. H., and Way, H. K. (1962). Pharyngeal proprioceptors in the dogfish *Squalus acanthias* L. *J. Exp. Biol.* **39**, 243–250.

Satchell, G. H., Hanson, D., and Johansen, K. (1970). Differential blood flow through the afferent branchial arteries of the skate, *Raja rhina*. *J. Exp. Biol.* **52**, 721–726.

Saunders, R. L., and Sutterlin, A. M. (1971). Cardiac and respiratory responses to hypoxia in the sea raven, *Hemitripterus americanus*, and an investigation of possible control mechanisms. *J. Fish. Res. Board Can.* **28**, 491–503.

Short, S., Butler, P. J., and Taylor, E. W. (1977). The relative importance of nervous, humoral and intrinsic mechanisms in the regulation of heart rate and stroke volume in the dogfish *Scyliorhinus canicula*. *J. Exp. Biol.* **70**, 77–92.

Smith, D. G. (1977). Sites of cholinergic vasoconstriction in trout gills. *Am. J. Physiol.* **233**, R222–R229.

Smith, D. G., and Chamley-Campbell, J. (1981). Localization of smooth-muscle myosin in branchial pillar cells of snapper (*Chrysophys auratus*) by immunofluorescence histochemistry. *J. Exp. Zool.* **215**, 121–24.

Smith, D. G., and Gannon, B. J. (1978). Selective control of branchial arch perfusion in air-breathing Amazonian fish *Hoploerythrinus unitaeniatus*. *Can. J. Zool.* **56**, 959–964.

Smith, D. G., Berger, P. J., and Evans, B. K. (1981). Baroreceptor control of the heart in the conscious toad, *Bufo marinus*. *Am. J. Physiol.* **241**, R307–R311.

Smith, F. M., and Jones, D. R. (1978). Localization of receptors causing hypoxic bradycardia in trout (*Salmo gairdneri*). *Can. J. Zool.* **56**, 1260–1265.

Stannius, H. (1849). "Das peripherische Nervensystem der Fische." Stiller, Rostock.

Steen, J. B., and Kruysse, A. (1964). The respiratory function of teleostean gills. *Comp. Biochem. Physiol.* **12**, 127–142.

Stevens, E. D., Bennion, G. R., Randall, D. J., and Shelton, G. (1972). Factors affecting arterial pressure and blood flow from the heart in intact, unrestrained lingcod, *Ophiodon elongatus*. *Comp. Biochem. Physiol. A* **43A**, 681–695.

Sutterlin, A. M., and Saunders, R. L. (1969). Proprioceptors in the gills of teleosts. *Can. J. Zool.* **47**, 1209–1212.

Trendelenburg, U. (1963). Supersensitivity and subsensitivity to sympathicomimetic amines. *Pharmacol. Rev.* **15**, 225–276.

Trendelenburg, U. (1972). Factors influencing the concentration of catecholamines at the receptors. *In* "Handbuch der experimentellen Pharmakologie" (H. Blaschko and E. Muscholl, eds.), Vol. 33, pp. 726–761. Springer-Verlag, Berlin and New York.

Vasilevskaya, N. E., and Polyakova, N. N. (1979). Comparison of single unit responses in the vagal and facial lobes of the carp medulla to chemoreceptor stimulation. *Neurophysiology (Engl. Transl.)* **10**, 613–621.

Vasilevskaya, N. E., and Polyakova, N. N. (1981). Unit responses in the vagal lobe of the carp brain to stimulation of chemoreceptors of the skin, mouth and gills. *Neurophysiology (Engl. Transl.)* **12**, 397–404.

Vogel, W., Vogel, V., and Schlote, W. (1974). Ultrastructural study of arteriovenous anastomoses in gill filaments of *Tilapia mossambica*. *Cell Tissue Res.* **155**, 491–512.

Wahlqvist, I. (1980). Effects of catecholamines on isolated systemic and branchial vascular beds of the cod, *Gadus morhua*. *J. Comp. Physiol.* **137**, 139–143.

Wahlqvist, I. (1981). Branchial vascular effects of catecholamines released from the head kidney of the Atlantic cod, *Gadus morhua*. *Mol. Physiol.* **1**, 235–241.

Wahlqvist, I., and Nilsson, S. (1980). Adrenergic control of the cardio-vascular system of the Atlantic cod, *Gadus morhua*, during "stress." *J. Comp. Physiol.* **137**, 145–150.

Walker, E. R., Fidler, S. F., and Hinton, D. E. (1981). Morphology of the buccopharyngeal portion of the gill in the fathead minnow *Pimephales promelas* (Rafinesque). *Anat. Rec.* **200**, 67–81.

Wikgren, B.-J. (1953). Osmotic regulation in some aquatic animals with special reference to the influence of temperature. *Acta Zool. Fenn.* **71**, 1–102.

Wilkes, P. R. H., Walker, R. L., McDonald, D. G., and Wood, C. M. (1981). Respiratory, ventilatory, acid-base and ionoregulatory physiology of the white sucker *Catostomus commersoni:* The influence of hyperoxia. *J. Exp. Biol.* **91**, 239–254.

Wood, C. M. (1974). A critical examination of the physical and adrenergic factors affecting blood flow through the gills of the rainbow trout. *J. Exp. Biol.* **60**, 241–265.

Wood, C. M. (1975). A pharmacological analysis of the adrenergic and cholinergic mechanisms regulating branchial vascular resistance in the rainbow trout (Salmo gairdneri). Can. J. Zool. 53, 1569–1577.

Wood, C. M., and Randall, D. J. (1973). The influence of swimming activity on sodium balance in the rainbow trout (Salmo gairdneri). J. Comp. Physiol. 82, 207–233.

Wood, C. M., and Shelton, G. (1975). Physical and adrenergic factors affecting systemic vascular resistance in the rainbow trout: A comparison with branchial vascular resistance. J. Exp. Biol. 63, 505–523.

Wood, C. M., and Shelton, G. (1980a). Cardiovascular dynamics and adrenergic responses of the rainbow trout in vivo. J. Exp. Biol. 87, 247–270.

Wood, C. M., and Shelton, G. (1980b). The reflex control of heart rate and cardiac output in the rainbow trout: Interactive influences of hypoxia, haemorrhage and systemic vasomotor tone. J. Exp. Biol. 87, 271–284.

Wood, C. M., Pieprzak, P., and Trott, J. N. (1979). The influence of temperature and anaemia on the adrenergic and cholinergic mechanisms controlling heart rate in the rainbow trout. Can. J. Zool. 57, 2440–2447.

Young, J. Z. (1931). On the autonomic nervous system of the teleostean fish Uranoscopus scaber. Q. J. Microsc. Sci. 74, 491–535.

Young, J. Z. (1933). The autonomic nervous system of selachians. Q. J. Microsc. Sci. 75, 571–624.

Young, J. Z. (1936). The innervation and reactions to drugs of the viscera of teleostean fish. Proc. R. Soc. (London) Ser. B 120, 303–318.

Young, S. (1972). Electromyographie activity during respiration and coughing in the tench (Tinca tinca L.). J. Physiol. (London) 227, 18–19.

# 4

# MODEL ANALYSIS OF GAS TRANSFER IN FISH GILLS

*JOHANNES PIIPER*

Abteilung Physiologie
Max-Planck-Institut für Experimentelle Medizin
Göttingen, Federal Republic of Germany

*PETER SCHEID*

Institut für Physiologie
Ruhr-Universität
Bochum, Federal Republic of Germany

FISH PHYSIOLOGY, VOL. XA

## I. INTRODUCTION

Over the past few decades, the analysis of gas exchange in mammalian lungs has been elaborated in great detail (for recent reviews, see Rahn and Farhi, 1964; Farhi, 1966; West, 1977a,b, 1980; Piiper and Scheid, 1981; Haab, 1982). In this chapter we will try to apply to fish gills the basic principles and methods that have proved successful in the analysis of gas exchange in mammalian lungs. In doing so, we will focus attention on the model that applies to fish gills, that is, the countercurrent model, as opposed to the ventilated pool system of mammalian lungs. Another basic difference between gills and lungs—water versus air as the external respiratory medium (see Rahn, 1966; Dejours, 1981)—will not be treated specifically in this chapter.

It will be shown that the countercurrent system provides the potential for the highest possible gas exchange efficiency, in that a complete equilibration of (arterial) blood to the inspired water, and of expired water to venous blood, is possible. Four basic factors may disturb this ideal situation, the first three of which—water–blood mismatch, diffusion resistance, and unequal distribution—may be described as functional shunts and are thus not easy to separate from the fourth factor, true shunts.

Reference to experimental data will be made only for exemplification, since the experimental results on gas exchange in fish are covered by other chapters in this series (Randall, 1970; Jones and Randall, 1978; Chapter 5, this volume).

The topic of this chapter is based on the concepts introduced by Hughes and Shelton (1962), which have been applied to analysis of experimental data by Randall *et al.* (1967) and reviewed by Randall (1970). It has been developed in a number of studies (Piiper and Baumgarten-Schumann, 1968b; Scheid and Piiper, 1971, 1976) and has also been reviewed (Piiper and Scheid, 1983).

## II. THE COUNTERCURRENT MODEL

It follows from the anatomic arrangement of the gill elements and their blood vessels that water flow through the interlamellar space and blood flow in the lamellae are in opposite directions (see Chapter 1, this volume). On the basis of this anatomic evidence, the countercurrent system is now generally accepted as the appropriate model for branchial gas exchange.

However, the physiological and functional evidence for the presence of the countercurrent system in gills is less conclusive. In the classical experiment of Hazelhoff and Evenhuis (1952), the substantially reduced $O_2$ extraction on reversal of water flow in gills of anesthetized tench has been attributed to the experimental conversion of the countercurrent into a cocurrent system with its inherently lower gas transfer efficiency. However, despite the authors' assertion that no change in the position of the gill lamellae was observable, disturbances of the distribution of water flow in the gills may have been present, which could lead to reduced gas transfer efficiency.

Higher $P_{O_2}$ in arterial blood as compared to exhaled water cannot be easily explained by a cocurrent system but is expected to occur with the countercurrent system. However, this behavior may also be produced by a crosscurrent or serial multicapillary system, which is believed to be operative in avian lungs (Scheid and Piiper, 1970; Scheid, 1979). Indeed this system has been taken into consideration in connection with fish gills (Piiper and Schumann, 1967). However, the gas transfer efficiency of the crosscurrent system is lower than that of the countercurrent system. The highly positive arterial–expired water $P_{O_2}$ differences reported in some studies are higher than possible with a crosscurrent system even under optimum conditions (Baumgarten-Schumann and Piiper, 1968) and therefore are explainable only by a countercurrent system.

This section will discuss the basic countercurrent model for branchial gas transfer, with its relevant quantities and relationships (Piiper and Scheid, 1972, 1975). Oxygen ($O_2$) will be used as the reference gas unless stated otherwise.

## A. Model and Equations

The countercurrent model is depicted in Fig. 1. A water flow ($\dot{V}_w$) and a blood flow ($\dot{V}_b$), both continuous and constant, are arranged in countercurrent manner. Gas transfer takes place through a barrier that is homogeneous and of constant thickness and width, its diffusive conductance, or diffusing capacity (transfer factor), being $D$. In water and blood, there are no gradients perpendicular to flow direction (because of complete mixing in each cross-sectional element). The effective solubilities or capacity coefficients of the media are $\beta_w$ for water and $\beta_b$ for blood.

Gas transfer ($O_2$ uptake or $CO_2$ output) per unit length, $d\dot{M}/dx$, is determined by the following simultaneous differential equations:

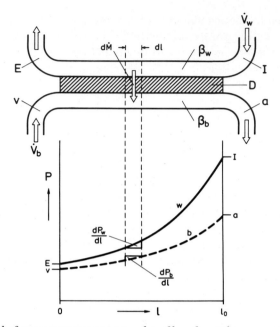

**Fig. 1.** Model of countercurrent system and profiles of partial pressure in water ($P_w$) and blood ($P_b$) along the contact length ($l$) of a substance, like $O_2$, that is taken up from water (w) into blood (b). $d\dot{M}$, infinitesimal uptake rate along the length element $dl$, across the element of the diffusing capacity, $dD$. I and E, inspired and expired water; a and v, arterial and venous blood; $\dot{V}$, flow rate; $\beta$, capacitance coefficient.

$$d\dot{M}/dl = \dot{V}_w\beta_w(dP_w/dl) \tag{1}$$

$$d\dot{M}/dl = \dot{V}_b\beta_b(dP_b/dl) \tag{2}$$

$$d\dot{M}/dl = [P_w(l) - P_b(l)](dD/dl) \tag{3}$$

These differential equations can be integrated to obtain the partial pressure profiles, $P_w(l)$ and $P_b(l)$, along the contact length, $l$. Using the boundary conditions (see Fig. 1),

$$\begin{aligned} P_w(l=l_0) &= P_I \\ P_b(l=0) &= P_v \end{aligned} \tag{4}$$

one obtains

$$\frac{P_I - P_w(l)}{P_I - P_v} = \frac{\exp[-Z(l/l_0)] - \exp(-Z)}{X - \exp(-Z)} \tag{5}$$

$$\frac{P_b(l) - P_v}{P_I - P_v} = \frac{X\{1 - \exp[-Z(l/l_0)]\}}{X - \exp(-Z)} \tag{6}$$

in which the following symbols for conductance ratios ($\dot{V}_w\beta_w$, ventilatory conductance; $\dot{V}_b\beta_b$, perfusive conductance; see Piiper and Scheid, 1983), have been used:

$$X = \frac{\dot{V}_w\beta_w}{\dot{V}_b\beta_b} \tag{7}$$

$$Y = \frac{D}{\dot{V}_b\beta_b} \tag{8}$$

$$Z = Y(1 - 1/X) \tag{9}$$

Although Eqs. (5) and (6) are valid for all values of $X$ and $Y$, the case in which water and blood flow conductances are equal, and hence $X = 1$, yields particularly simple profiles as listed in the second column of Table IA. In this case, the partial pressures in both water and blood constitute linear and parallel lines, of slope $Y$ (when plotted against $l/l_0$). The difference between $P_w$ and $P_b$ is thus constant along the contact length at a value that depends only on $Y$:

$$\frac{P_w(l) - P_b(l)}{P_I - P_v} = \frac{1}{1 + Y} \tag{10}$$

with the limits of 1 at $Y = 0$, and 0 at $Y\rightarrow\infty (e.g., D\rightarrow\infty)$.

**Table I**

Partial Pressures in Countercurrent System for Finite and Infinite Values of Diffusing Capacity $D$, and for Matched ($X = 1$) and Nonmatched Convective Conductances[a,b]

| | | Finite $D$ | | $D \rightarrow \infty$ | | |
|---|---|---|---|---|---|---|
| | | $X \neq 1$ | $X = 1$ | $X > 1$ | $X = 1$ | $X < 1$ |
| **A.** | $\dfrac{P_I - P_w(l)}{P_I - P_v}$ | $\dfrac{\exp[-Z(l/l_0)] - \exp(-Z)}{X - \exp(-Z)}$ | $\dfrac{Y(1-l/l_0)}{1 + Y}$ | $0$ | $1 - l/l_0$ | $1$ |
| | $\dfrac{P_b(l) - P_v}{P_I - P_v}$ | $\dfrac{X\{1 - \exp[-Z(l/l_0)]\}}{X - \exp(-Z)}$ | $\dfrac{Y(l/l_0)}{1 + Y}$ | $1$ | $l/l_0$ | $0$ |
| **B.** | $\dfrac{P_I - P_E}{P_I - P_v}$ | $\dfrac{1 - \exp(-Z)}{X - \exp(-Z)}$ | $\dfrac{Y}{1 + Y}$ | $\dfrac{1}{X}$ | $1$ | $1$ |
| | $\dfrac{P_a - P_v}{P_I - P_v}$ | $\dfrac{X[1 - \exp(-Z)]}{X - \exp(-Z)}$ | $\dfrac{Y}{1 + Y}$ | $1$ | $1$ | $X$ |
| | $\dfrac{P_E - P_a}{P_I - P_v}$ | $\dfrac{X \exp(-Z) - 1}{X - \exp(-Z)}$ | $\dfrac{1 - Y}{1 + Y}$ | $-\dfrac{1}{X}$ | $-1$ | $-X$ |

[a](A) Partial pressure profiles along gas exchange contact ($l$) in water ($P_W$) and blood ($P_b$). (B) Values of partial pressures in inflowing ($P_I$) and outflowing water ($P_E$) and in venous ($P_v$) and arterial blood ($P_a$).
[b]$X = \dot{V}_w\beta_w/\dot{V}_b\beta_b$; $Y = D/\dot{V}_b\beta_b$; $Z = Y[1 - 1/X]$.

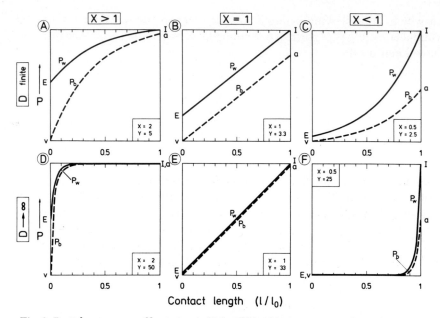

**Fig. 2.** Partial pressure profiles in water ($P_w$) and blood ($P_b$) at various values of the conductance ratio, $X = (\dot{V}_w\beta_w)/(\dot{V}_b\beta_b)$, both with finite $D$ and with $D$ approximating infinity [expressed by the conductance ratio, $Y = D/(\dot{V}_b\beta_b)$]. E, I, a, v, as in Fig. 1.

In Table IA are listed the partial pressure profiles for the general case, that is, $X \neq 1$ and finite $D$, and for the special cases, $X = 1$ and infinite $D$ values. These profiles are also illustrated in Fig. 2. The particular symmetry of the countercurrent system becomes evident from these curves.

Of particular interest are the values of $P_w$ and $P_b$ at the exit from the system, that is, $P_E$ and $P_a$. These values can be obtained from Eqs. (5) and (6) and may be expressed as the relative partial pressure differences,

$$\frac{P_I - P_E}{P_I - P_v} = \frac{1 - \exp(-Z)}{X - \exp(-Z)} \tag{11}$$

$$\frac{P_a - P_v}{P_I - P_v} = \frac{X[1 - \exp(-Z)]}{X - \exp(-Z)} \tag{12}$$

$$\frac{P_E - P_a}{P_I - P_v} = \frac{X \exp(-Z) - 1}{X - \exp(-Z)} \tag{13}$$

These equations are listed in Table IB, together with the expressions obtained for $X = 1$.

Of particular importance is the difference $P_E - P_a$, which can be negative in the countercurrent system, indicating an overlap of partial pressure ranges in blood and water. The high gas exchange efficiency of the system is directly related to this overlap, which can be described by the "overlap coefficient" $u$ as

$$u = (P_a - P_E)/(P_I - P_v) \qquad (14)$$

This coefficient can assume values between $+1$, for complete overlap ($P_a = P_I$ and $P_E = P_v$) and $-1$ for zero gas transfer ($P_a = P_v$, $P_E = P_I$). It can be seen from Table IB that this overlap becomes more extensive, thus increasing toward $+1$, with increasing $D$. However, the matching of water and blood conductances is critical too, in that, for a given $D$ value, the largest overlap is obtained when $X = 1$, that is, when both convective conductances are perfectly matched. This is also evident from Fig. 2, which shows that the largest overlap possible ($u = +1$, $P_a = P_I$, and $P_E = P_v$), is obtained when $D \to \infty$ and $X = 1$.

## B. Countercurrent versus Cocurrent Models

The high intrinsic efficiency of the countercurrent model, shown by the crossing over of water and blood partial pressures, cannot be achieved by a cocurrent model (Fig. 3). In the cocurrent model, $P_w$ and $P_b$ approach a common asymptotic value, $P_{eq}$, from each side, and at most an equality of $P_E$ and $P_a$ can be reached (see Appendix).

The position of $P_{eq}$ between $P_I$ and $P_v$ is given by the relation

$$\frac{P_{eq} - P_v}{P_I - P_v} = X \qquad (15)$$

The equilibration along the $l$ axis of the model is given by the relationship (see Appendix)

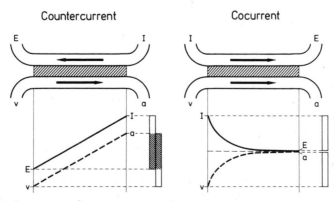

Fig. 3. Comparison of countercurrent with cocurrent system. Symbols as in Fig. 1.

$$\frac{P_w - P_{eq}}{P_I - P_{eq}} = \frac{P_{eq} - P_b}{P_{eq} - P_v} = \exp(-Z^* \, l/l_0) \tag{16}$$

with the abbreviation

$$Z^* = Y(1 + 1/X) \tag{17}$$

The partial pressure profiles for countercurrent and cocurrent models for same $X$ (=1) and $Y$ are comparatively shown in Fig. 3.

The equations for $P_E$ and $P_a$ for the cocurrent model corresponding to the Eqs. (11) and (13) for the countercurrent model may be obtained from Eqs. (A4) and (A5) of the Appendix:

$$\frac{P_I - P_E}{P_I - P_v} = [1 - \exp(-Z^*)]/(X + 1) \tag{18}$$

$$\frac{P_a - P_v}{P_I - P_v} = [1 - \exp(-Z^*)]X/(X + 1) \tag{19}$$

$$\frac{P_E - P_a}{P_I - P_v} = \exp(-Z^*) \tag{20}$$

It is evident that $P_E - P_a$ is always positive, approaching zero with $Z^* \to \infty$ $(D \to \infty)$.

## C. Ventilation–Perfusion Conductance Ratio and Functional Shunt

The gas exchange efficiency of the countercurrent model depends strongly on the value of $X = (\dot{V}_w \beta_w)/(\dot{V}_b \beta_b)$, which is the ventilatory:perfusive conductance ratio. The dependence is particularly pronounced with little or no diffusion limitation $(D \to \infty)$.

Even with no diffusion limitation $(D \to \infty)$, complete blood–gas equilibration—that is, $P_E = P_v$ and at the same time $P_a = P_I$—is only achieved when $X = 1$ (Fig. 2E). This ideal condition constitutes that of maximum efficiency. When there is excess perfusive conductance, and hence $X < 1$, $P_E = P_v$ but $P_a < P_I$, despite the absence of diffusion limitation (Fig. 2F).

The same result of partial pressures is obtained in a model in which the excess perfusive conductance is diverted from the gas exchange zone to pass through a blood shunt (see Fig. 4A), which may be termed "mismatch blood shunt" to distinguish it from other shunts (see later).

In analogy, excessive ventilatory conductance, resulting in $X > 1$, is equivalent to a "mismatch water shunt" (Fig. 4C).

The characteristic feature of shunt, produced either by mismatch or by true shunt (see later), is that it affects only the partial pressure in outflowing water (mismatch with $X > 1$ or water shunt) or in arterial blood (mismatch

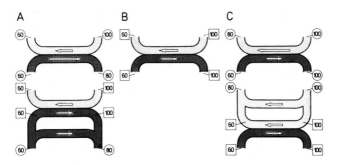

**Fig. 4.** (A) and (C) Equivalence between the effects of conductance mismatch (i.e., $X \neq 1$) and shunt. (B) The ideal system with $X = 1$, assuming $P_I = 100$ and $P_v = 60$ (units). (A) Mismatch is produced by doubling perfusive conductance ($X < 1$) thereby lowering $P_a$ to 80. The same effect is obtained when the extra blood flow is chaneled through a blood shunt, whereby ideal matching conditions in the gas exchange compartment are restored. (C) Similarly, a mismatch is created by doubling water flow conductance ($X > 1$; for details, see text).

with $X < 1$ or blood shunt). Diffusion limitation or unequal distribution generally affect partial pressures at both sites at the same time.

It follows from these considerations on the mismatch shunts that, starting from the ideal match (i.e., $X = 1$), an increase in either water or blood flow does not increase the rates of gas transfer. In particular, when ventilation is increased, leading from $X = 1$ to $X > 1$, blood oxygenation is unaffected. If, on the other hand, blood flow is increased, leading from $X = 1$ to $X < 1$, arterial oxygenation is diminished, but the total $O_2$ supplied by the arterial blood remains unaffected due to the simultaneous increase in blood flow.

Such changes are specific for the countercurrent model (Piiper and Scheid, 1983). In the ventilated pool model of alveolar lungs, increased ventilation always increases arterial $P_{O_2}$ (and decreases $P_{CO_2}$), and increased cardiac output means little change in arterial $P_{O_2}$ but improves oxygen supply to the tissues. Thereby, $P_{vO_2}$ in these conditions will increase. With finite diffusion resistance (finite $D$) the fall of arterial $P_{O_2}$ on increase of $P_{vO_2}$ becomes more important, and with high degree of diffusion limitation both changes become of equal magnitude.

It is possible that the marked variations in arterial $P_{O_2}$ observed in fish with rather constant ventilation and constant expired $P_{O_2}$ are due to changes in gas transfer efficiency caused by transitory changes in the cardiac output. (Changes in functional blood shunting of other origin may also be involved; see later.)

## D. Diffusion Limitation

The effect of diffusion limitation is to reduce the gas transfer efficiency. Quantitatively, the parameters determining the extent of diffusion limitation

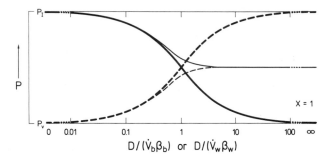

**Fig. 5.** Dependence of $P_a$ (broken lines) and $P_E$ (solid lines) on diffusion limitation, expressed by the conductance ratios, $D/(\dot{V}_b\beta_b)$ or $D/(\dot{V}_w\beta_w)$. Ideal matching ($X = 1$). Heavy lines, counter-current system; thin lines, cocurrent system.

are the diffusive:ventilatory and diffusive:perfusive conductance ratios, $D/(\dot{V}_w\beta_w)$ and $D/(\dot{V}_b\beta_b)$.

Figure 5 shows the dependence of $P_a$ and $P_E$ on the diffusion limitation for matched countercurrent and cocurrent systems, that is, at $X = 1$. The following features are noted for the countercurrent system:

1. $P_a$ varies between the limits of $P_v$, attained with infinite diffusion limitation [i.e., with $D/(\dot{V}_b\beta_b) = 0$], and $P_I$, attained at zero diffusion limitation [i.e., at $D/(\dot{V}_b\beta_b) \to \infty$].
2. Similarly, $P_E$ varies between $P_I$ and $P_v$ in the same range.
3. An overlap in water and blood partial pressures is attained only with $D/(\dot{V}_b\beta_b) > 1$.
4. A difference between the corresponding partial pressures in the countercurrent and cocurrent systems becomes evident only for $D/(\dot{V}_b\beta_b) > 0.5$.

It may thus be concluded that the particularly high gas exchange efficiency can only be observed with low diffusion limitation, $D/(\dot{V}_b\beta_b) > 1$. Below this value, gas exchange is prevalently diffusion limited with no intrinsic efficiency differences between the systems.

The diffusive conductance, designated diffusing capacity ($D$) or transfer factor ($T$), is usually the unknown variable in an experimental branchial gas exchange analysis. It can be determined from the gas transfer rate $\dot{M}$ and the partial pressures in inflowing and outflowing water and blood, respectively ($P_I$, $P_E$, $P_v$, $P_a$),

$$D = \frac{\dot{M}}{(P_I - P_a) - (P_E - P_v)} \ln\left(\frac{P_I - P_a}{P_E - P_v}\right) \qquad (21)$$

This equation can be obtained by rearranging Eqs. (11) to (13) and using the mass balance,

$$\dot{M} = \dot{V}_b \beta_b (P_a - P_v) = \dot{V}_w \beta_w (P_I - P_E) \tag{22}$$

Since $D$ is, by definition, transfer rate, $\dot{M}$, divided by the mean effective water–blood partial pressure difference, $\overline{P_w - P_b}$, one obtains from Eq. (20)

$$\overline{(P_w - P_b)} = \frac{(P_I - P_a) - (P_E - P_v)}{\ln[(P_I - P_a)/(P_E - P_v)]} \tag{23}$$

which can also be obtained from Eqs. (5) and (6) by integration along the contact length, $l$.

For $X = 1$, the equations listed in Table IB may be solved to obtain an equation analogous to Eq. (21):

$$D = \frac{\dot{M}}{(P_I - P_a)} = \frac{\dot{M}}{(P_E - P_v)} \tag{24}$$

For the mean water–blood partial pressure difference one obtains therefrom,

$$(\overline{P_w - P_b}) = P_I - P_a = P_E - P_v \tag{25}$$

It is thus only in the special case of perfect matching (i.e., $X = 1$) that the mean water–blood partial pressure difference, from which $D$ has to be calculated, equals the partial pressure difference at the ends of the system.

It follows from Fick's diffusion law that $D$ is a function of diffusion coefficient ($d$) and solubility ($\alpha$) of the gas in the barrier, of its surface area ($A$) and thickness ($l$):

$$D = d\alpha A/l \tag{26}$$

This relationship is useful for comparison of the behavior of different gases (see later).

## III. COMPOUND MODELS FOR FUNCTIONAL INHOMOGENEITIES

### A. Ventilation–Perfusion Inhomogeneity

Since the gills are composed of numerous gas exchange "units" (lamellae and interlamellar spaces) arranged in parallel, unequal distribution between these units of ventilation and perfusion may easily occur. The simplest model is a two-unit parallel model without diffusion limitation, depicted in Fig. 6A.

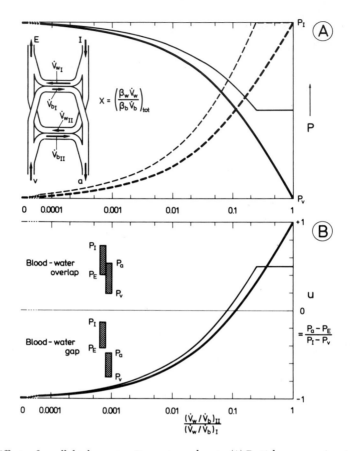

**Fig. 6.** Effects of parallel inhomogeneity on gas exchange. (A) Partial pressures in arterial blood (broken lines, $P_a$) and expired water (solid lines, $P_E$), at constant $P_I$ and $P_v$ are plotted at different degrees of parallel inhomogeneity, expressed as the ratio $(\dot{V}_w/\dot{V}_b)_{II}/(\dot{V}_w/\dot{V}_b)_I$ (see inset). For both (A) and (B): heavy lines for overall matching ($X = 1$); thin lines for $X = 2$. (B) Overlap parameter $u = (P_a - P_E)/(P_I - P_v)$ for both cases, $X = 1$ and $X = 2$. (For details, see text.)

It is evident from Fig. 6A that, in general, $\dot{V}_w/\dot{V}_b$ inhomogeneity reduces gas exchange efficiency in that $P_a$ levels stay closer to $P_v$, and $P_E$ to $P_I$. However, the extent of this reduction depends on the matching of total water to total blood flow conductances, $X = (\dot{V}_w\beta_w)/(\dot{V}_b\beta_b)_{tot}$. If $X = 1$ (Fig. 6A), the reduction in efficiency starts with the smallest level of $\dot{V}_w/\dot{V}_b$ inhomogeneity. If, however, $X \neq 1$ (e.g., $X = 2$ in Fig. 6A), then the partial pressures in blood and water, and hence the gas exchange efficiency, are independent within a certain range of $\dot{V}_w/\dot{V}_b$ inhomogeneities.

The overlap coefficient $u$ (Fig. 6B) thus depends on both the total matching $(X)$ as well as the degree of inhomogeneity. Whereas with no or little inhomogeneity, $u$, and hence the efficiency, with ideal overall matching $(X = 1)$ exceeds that with no ideal matching, the situation reverses with high degrees of inhomogeneities when the nonmatched model exerts a higher efficiency than the matched system.

The deviation of $P_a$ and $P_E$ from complete equilibration, and hence $u < +1$, may be described with a simple model involving a single, perfectly matched gas exchange unit and one shunt compartment each on the water and the blood side. (Note that mismatch in the absence of unequal distribution may be described by a single, perfectly matched gas exchange unit, with only one shunt on either water or blood side, depending on whether $X > 1$ or $X < 1$.) The shunt compartments are quantified as follows:

1. *Water shunt.* The deviation of $P_E$ from its ideal value (i.e., $P_v$) is attributed to water shunt, $\dot{V}_{w,s}$, whose fraction $\dot{V}_{w,s}/\dot{V}_{w,tot} = S_w$ is hence given by

$$S_w = \frac{P_E - P_v}{P_I - P_v} \tag{27}$$

2. *Blood shunt.* Similarly, deviation of $P_a$ from $P_I$ is attributed to a blood shunt, $\dot{V}_{b,s}$, with fraction $S_b = \dot{V}_{b,s}/\dot{V}_{b,tot}$

$$S_b = \frac{P_I - P_a}{P_I - P_v} \tag{28}$$

3. *Matched gas exchange unit.* The conductances of flows through the gas exchange unit, $\dot{V}_{tot} - \dot{V}_s$, for both water and blood can easily be shown to be matched, since $\dot{V}_{w,tot}\beta_w(P_I - P_E) = \dot{V}_{b,tot}\beta_b(P_a - P_v)$.

The following relationship results from Eqs. (14), (27), and (28):

$$u = 1 - S_w - S_b \tag{29}$$

This shunt model for unequal distribution is capable of explaining the fact that a low degree of unequal distribution does not affect partial pressures, $P_a$ and $P_E$, and thus not gas exchange, provided the total flow conductances are not matched (see Fig. 6 with $X = 2$). It was shown in the homogeneous model that with a mismatch at $X > 1$, water can be diverted from the gas exchange unit (into a shunt compartment) without affecting gas exchange. In the heterogeneous model with overall $X > 1$, water can similarly be diverted from one unit to the other with no effect on gas exchange, as long as $X$ for each subunit remains $>1$.

## B. Inhomogeneities Involving Diffusing Capacity

Besides $\dot{V}_w$ and $\dot{V}_b$, $D$ as well may be unequally distributed. This inequality also leads to decrease of gas exchange efficiency compared to equal distribution of the same $D$. A detailed treatment of this case does not lead to general insights beyond the ones discussed so far.

## C. Functional Shunt versus True Shunt

All disturbances of gas exchange occurring when the countercurrent model departs from the ideal state—that is, perfect matching $(\dot{V}_w\beta_w)/(\dot{V}_b\beta_b)$ = 1, lack of diffusion limitation ($D \rightarrow \infty$), and equal distribution—lead to decreased efficiency. This decreased efficiency may be expressed in terms of functional shunt, which may be due to the following basic mechanisms (Fig. 7):

1. *Mismatch shunt.* The model is ideal except for $(\dot{V}_w\beta_w)/(\dot{V}_b\beta_b)$ unequal unity. The situation in which $(\dot{V}_w\beta_w)/(\dot{V}_b\beta_b) > 1$ is equivalent to mismatch water shunt, whereas $(\dot{V}_w\beta_w)/(\dot{V}_b\beta_b) < 1$ corresponds to mismatch blood shunt. Hence, either water or blood shunt occurs.
2. *Diffusion shunt.* The model is ideal except for finite $D$, which results in shunt. Water shunt and blood shunt occur simultaneously and are of equal magnitude.
3. *Distribution shunt.* The model is ideal except for unequal distribution among parallel units. Water and blood shunts of varied proportions arise, depending on the pattern of inhomogeneity. In general, both shunts occur simultaneously.
4. *True water and/or blood shunts* exist (see later).

It is thus evident that these four basic mechanisms that lead to deviations in the countercurrent system from the ideal state result in somewhat different patterns of equivalent shunt.

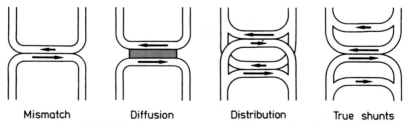

|   Mismatch   |   Diffusion   |   Distribution   |   True  shunts   |

**Fig. 7.** Models for functional shunts and for true shunt (see text).

## IV. NONLINEAR EQUILIBRIUM CURVES

### A. Capacitance Coefficients

In the preceding sections, the capacitance coefficient (or effective solubilities) in water and blood, $\beta_w$ and $\beta_b$, were considered as constant. It is, however, well established that this is generally not the case, since the respective equilibrium curves—that is, the plots of content ($C$) against partial pressure ($P$)—are nonlinear (Fig. 8). The capacitance coefficients may therefore be defined as the (partial pressure-dependent) slopes of these equilibrium curves:

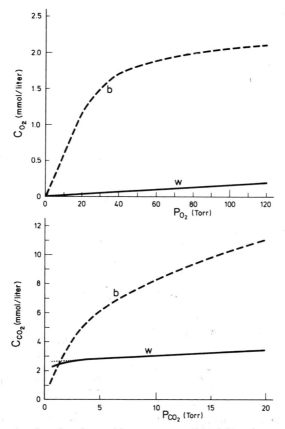

**Fig. 8.** Oxygen and carbon dioxide equilibrium curves of blood (b) and water (w). (Data for the elasmobranch, *Scyliorhinus stellaris*, after Piiper and Baumgarten-Schumann, 1968a.)

$$\beta_b = dC_b/dP_b \tag{30}$$

$$\beta_w = dC_w/dP_w \tag{31}$$

### 1. Effective Solubility of $O_2$ in Blood

Effective solubility of $O_2$ in blood is determined by both physical solubility ($\alpha$) and chemical binding to hemoglobin (Hb). For mammals, physical solubility is small compared with Hb binding. However, when Hb concentration is low and when $\alpha$ is increased at low temperature, physical binding may become significant in $\beta_{bO_2}$. An extreme is the hemoglobin-free blood of the Antarctic icefish (Chaenichthyidae), where $O_2$ transport is exclusively by physical solubility (analyzed by Holeton, 1972).

The $O_2$ bonding to hemoglobin has been investigated in great detail in recent years (reviewed by Bartels and Baumann, 1977; with stress on evolutionary aspects, by Wood and Lenfant, 1979; for fish blood, by Riggs, 1970). In this section, only the relationships between the customarily used blood $O_2$ parameters and $\beta_{b,O_2}$ will be summarized.

Hemoglobin-bound $O_2$ carriage in blood is conventionally characterized

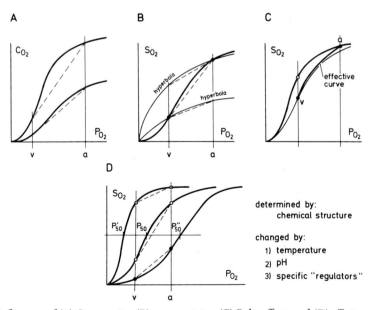

**Fig. 9.** Influences of (A) $O_2$ capacity, (B) cooperativity, (C) Bohr effect, and (D) affinity on the shape and the location of the $O_2$ dissociation curve, and thereby on $\beta_{bO_2}$ (for details, see text). (Reproduced from Piiper and Scheid, 1981, with permission.)

by (a) half-saturation pressure ($P_{50}$), or affinity, and (b) Hill coefficient (n) or heme–heme cooperativity. The value of $\beta_{bO_2}$ at any $P_{O_2}$ changes with each of these parameters (Fig. 9).

An increase in $O_2$ capacity leads to an increase in $\beta_{bO_2}$ (Fig. 9A). For the same $P_{50}$, an increase in Hill's n, and hence an increase in the sigmoidicity of the $O_2$ dissociation curve, results in increased $\beta_{bO_2}$ at $P_{50}$ and in decreased $\beta_{bO_2}$ at higher or lower $P_{O_2}$ values (Fig. 9B). The value of $P_{50}$ determines the $P_{O_2}$ range in which $\beta_{bO_2}$ reaches its maximum (Fig. 9D).

There exists an important influence on $O_2$ capacity and affinity of a number of variables. Notable are (a) $H^+$ ion concentration (Bohr and Root effects), (b) "allosteric modifiers," ATP and GTP, and (c) temperature.

The Bohr effect results in a higher $\beta_b$ value of the physiological dissociation curve (Fig. 9C). For gill $O_2$ transfer, the Root effect (i.e., the decrease in $O_2$ capacity with decreasing pH) exerts an effect similar to the Bohr effect. However, the significance of the Root effect appears to be linked to $O_2$ secretion into the swimbladder or the retina.

## 2. Effective Solubility of $CO_2$ in Blood

The major part of $CO_2$ in blood is transported as $HCO_3^-$, even though $\alpha$ for $CO_2$ is about 20 times larger than $\alpha$ for $O_2$. Reversible formation of $HCO_3^-$ is dependent on the buffer power of nonbicarbonate buffers, Hb being the most important nonbicarbonate buffer in most fishes. Hence, $\beta_{bCO_2}$ increases with the nonbicarbonate buffering power. Likewise, increasing $HCO_3^-$ at constant $P_{CO_2}$ increases $\beta_{bCO_2}$. For the quantitative relationships, see Albers (1970) and Burton (1973).

The value of $\beta_{bCO_2}$ is continually increasing with decreasing $P_{CO_2}$. Hence, water-breathing fish, which display low values of blood $P_{CO_2}$ (cf. Piiper and Scheid, 1977) are endowed with comparatively large values of $\beta_{bCO_2}$.

The Haldane effect—that is, the decrease of $CO_2$ content with increasing $O_2$ saturation at constant $P_{CO_2}$—leads to an increase in $\beta_{bCO_2}$ for the physiological $CO_2$ dissociation curve, analogous to the influence of the Bohr effect on $\beta_{bO_2}$.

The role of nonbicarbonate buffers in $CO_2$ binding has been theoretically analyzed by Burton (1973) and by others. It should be borne in mind that the derived, relatively simple relationship between $\beta_{CO_2}$ and buffer power applies to solutions containing only one phase (e.g., water; see later). Therefore, calculation of $\beta_{CO_2}$ for blood from buffer values obtained for "true plasma" (i.e., plasma in exchange contact with erythrocytes) is not easy because of the coexistence of two phases—that is, plasma and erythrocytes—with different pH and buffer value, and because of interchange of ions.

## 3. Effective Solubility of $O_2$ in Water

For $O_2$ in water (fresh water or seawater), the solubility is independent of partial pressure, and hence the curve of content versus partial pressure is linear (Fig. 8). The solubility decreases with increasing temperature and salinity (e.g., Rahn, 1966).

## 4. Effective Solubility of $CO_2$ in Water

For $CO_2$ in carbonated water (e.g., seawater or "hard" fresh water), the dissociation curve is curved at low $P_{CO_2}$ and high pH (Dejours et al., 1968; Piiper and Baumgarten-Schumann, 1968a; Dejours, 1978), as a result of the reversible formation of carbonate from $HCO_3^-$ at decreasing $P_{CO_2}$ according to the following coupled reaction equations:

$$CO_2 + H_2O \leftrightharpoons H^+ + HCO_3^- \tag{32}$$

$$CO_3^{2-} + H^+ \leftrightharpoons HCO_3^- \tag{33}$$

$$B^- + H^+ \leftrightharpoons HB \tag{34}$$

in which $HB/B^-$ constitutes the nonbicarbonate buffer system. Both the steepness of the water dissociation curve, $\beta_{wCO_2}$, and its curvature depend on the relative amounts of $HCO_3^-$ and nonbicarbonate buffers.

## 5. Effects of Ammonia

Teleost fish are ammonotelic, the end product of their protein catabolism being ammonia, $NH_3$. Almost all ammonia formed at the pH of body fluids appears as $NH_4^+$, and ammonia reacts thus as a base. There is a significant rate of $NH_3$ production, $\dot{M}_{NH3}$. In rainbow trout under normal resting conditions, Cameron and Heisler (1983) found a mean $\dot{M}_{NH3}/\dot{M}_{O2}$ ratio of 0.12.

Since transport in blood is by $NH_4^+$ but excretion across the gills mainly as $NH_3$, the ammonia mechanism leads to a reduction (or reversal) of the (normally positive) pH differences between arterial and venous blood, and inspired and expired water. This has consequences for $\beta_{CO_2}$ in the following senses:

1. $\beta_{CO_2}$ is increased for both water and blood. In blood, the effect of ammonia is similar to the Haldane effect. When $NH_3$ is excreted, the remaining $H^+$ ions lead to formation of $CO_2$ from $HCO_3^-$, potentially without a change in $P_{CO_2}$

$$NH_4^+ + HCO_3^- \leftrightharpoons NH_3 + CO_2 + H_2O \tag{35}$$

2. The Bohr effect, and hence its enhancing influence on blood $\beta_{O_2}$, is decreased.

## 6. Effective Solubility Relationships: $CO_2/O_2$ and Water/Blood

Since $\beta_{wCO_2}/\beta_{wO_2}$ is in general unequal to $\beta_{bCO_2}/\beta_{bO_2}$, it is evident that the ideal matching condition of $(\dot{V}_w\beta_w)/(\dot{V}_b\beta_b) = 1.0$ can be reached either for $O_2$ or for $CO_2$, but not for both at the same time (cf. Piiper and Scheid, 1972, 1975, 1977).

## B. Diffusing Capacity

With a curvilinear $O_2$ equilibrium curve, Eq. (21) cannot be used to calculate $D_{O_2}$ from partial pressures in blood and water. However, the differential Eqs. (1) to (3) remain valid, in which $\beta_b$ becomes a function of the blood partial pressure, $P_b$. If, in the range between $P_a$ and $P_b$, the equilibrium curve can be approximated by a sufficiently simple expression, yielding $C_b$ as a sufficiently simple function of $P_b$, then integration of Eqs. (1) to (3) may still be possible to yield an analytic expression for $D_{O_2}$, analogous to Eq. (21).

However, numerical or graphical procedures will generally have to be employed to obtain $D_{O_2}$ from measurements of $\dot{M}$ and partial pressures. It is appropriate in this case to use Eqs. (30) and (31) to replace $dP$ by $dC$ in Eqs. (1) and (2). In practice, the blood–water contact is subdivided into $N$ parts of equal $O_2$ transfer rate (Piiper and Baumgarten-Schumann, 1968b; Piiper et al., 1977):

$$\Delta \dot{M}_{O2} = \dot{M}_{O2}/N \tag{36}$$

According to Eqs. (1) and (2), $O_2$ content in water and blood thus changes from compartment to compartment in $N$ equal steps, between $P_E$ and $P_I$, and between $P_a$ and $P_v$ (Fig. 10). The water and blood partial pressures in any compartment, $n$, $(P_w)_n$, and $(P_b)_n$ can thus be read from the equilibrium curves, and the diffusing capacity of this compartment, $\Delta D_n$, can be calculated from Eq. (3) as

$$\Delta D_n = \frac{\Delta \dot{M}_{O2}}{(P_w - P_b)_n} \tag{37}$$

Summing up all the elements then gives

$$D = \sum_{n=1}^{N} \Delta D_n = \frac{\dot{M}_{O2}}{N} \sum_{n=1}^{N} \frac{1}{(P_w - P_b)_n} \tag{38}$$

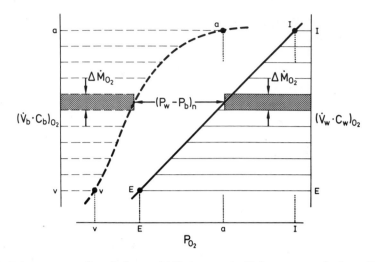

**Fig. 10.** Bohr integration for calculation of diffusing capacity $D$, from measured values of $P_{O_2}$ in arterial (a) and venous blood (v) and in inspired (I) and expired water (E). On the left ordinate is plotted the perfusive transport rate $(\dot{V}_b C_b)_{O_2}$, between a and v; on the right ordinate, the ventilatory transfer rate $(\dot{V}_w C_w)_{O_2}$, between I and E. Both transfer rates are plotted so that corresponding values, pertaining to the same element along the contact length in the countercurrent system, are at the same ordinate level. Thereby, the water–blood partial pressure difference in any element, $(P_w - P_b)_n$, corresponds to the horizontal distance between both curves.

The same procedure may be used for calculation of $D_{CO_2}$ (Piiper and Baumgarten-Schumann, 1968b). The use of the correct water equilibrium curve is important for bicarbonate-containing water with low (normal) inspired $P_{CO_2}$ (see earlier).

## C. Ventilation–Perfusion Inhomogeneity with Curvilinear Equilibrium Curves

The analysis of the effects of inhomogeneity is difficult with curvilinear equilibrium curves, because (*a*) the equilibration is determined by partial pressures, whereas (*b*) the shunt effects are quantified in terms of content. Considering the equilibration of blood with water, ideally $P_a$ can attain $P_I$. The content in blood equilibrated with $P_I$ is $C_{I'}$. Then the effective shunt $(S_b)$ is calculated in analogy with Eq. (28) as

$$S_b = \frac{C_{I'} - C_a}{C_{I'} - C_v} \qquad (39)$$

The complementary index, $1 - S_b$, has been termed effectiveness of $O_2$ uptake by blood or of $CO_2$ release from blood by Randall *et al.* (1967).

When, conversely, the equilibration of water is considered with blood, $P_E$ could ideally approach $P_v$. The corresponding content in water is $C_{v'}$. Hence, in analogy with Eq. (27) for water shunt,

$$S_w = \frac{C_E - C_{v'}}{C_I - C_{v'}} \qquad (40)$$

For $O_2$ the partial pressures could be used in Eq. (40), since all content values are located on the linear water equilibrium curve. For $CO_2$, however, the curvature of this curve (see Section IV,A) should be taken into account.

In interpreting $S_b$ and $S_w$ values, it should be borne in mind that eliminating diffusion limitation and unequal distribution from the system would not mean that the ideal state with $P_a = P_I$ and $P_E = P_v$ is reached. This is only the case with perfect matching, that is, when $(\dot{V}_w \beta_w)/(\dot{V}_b \beta_b) = 1$. Otherwise, either a mismatch water shunt, with $P_E \neq P_v$, or a mismatch blood shunt, with $P_a \neq P_I$, occurs (see Section II,C).

## V. DIFFUSION RESISTANCE OF INTERLAMELLAR WATER

Since the width of the interlamellar spaces (20–100 μm) is much greater than the thickness of the water–blood tissue barrier (0.2–10 μm) (cf. Piiper, 1971), a sizeable and possibly a major part of the total diffusive resistance, $R$ ($= 1/D$), to water–blood gas transfer should reside in the interlamellar water. Models will be presented in this section that allow estimation of the diffusion resistance offered by interlamellar water to $O_2$ uptake. The analysis is, however, similarly valid for any gas, including $CO_2$.

### A. Simplified Model

The role of diffusion in interlamellar water in limiting branchial gas transfer has been studied by Scheid and Piiper (1971) using a highly simpliifed gill model (Fig. 11). This model represents an interlamellar space, bounded by two lamellae of trapezoidal shape, the length at the edge being a fraction, $\lambda$ ($<1$), of the length at the base ($l_0$). Slit width is $b_0$, and lamellar height, $h_0$.

The $O_2$ partial pressure in water at the lamellar surface is assumed to be constant at $P_0$ throughout the lamella (in reality, this $P_{O_2}$ decreases from the water inflow to its outflow end). Water flow is assumed to be laminar (with parabolic velocity profile) and inversely proportional to slit length, which decreases linearly from the base to the free edge as a result of the trapezoidal shape of the lamella. Thus, there are velocity gradients in both $b$ and $h$

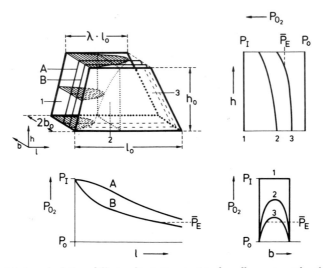

**Fig. 11.** Model for calculating diffusion limitation in interlamellar water. The three profiles reveal the gradients in $P_{O_2}$ occurring along the three coordinates of the model: height, $h$; length, $l$; width, $b$. The curves reflect $P_{O_2}$ profiles at sites indicated in the model (for details, see text). (After Piiper, 1982.)

directions (see Fig. 11). Oxygen is assumed to diffuse in the $b$ direction only, that is, perpendicularly to the lamellar surface.

The $P_{O_2}$ profiles of the model can be calculated from the differential equation ($P = P_{O_2}$),

$$\frac{\partial P}{\partial l} = \frac{d}{v(b,h)} \frac{\partial^2 P}{\partial b^2} \tag{41}$$

in which $d$ represents the diffusion coefficient of $O_2$ in water and $v$, the linear water velocity (varying with $b$ and $h$). The equation, which is derived from Fick's second law of diffusion ($\partial P/\partial t = d\partial^2 P/\partial b^2$) by introducing $v = dl/dt$, can be numerically integrated for the appropriate boundary and initial conditions.

The resulting $P_{O_2}$ profiles, schematically shown in Fig. 11, display $P_{O_2}$ gradients in all three space dimensions: (1) $\partial P/\partial l$, reflecting the progressive $O_2$ depletion, (2) $\partial P/\partial b$, due to resistance to diffusion, and (3) $\partial P/\partial h$, due to the base-to-apex velocity gradient.

Because of the gradients (2) and (3), $P_{O_2}$ in outflowing water ($P_E$) varies across the cross section. The decisive variable for the overall $O_2$ uptake is the $P_{O_2}$ in the (flow-weighted) mixed expired water ($\bar{P}_E$) obtained by double integration of $P_E$ with respect to $b$ (in the limits $-b_0 < b < +b_0$) and $h$ (in the limits $0 < h < h_0$):

$$\overline{P_{\mathrm{E}}} = \frac{1}{2b_0 h_0 \bar{v}} \int_{-b_0}^{+b_0} \int_0^{h_0} P_{\mathrm{E}} v \, db \, dh \qquad (42)$$

where $\bar{v}$ is the average water velocity.

The inefficiency of $O_2$ equilibration, $\varepsilon$, may be expressed by

$$\varepsilon = \frac{\overline{P_{\mathrm{E}}} - P_0}{P_{\mathrm{I}} - P_0} \qquad (43)$$

Since, in the absence of diffusion limitation in water, $\overline{P_{\mathrm{E}}}$ must equal $P_0$, $\varepsilon$ constitutes an equivalent shunt, which may be termed "water diffusion shunt" in distinction from the "diffusion shunt" due to membrane diffusion (Section III,C).

By integrating Eq. (41) it can be shown (Scheid and Piiper, 1971) that $\varepsilon$ is determined by the dimensionless parameter,

$$\varphi = \frac{b_0^2 \bar{v}}{l_0 d} \qquad (44)$$

and by the tapering factor, $\lambda$ (see Fig. 11). The parameter, $\varphi$, contains two geometric variables (the half-width of the interlamellar space, $b_0$, and its base length, $l_0$), the mean water velocity, $\bar{v}$, and the diffusion coefficient of $O_2$ in water, $d$. $\varphi$ is proportional to the mean water velocity, $\bar{v}$, and thereby to the water flow rate, $\dot{V}_{\mathrm{w}}$. With increasing $\varphi$ (and decreasing $\lambda$), the inefficiency $\epsilon$ increases (Fig. 12).

Estimates using experimental values for ventilation and morphometric data (e.g., $\lambda$ ranging from 0.5 to 1.0) for teleost fish gills yield $\varphi$ values for $O_2$ equilibration ranging from 0.02 to about 10 (Scheid and Piiper, 1971). The corresponding $\epsilon$ values range from close to zero to about 80%. Typical values for basal resting conditions are 0–5%; for increased ventilation due to hypoxia or activity, they are 10–50%.

Since the mechanism involved is diffusion limitation, it is appropriate to transform the inefficiency index, $\epsilon$, into an equivalent or apparent diffusing capacity,

$$D_{\mathrm{app}} = \dot{V}_{\mathrm{w}} \alpha \ln(\epsilon) \qquad (45)$$

$D_{\mathrm{app}}$ is the diffusing capacity of an equivalent model in which water is mixed in each cross section (in $b$ and $h$ directions, Fig. 11), and diffusion limitation is offered by a stagnant water layer of appropriate thickness bordering upon the interlamellar surface.

Remarkably, $D_{\mathrm{app}}$ increases considerably with increasing $\varphi$ and thus with increasing mean water flow velocity, $\bar{v}$ (Fig. 13), or increasing ventilation, $\dot{V}_{\mathrm{w}}$. This feature may be important for facilitating $O_2$ uptake in conditions of increased $O_2$ requirement (swimming, hypoxia).

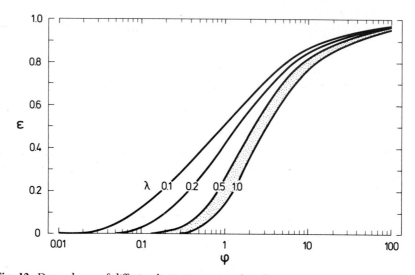

**Fig. 12.** Dependence of diffusion limitation in interlamellar water, expressed as inefficiency coefficient, $\epsilon = (\bar{P}_E - P_0)/(P_I - P_0)$, on geometric and physical parameters, lumped into the dimensionless parameter, $\varphi = (b_0^2 \bar{v})/(l_0 d)$. Curves at different values of the shape parameter, $\lambda$ (see Fig. 11). The region of $\epsilon$ values with experimental data of $\lambda$ (between 0.5 and 1.0) is indicated by stippling. (After Piiper, 1982.)

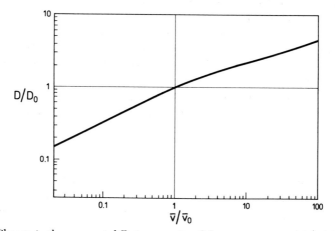

**Fig. 13.** Change in the apparent diffusing capacity of the stagnant water layer $(D)$ with mean linear velocity of interlamellar water, $\bar{v}$, $\bar{v}_0$ denotes mean linear velocity in a system in which $\varphi_0 = (b^2\bar{v}_0)/(dl_0) = 1$; $D_0$, apparent diffusing capacity of water at $\varphi_0$. (After Scheid and Piiper, 1971.)

## B. Comprehensive Models

Since the lamellar surface $P_{O_2}$ was assumed to be constant (at $P_0$) in the simplified model of Fig. 11, diffusion resistance offered by water was artificially isolated from other resistances to gas transfer, notably from those due to membrane diffusion and to blood perfusion. If it is assumed that water diffusion and membrane diffusion resistances may be considered as being arranged in series, then $D_{app}$ of Eq. (45) can be compared with membrane $D$ to estimate their relative magnitudes. This has been performed by Scheid and Piiper (1971, 1976).

A more comprehensive model has been subsequently investigated in which diffusion both across a tissue barrier and in flowing water are considered, and blood perfusion is finite (J. Piiper, C. Hook, and P. Scheid, unpublished). According to preliminary results of calculations, the effects of diffusion limitation in interlamellar water are slightly less than calculated for the simplified model, because the water and membrane diffusion resistances are not strictly in series (or additive) in the comprehensive model. However, the simplified model seems to yield sufficiently good estimates of the order of magnitude of diffusion limitation in interlamellar water.

## VI. FACTORS NOT INCLUDED IN MODELS

A number of additional factors that are expected to affect gas transfer in fish gills are not explicitly included in the models. Some of the more important factors will be considered in this section.

## A. Diffusion in Blood

Part of the diffusion path for gas transfer resides within blood, that is, in plasma, red cell membrane, and red cell interior. The role of intravascular diffusion in limiting branchial gas exchange has not been estimated. However, since there is diffusion limitation in interlamellar water, the fractional effect of intravascular diffusion in overall gas exchange may be less than in the lungs of air breathers. Also, the reaction kinetics of hemoglobin oxygenation may exert a limiting effect, particularly at the temperatures of fish, which are usually much lower than those in the mammalian body.

In all vertebrates, red cell carbonic anhydrase appears to play a major role in accelerating the hydration–dehydration reaction, $CO_2 + H_2O \rightleftharpoons H_2CO_3$ (reviewed by Maren, 1967). Haswell, Randall, and co-workers have since postulated, on the basis of experimental results, that the membrane of fish red cells was impermeable to bicarbonate, so that the intraerythrocyte carbonic anhydrase and buffering power could not be utilized for $CO_2$ transport and exchange (Haswell and Randall, 1976; summarized by Haswell *et al.*, 1980). This would mean a strong reaction–diffusion limitation of $CO_2$ transfer corresponding to a low apparent $D_{CO_2}$ value in our simplified model.

Estimates of $D_{CO_2}$ for *Scyliorhinus stellaris*, however, have yielded unexpectedly high values (Piiper and Baumgarten-Schumann, 1968b). Moreover, later reinvestigation of the experimental methods and data of Haswell *et al.* (1980) has invalidated the interpretations (Heming and Randall, 1982), and the prominent role of red cells in circulatory transport of $CO_2$ has been demonstrated (Obaid *et al.*, 1979; Perry *et al.*, 1983). Thus, at present, it is generally believed that the interaction of red cells and plasma in $CO_2$ transport in fish blood is similar to that in mammals, probably leaving little space for the limiting role of red cell–plasma bicarbonate transfer in $CO_2$ transport (see Chapter 5).

A rather comprehensive model (including countercurrent blood and water flow, diffusion and hydration–dehydration kinetics of the $CO_2/H_2CO_3$ system, and $HCO_3^-$ equilibration kinetics between red cells, plasma, and gill water) has been tentatively used by Cameron and co-workers (Cameron and Polhemus, 1974; Cameron, 1978) for analysis of $CO_2$ transfer in teleost gills, with particular emphasis on the role of intraerythrocyte carbonic anhydrase. Cameron (1978) also specifically mentions the possible limitation imposed by the (uncatalyzed) hydration of $CO_2$ in gill water.

## B. Water Shunting between Filaments

It is quite conceivable that gill filaments may move apart, giving rise to (*a*) water shunted outside the interlamellar spaces of adjacent filaments and to (*b*) water shunted between the free ends of filaments. This is expected to happen in particular with enhanced inspiratory enlargements of the branchial apparatus, accompanying increased ventilation. Indeed, the high $O_2$ extraction from water may be found surprising, given the anatomic potential for disruption of the "branchial sieve."

For gas transfer, such "true water shunt" would have the same effect as interlamellar diffusion limitation ("water diffusion shunt"), both contributing to inefficiency of branchial gas exchange.

## C. Extralamellar Blood Shunting

The vascular connections beside the main route, that is, afferent arteries–lamellar lacunae–efferent arteries, have been receiving considerable

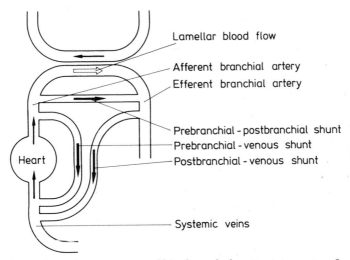

**Fig. 14.** Schema of vascular arrangement of blood vessels that give rise to various forms of true shunt (for details, see text).

attention (e.g., Laurent and Dunel, 1976; Vogel, 1978). According to their significance for gas transport, three categories of extralamellar shunts may be distinguished (Fig. 14).

## 1. PREBRANCHIAL–POSTBRANCHIAL SHUNT

In this category, blood is flowing in the very basis of the lamella or in the filament itself, bypassing the lamella (Steen and Kruysse, 1964). This is a "true shunt," parallel to lamellar circulation, and it leads to decreased arterialization.

## 2. PREBRANCHIAL–VENOUS SHUNT

Blood flowing through anastomotic vessels from the afferent arteries to the central venous sinus of the filament drains into the systemic venous system. The effect is to bypass lamellae, but there is no decrease in arterialization.

## 3. POSTBRANCHIAL–VENOUS SHUNT

Arterialized blood here passes through anastomotic connections from postbranchial arteries to the central venous sinus of the filament, probably subserving $O_2$ supply to branchial tissues. This shunt decreases $O_2$ availability to body tissues but leads to no venous admixture to arterialized blood.

## D. Pulsatile Water and Blood Flow

In the models, continuous, constant water and blood flows were assumed. Although in fishes the respiratory water flow through the interlamellar spaces may be smoothed by virtue of well-balanced pressure–suction pumping mechanisms, water flow remains pulsatile to some degree. The same pertains to the imperfectly smoothed cardiac ejection of blood.

The effects of this pulsatility are not easy to assess and depend mainly on two factors: (1) the period of the pulsatile flow relative to the contact time in the lamella and (2) the phase relation between water flow and blood flow pulsations. In general, this pulsatility may lead to effects that are similar to those of mismatching, particularly when slow (compared with contact time) flow oscillations are out of phase. The pulsatility in general is hence expected to reduce gas exchange efficiency. Since the respiratory and cardiac frequencies in fishes are in many cases similar, coupling of these rhythms has been postulated and repeatedly investigated. However, only a feeble coupling, a slight relative coordination, seems to be present (Hughes, 1972; Taylor and Butler, 1971).

## E. Unsteady State (Irregular, Intermittent Breathing)

Although breathing and heart rates in fishes as a group are usually more constant and stable as compared to amphibians and reptiles (Shelton and Boutilier, 1982), irregularities and intermittence of these rhythms, and hence of water and blood flows, do occur. When of relatively short duration, the effects of periodicities are similar to those of pulsatilities of water and blood flow, that is, they produce temporal inhomogeneity with reduced gas exchange efficiency.

Variations with long duration produce changes the kinetics of which are determined generally by conductance:capacitance ratios of the limiting processes; therefore, they are not easily analyzed. However, since steady state, for which the models have been developed, is only a limiting case of transient and unsteady state, it should be possible to establish adequate models and approaches in future.

## F. Oxygen Consumption and Uptake

Fish gills are important organs not only for (passive) gas exchange, but also for (active) ion transfer. They thus have a relatively high metabolic rate and $O_2$ consumption (Johansen and Petterson, 1981). Gill blood flow (cardiac output) is overestimated when calculated from the arteriovenous $O_2$ content difference and total $O_2$ uptake (Fick's principle), since part of this uptake is consumed in the gills.

Gas exchange occurs not only in gills, but to some extent also across the body surface; this is particularly the case in some fishes, such as eels, in which the skin $O_2$ uptake in water has been estimated to be a significant fraction of the total $O_2$ uptake (Berg and Steen, 1965). However, in other fishes there is also measurable cutaneous gas exchange (Kirsch and Nonnotte, 1977; Nonnotte and Kirsch, 1978). When this $O_2$ uptake is included in calculating gill blood flow by the Fick principle, gill blood flow is overestimated. Thus, in respect of calculating gill blood flow, the effect of extra-branchial gas exchange is similar to that of branchial $O_2$ consumption.

## VII. INTERPRETATION OF EXPERIMENTAL DATA

An important result of the simplified functional fish gill model is its ambiguity with respect to the modifying mechanisms: reduction of effectiveness from the ideal, maximum value may be due to (a) diffusion limitation (in water, in water–blood barrier, in blood), (b) unequal distribution of $\dot{V}_w$ to $\dot{V}_b$ either between parallel units (e.g., of lamellae, filaments, or gill slits) or due to pulsatile water and blood flows, or (c) various combinations of these factors and/or other factors.

This conclusion is highly unsatisfactory. What can be done to reduce the ambiguity of interpretations, for example, to establish that it is more diffusion limitation or more parallel inhomogeneity that limits gas exchange efficiency in a particular example? Additional relevant information is necessary to this end, which may be obtained from the following approaches based on experience obtained with analysis of alveolar gas exchange in mammals.

1. The relevant physicochemical properties of water and blood may be systematically varied [$\beta_{bO_2}$ (e.g., by hypoxia or experimental anemia), $\beta_{bCO_2}$ (e.g., by hypercapnia), $\beta_{wO_2}$ (e.g., by addition of hemoglobin), $\beta_{wCO_2}$ (e.g., by artificial buffering)]. Obviously, such variations are expected to induce other (adaptive) changes in gas transfer variables ($\dot{V}_w$, $\dot{V}_b$, distribution), which must be taken into account.

2. Simultaneous analysis of several gases significantly sharpens the analysis. Thus, simultaneous analysis of the behavior of $O_2$ and $CO_2$ may reveal important features (comparison of $D_{O_2}$ and $D_{CO_2}$, calculation of functional shunts using $CO_2$ and $O_2$). An important gas for assessment of diffusion in mammalian lungs is CO. To our knowledge, there is only one single pioneer determination of $D_{CO}$ in fish gills (Fisher et al., 1969). The counterpart of

CO are inert gases, which are expected to be at least diffusion limited (high values of $D/(\dot{V}_b\beta_b)$ and, therefore, may permit more precise definitions of distribution inequalities.

3. Local measurements of $P_{O_2}$ in gill water reveal inhomogeneities (Hughes, 1973). Also, local determinations of water flow and its time patterns may yield useful information.

4. Comparison of morphological and morphometric data with functional measurements is a particularly promising approach. Results have been obtained from comparison of functional $D_{O_2}$ and morphometric data on tissue–blood barrier and on interlamellar space (Scheid and Piiper, 1976). Highly interesting results were obtained by comparing the kinetics of oxygenation and deoxygenation of stagnant red cells in isolated lamellae with thickness and surface area of the water–blood barrier (Hills et al., 1982).

Thus, it appears that combining a number of methods and approaches may enable us to achieve a sharper picture of gas exchange processes in fish gills.


## APPENDIX: COCURRENT SYSTEM


Assume that in the countercurrent system water flow direction is reversed. Hence, a cocurrent system would result, which can be described by differential equations analogous to Eqs. (1) to (3) of the countercurrent system:

$$\frac{d\dot{M}}{dl} = -\dot{V}_w\beta_w\frac{dP_w}{dl} \tag{A1}$$

$$\frac{d\dot{M}}{dl} = \dot{V}_b\beta_b\frac{dP_b}{dl} \tag{A2}$$

$$\frac{d\dot{M}}{dl} = (P_w - P_b)\frac{dD}{dl} \tag{A3}$$

Integration of these differential equations with the boundary conditions

$$\begin{aligned} l = 0: \ &P_w = P_I; \ P_b = P_v \\ l = l_0: \ &P_w = P_E; \ P_b = P_a \end{aligned} \tag{A4}$$

yields the following partial pressure profiles:

$$\frac{P_I - P_w(l)}{P_I - P_v} = \frac{1 - \exp(-Z^*l/l_0)}{1 + X} \tag{A5}$$

$$\frac{P_b(l) - P_v}{P_I - P_v} = \frac{[1 - \exp(-Z^* \cdot l/l_0)]X}{1 + X} \qquad \text{(A6)}$$

in which

$$Z^* = Y(1 + 1/X) \qquad \text{(A7)}$$

and $X$ and $Y$ are given by Eqs. (7) and (8).

It is evident from Eqs. (A4) and (A5) that at any length, $l/l_0$, the ratio

$$\frac{P_b(l) - P_v}{P_I - P_w(l)} = X \qquad \text{(A8)}$$

is constant. This holds true in particular for the final equilibrium value, $P_{eq}$, which both $P_w$ and $P_b$ approach, irrespective of whether it is reached or not:

$$\frac{P_{eq} - P_v}{P_I - P_{eq}} = X \qquad \text{(A9)}$$

With the use of $P_{eq}$, the partial pressure profiles of Eqs. (A4) and (A5) assume a simple form,

$$\frac{P_w(l) - P_{eq}}{P_I - P_{eq}} = \frac{P_{eq} - P_b(l)}{P_{eq} - P_v} = \exp(-Z^* l/l_0) \qquad \text{(A10)}$$

which shows that both $P_w(l)$ and $P_b(l)$ approach in an exponential course the final equilibrium value, and that the relative degree of equilibration, which is given by the expressions in Eq. (A9), at any length, $l/l_0$, is the same for both.

## REFERENCES

Albers, C. (1970). Acid-base balance. *In* "Fish Physiology" (W. S. Hoar and D. J. Randall, eds.), Vol. 4, pp. 173–208. Academic Press, New York.

Bartels, H., and Baumann, R. (1977). Respiratory function of hemoglobin. *Int. Rev. Physiol.* **14**, 107–134.

Baumgarten-Schumann, D., and Piiper, J. (1968). Gas exchange in the gills of resting unanesthetized dogfish (*Scyliorhinus stellaris*). *Respir. Physiol.* **5**, 317–325.

Berg, T., and Steen, J. B. (1965). Physiological mechanisms for aerial respiration in the eel. *Comp. Biochem. Physiol.* **15**, 469–484.

Burton, R. F. (1973). The roles of buffers in body fluids: Mathematical analysis. *Respir. Physiol.* **18**, 34–42.

Cameron, J. N. (1978). Regulation of blood pH in teleost fish. *Respir. Physiol.* **33**, 129–144.

Cameron, J. N., and Heisler, N. (1983). Studies of ammonia in the rainbow trout: Physicochemical parameters, acid-base behavior, and respiratory clearance. *J. Exp. Biol.* **105**, 107–125.

Cameron, J. N., and Polhemus, J. A. (1974). Theory of $CO_2$ exchange in trout gills. *J. Exp. Biol.* **60**, 183–194.

Dejours, P. (1978). Carbon dioxide in water- and air-breathers. *Respir. Physiol.* **33**, 121–128.

Dejours, P. (1981). "Principles of Comparative Respiratory Physiology," 2nd ed. North-Holland Publ., Amsterdam.

Dejours, P., Armand, F., and Verriest, G. (1968). Carbon dioxide dissociation curves of water and gas exchange of waterbreathers. *Respir. Physiol.* **5**, 23–33.

Farhi, L. E. (1966). Ventilation-perfusion relationship and its role in alveolar gas exchange. *In* "Advances in Respiratory Physiology" (C. G. Caro, ed.), pp. 148–197. Arnold, London.

Fisher, T. R., Coburn, R. F., and Forster, R. E. (1969). Carbon monoxide diffusing capacity in the bullhead catfish. *J. Appl. Physiol.* **26**, 161–169.

Haab, P. (1982). Systématisation des échanges gazeux pulmonaires. *J. Physiol. (Paris)* **78**, 108–118.

Haswell, M. S., and Randall, D. J. (1976). Carbonic anhydrase inhibitor in trout plasma. *Respir. Physiol.* **28**, 17–27.

Haswell, M. S., Randall, D. J., and Perry, S. F. (1980). Fish gill carbonic anhydrase: Acid-base regulation or salt transport? *Am. J. Physiol.* **238**, R240–R245.

Hazelhoff, E. H., and Evenhuis, H. H. (1952). Importance of the "counter current principle" for the oxygen uptake in fishes. *Nature (London)* **169**, 77.

Heming, T. A., and Randall, D. J. (1982). Fish erythrocytes are bicarbonate permeable: Problems with determining carbonic anhydrase using the modified boat technique. *J. Exp. Zool.* **219**, 125–128.

Hills, B. A., Hughes, G. M., and Koyama, T. (1982). Oxygenation and deoxygenation kinetics of red cells in isolated lamellae of fish gills. *J. Exp. Biol.* **98**, 269–275.

Holeton, G. F. (1972). Gas exchange in fish with and without hemoglobin. *Respir. Physiol.* **14**, 142–150.

Hughes, G. M. (1972). The relationship between cardiac and respiratory rhythms in the dogfish, *Scyliorhinus canicula* L. *J. Exp. Biol.* **57**, 415–434.

Hughes, G. M. (1973). Comparative vertebrate ventilation and heterogeneity. *In* "Comparative Physiology" (L. Bolis, K. Schmidt-Nielsen, and S. H. P. Maddrell, eds.), pp. 187–220. North-Holland Publ., Amsterdam.

Hughes, G. M., and Shelton, G. (1962). Respiratory mechanisms and their nervous control in fish. *Adv. Comp. Physiol. Biochem.* **1**, 275–364.

Johansen, K., and Pettersson, K. (1981). Gill $O_2$ consumption in a teleost fish, *Gadus morhua*. *Respir. Physiol.* **44**, 277–284.

Jones, D. R., and Randall, D. J. (1978). The respiratory and circulatory systems during exercise. *In* "Fish Physiology" (W. S. Hoar and D. J. Randall, eds.), Vol. 7, pp. 425–501. Academic Press, New York.

Kirsch, R., and Nonnotte, G. (1977). Cutaneous respiration in three freshwater teleosts. *Respir. Physiol.* **29**, 339–354.

Laurent, P., and Dunel, S. (1976). Functional organization of the teleost gill. I. Blood pathways. *Acta Zool. (Stockholm)* **57**, 189–209.

Maren, T. H. (1967). Carbonic anhydrase: Chemistry, physiology, and inhibition. *Physiol. Rev.* **47**, 595–781.

Nonnotte, G., and Kirsch, R. (1978). Cutaneous respiration in seven sea-water teleosts. *Respir. Physiol.* **35**, 111–118.

Obaid, A. L., McElroy Critz, A., and Crandall, E. D. (1979). Kinetics of bicarbonate/chloride exchange in dogfish erythrocytes. *Am. J. Physiol.* **237**, R132–R138.

Perry, S. F., Davie, P. S., Daxboeck, C., and Randall, D. J. (1983). A comparison of $CO_2$ excretion in a spontaneously ventilating, blood-perfused trout preparation and saline-perfused gill preparations: Contribution of the branchial epithelium and red blood cell. *J. Exp. Biol.* (in press).

Piiper, J. (1971). Gill surface area: Fishes. *In* "Respiration and Circulation" (P. L. Altman and D. S. Dittmer, eds.), pp. 119–121. Fed. Am. Soc. Exp. Biol., Bethesda, Maryland.

Piiper, J. (1982). Diffusion in the interlamellar water of fish gills. *Fed. Proc., Fed. Am. Soc. Exp. Biol.* **41**, 2140–2142.

Piiper, J., and Baumgarten-Schumann, D. (1968a). Transport of $O_2$ and $CO_2$ by water and blood in gas exchange of the dogfish (*Scyliorhinus stellaris*). *Respir. Physiol.* **5**, 326–337.

Piiper, J., and Baumgarten-Schumann, D. (1968b). Effectiveness of $O_2$ and $CO_2$ exchange in the gills of the dogfish (*Scyliorhinus stellaris*). *Respir. Physiol.* **5**, 338–349.

Piiper, J., and Scheid, P. (1972). Maximum gas transfer efficacy of models for fish gills, avian lungs and mammalian lungs. *Respir. Physiol.* **14**, 115–124.

Piiper, J., and Scheid, P. (1975). Gas transport efficacy of gills, lungs and skin: Theory and experimental data. *Respir. Physiol.* **23**, 209–221.

Piiper, J., and Scheid, P. (1977). Comparative physiology of respiration: Functional analysis of gas exchange organs in vertebrates. *Int. Rev. Physiol.* **14**, 219–253.

Piiper, J., and Scheid, P., eds. (1981). "Gas Exchange Function of Normal and Diseased Lungs," Prog. Respir. Res., Vol. 16. Karger, Basel.

Piiper, J., and Scheid, P. (1983). Physical principles of respiratory gas exchange in fish gills. *In* "Gills" (D. F. Houlihan, J. C. Rankin, and T. J. Shuttleworth, eds.), pp. 45–62. Cambridge Univ. Press, London and New York.

Piiper, J., and Schumann, D. (1967). Efficiency of $O_2$ exchange in the gills of the dogfish, Scyliorhinus stellaris. *Respir. Physiol.* **2**, 135–148.

Piiper, J., Meyer, M., Worth, H., and Willmer, H. (1977). Respiration and circulation during swimming activity in the dogfish *Scyliorhinus stellaris*. *Respir. Physiol.* **30**, 221–239.

Rahn, H. (1966). Aquatic gas exchange: Theory. *Respir. Physiol.* **1**, 1–12.

Rahn, H., and Farhi, L. E. (1964). Ventilation perfusion and gas exchange—the $\dot{V}_A/\dot{Q}$ concept. *In* "Handbook of Physiology" (W. O. Fenn and H. Rahn, eds.), Sect. 3, Vol. I, pp. 735–766. Am. Physiol. Soc., Washington, D.C.

Randall, D. J. (1970). Gas exchange in fish. *In* "Fish Physiology" (W. S. Hoar and D. J. Randall, eds.), Vol. 4, pp. 253–292. Academic Press, New York.

Randall, D. J., Holeton, G. F., and Stevens, E. D. (1967). The exchange of oxygen and carbon dioxide across the gills of rainbow trout. *J. Exp. Biol.* **46**, 339–348.

Riggs, A. (1970). Properties of fish hemoglobins. *In* "Fish Physiology" (W. S. Hoar and D. J. Randall, eds.), Vol. 4, pp. 209–252. Academic Press, New York.

Scheid, P. (1979). Mechanisms of gas exchange in bird lungs. *Rev. Physiol., Biochem. Pharmacol.* **86**, 137–186.

Scheid, P., and Piiper, J. (1970). Analysis of gas exchange in the avian lung: Theory and experiments in the domestic fowl. *Respir. Physiol.* **9**, 246–262.

Scheid, P., and Piiper, J. (1971). Theoretical analysis of respiratory gas equilibration in water passing through fish gills. *Respir. Physiol.* **13**, 305–318.

Scheid, P., and Piiper, J. (1976). Quantitative functional analysis of branchial gas transfer: Theory and application to *Scyliorhinus stellaris* (Elasmobranchii). *In* "Respiration of Amphibious Vertebrates" (G. M. Hughes, ed.), pp. 17–38. Academic Press, New York.

Shelton, G., and Boutilier, R. G. (1982). Apnoea in amphibians and reptiles. *J. Exp. Biol.* **100**, 245–273.

Steen, J. B., and Kruysse, A. (1964). The respiratory function of teleostean gills. *Comp. Biochem. Physiol.* **12**, 127–142.

Taylor, E. W., and Butler, P. J. (1971). Some observations on the relationship between heart beat and respiratory movements in the dogfish (*Scyliorhinus canicula* L.). *Comp. Biochem. Physiol. A* **39A**, 297–305.

Vogel, W. O. P. (1978). Arteriovenous anastomoses in the afferent region of trout gill filaments (*Salmo gairdneri* Richardson, Teleostei). *Zoomorphologie* **90,** 205–212.

West, J. B., ed. (1977a). "Bioengineering Aspects of the Lung." Dekker, New York.

West, J. B. (1977b). "Ventilation/Blood Flow and Gas Exchange," 3rd ed. Blackwell, Oxford.

West, J. B., ed. (1980). "Pulmonary Gas Exchange," Vols. 1 and 2. Academic Press, New York.

Wood, S. C., and Lenfant, C. (1979). Oxygen transport and oxygen delivery. *In* "Evolution of Respiratory Processes: A Comparative Approach" (S. C. Wood and C. Lenfant, eds.), pp. 193–223. Dekker, New York.

# 5

# OXYGEN AND CARBON DIOXIDE TRANSFER ACROSS FISH GILLS

*DAVID RANDALL*

Department of Zoology
University of British Columbia
Vancouver, British Columbia, Canada

*CHARLES DAXBOECK*

Pacific Gamefish Foundation
Kailua-Kona, Hawaii

## I. INTRODUCTION

The gills of fish are the major site, though not the only one, for oxygen and carbon dioxide transfer. The skin and fins may also serve in this capacity, and many fish have evolved accessory air-breathing organs. These may be modifications of the skin, buccal, pharyangeal, or gill surface, or they may be regions of the gut or the swim bladder (Randall *et al.*, 1981). The gills are designed for the transfer of oxygen and carbon dioxide between water and blood. The blood then transports these gases to and from the tissues. The body surface in fish, however, is usually supplied with oxygen directly from

FISH PHYSIOLOGY, VOL. XA

the water. The gill epithelium and the buccal and opercular cavities may consume from 6 (Johansen and Pettersson, 1981) to 27% (Daxboeck et al., 1982) of the resting oxygen uptake of the fish, and the skin surface a further 13–35% (Kirsch and Nonnotte, 1977). Kirsch and Nonnotte (1977) concluded that the skin of fish does not play a role in oxygen transfer between water and blood, because all skin oxygen uptake can be accounted for by skin consumption. Smith et al. (1983) concluded that the swim bladder gas was probably the source of oxygen during apnea in eels, and they agreed with Kirsch and Nonnotte (1977) that cutaneous oxygen uptake supplies only skin tissue, even during these periods of apnea. The importance of the skin in gas transfer in species such as lungfish, although often considered to be significant, has yet to be demonstrated quantitatively. Thus, in resting fish, only 60–80% of the oxygen leaving the water may cross the gill epithelium and enter the blood, the remainder being utilized by the gill tissue and skin. During exercise, when there is a large increase in oxygen consumption, 90% of which is utilized by the working muscles (Randall and Daxboeck, 1982), the proportion of the total oxygen uptake utilized by the skin and gill tissue is clearly much less.

In general, the gills of fish are the major pathway for oxygen and carbon dioxide transfer between the environment and the body tissues. The oxygen stores within the body, with the exception of that in the swim bladder, are small. Assuming a fish could utilize all the store and no oxygen was available from the swim bladder, then the oxygen store will last no longer than about 3 to 5 min. Thus, fish must breathe continuously to take up the oxygen to supply the metabolic needs of the animal. Conversely, the oxygen uptake of the fish is generally a good indication of the level of aerobic metabolism at that time. An exception is when the swim bladder is utilized as an oxygen source during periods of reduced oxygen availability or during periods of apnea. The swim bladder, if full of oxygen, could supply the requirements of the fish for up to 2 hr. Generally, however, fish breathe continuously, supplying oxygen at approximately the same rate as it is utilized.

Carbon dioxide stores ($[CO_2] + [HCO_3^-] + [CO_3]$) in the body are large compared with the rate of production. At resting rates of $CO_2$ production, it would take the animal several hours to accumulate the equivalent of the body $CO_2$ stores. Thus, minor changes in the magnitude of the $CO_2$ stores, for example related to acidification of the body tissues, can have a marked effect on $CO_2$ excretion across the gills. Thus, because of the large and variable $CO_2$ stores in the body, the respiratory exchange ratio (RE = $CO_2$ excretion/oxygen uptake) need not and often does not reflect the respiratory quotient (RQ = $CO_2$ production by tissues/$O_2$ utilization by tissues).

The gills are relatively impermeable to ions, but because of their large area, there is a measurable ion flux across the respiratory epithelium. Carbon dioxide may be excreted as $HCO_3^-$ and $H^+$ with $Na^+$ and $Cl^-$ as the counterions. Sodium chloride fluxes, however, are small compared with $CO_2$ excretion rates. For example, $Na^+$ influx across the gills of rainbow trout (Wood and Randall, 1973) is only approximately 7% of the $CO_2$ excretion of the resting fish in fresh water and is a much smaller percentage in the active fish because of a sharp rise in $CO_2$ excretion with little change in $Na^+$ influx.

The gills are also involved in the transfer of ammonia between blood and water. The gills have the capacity to produce ammonia by deamination of adenylates and transamination of glutamine (Cowey and Sargent, 1979), but most of the ammonia is produced in the liver and transported in the blood to the gills. Most of the ammonia will diffuse across the gills as ammonia gas, but there is some evidence for $NH_4^+$ excretion coupled to $Na^+$ influx (see Chapter 8). The rate of ammonia excretion is high following feeding and low during starvation in fish. It can be calculated from the data of Brett and Zala (1975) for sockeye salmon, that the mean ratio of ammonia excretion to oxygen uptake is about 0.12. This ratio increases to 0.22 just after feeding, approaching a value of 0.27 expected for the oxidation of amino acids. The ratio falls to 0.08 during starvation, reflecting very low rates of ammonia excretion.

The respiratory surface area of fish increases with weight and the general level of activity of the species. Most fish have a smaller respiratory surface area than mammals, the exception being tuna (Table I). The respiratory area

**Table I**

Comparison of Respiratory Parameters in Humans and Several Species of Fish

| | | Fish | | |
|---|---|---|---|---|
| | Human (37°C) | Tuna *Thunnus albacares* | Trout *Salmo gairdneri* (10–15°C) | Catfish *Ictaurus nebulosus* |
| Body weight, $W$ (g) | 55,000 | 1450 | 200 | 50 |
| Respiratory surface area, $A$ (m²) | 63 | 13.4 | 0.2 | 0.008 |
| A/W (cm²/g) | 11.5 | 9.21 | 2.97 | 1.58 |
| Oxygen uptake, $\dot{M}_{O_2}$ (ml g⁻¹ hr⁻¹) | 0.23 | 0.48[a] | 0.04 | 0.046 (20°C) 0.014 (10°C) |
| $A/\dot{M}_{O_2}$ | 50 | 19 | 74 | 34 (20°C) 113 (10°C) |
| Diffusion distance medium–blood (μm) | 1 | 0.38–0.85 | 5 | 10 |

[a]From Stevens (1972); other data from Altman and Dittmer (1971).

per unit oxygen uptake is similar in fish and mammals, indicating that respiratory area is linked to the level of oxygen uptake by the animal.

## II. GILL VENTILATION ($\dot{V}_g$)

The gas exchange unit of the gills, the lamellae, are flattened, trapezoidal leaflets extending from both sides of the gill filament. The total surface area of all lamellae is generally considered to represent the functional gas transfer surface of the gill, because the epithelium overlaying other regions is too thick to permit much gas transfer between water and blood.

The lamellae are about 17 to 18 times as high as they are thick (Hughes and Morgan, 1973) and twice as long as high. These platelike structures are in parallel rows and, in total, form a sieve placed in the water flow (see Chapter 1, this volume). The gaps between lamellae—that is, the water pores within the gill sieve—have a rectangular cross section and are much longer than wide, forming a slitlike hole. In a 300- to 400-g trout this water slit is about 25 μm wide, the lamellae has a maximum height of about 450 μm, and the slit is about 900 μm long. The lamellae are somewhat wedge-shaped and the highest portion is near the front of the slit, with the sides tapering away toward the water outflow. The size of lamellae is larger and the pores smaller in fast-swimming fish compared with more sluggish forms (Hughes, 1966). In air-breathing forms the gills are often very reduced, and the lamellae are small and more widely spaced (see Chapter 1, this volume).

Water flow over the gills is generated by contractions of both the buccal force and opercular suction pumps (Shelton, 1970). Water flows in through the mouth into the buccal cavity, over the gills into the opercular cavity, and then out via the opercular clefts. The buccal chamber expands, drawing water through the open mouth, and at the same time the operculum swings out, enlarging the opercular cavity, drawing water across the gills. During this period the opercular cleft remains closed. The buccal pump then contracts, the mouth closes, and water is forced over the gills; the operculum then swings in, and the volumes of the opercular cavities are reduced. The opercular cleft, however, is open so water can escape through the cleft. The pressure in the buccal cavity remains above that in the opercular cavity for most of each breathing cycle; there is, however, a short period of pressure (Shelton, 1970) and possibly flow (Holeton and Jones, 1975) reversal across the gills of some fish. The general sequencing of the ventilatory movements has been described in detail for teleost fishes (Hughes and Shelton, 1958; Hughes, 1960b; Saunders, 1961; Hughes and Umezawa, 1968; Burggren, 1978) and the elasmobranchs (Hughes, 1960a; Hughes and Ballintijn, 1965).

An important exception to these general schemes is the parasitic adult sea lamprey, whose gills are ventilated by tidal water movements through the external openings of each gill pouch (Randall, 1972). Fish moving forward in the water need only to open their mouths to ventilate their gills. This "ram" ventilation is seen in many pelagic fish, like tuna, and in many species buccal movements cease once the animal has gained speed (reviewed by Jones and Randall, 1978). In these animals flow across the gills is presumably steady. If the mouth is at the front of the fish, forward motion will contribute to gill water flow even before breathing movements cease. Thus, the importance of ram ventilation will increase with swimming speed even in the presence of rhythmic breathing. Swimming speed is not the sole determinant of ram ventilation, however, because mouth gap is adjusted to control water flow over the gills at any given swimming speed (see Jones and Randall, 1978, for references).

The cost of breathing is probably in the region of 10% of oxygen uptake independent of the exercise level of the fish (see Jones and Randall, 1978) and the mode of ventilation, that is, whether water flow is maintained by ram ventilation or rhythmic contractions of the buccal and opercular cavities. Ram ventilation, however, confers some advantages on the fish. Rhythmic breathing movements induce turbulence over the body of the fish, whereas a more streamlined flow is maintained over the body of the fish during ram ventilation (Freadman, 1981). This change results in some hydrodynamic advantage to the fish, and a small but measurable reduction in oxygen consumption occurs when the animal switches from rhythmic to ram ventilation at cruising swimming speeds (Freadman, 1981).

A body of water is in contact with the lamellar surface for only a short period of time, being around 100 to 300 msec in many fish (Randall, 1982a). Gill ventilation is increased in many fish during hypoxia (Saunders, 1962; Holeton and Randall, 1967; Lomholt and Johansen, 1979; Steffensen et al., 1982), hypercapnia (Janssen and Randall, 1975; Randall et al., 1976), exercise (Jones and Randall, 1978; Freadman, 1979, 1981), and with an increase in temperature (Randall and Cameron, 1973). Any increase in gill ventilation will reduce the residence time for water at the gill surface. In resting, quiet conditions some fish show periodic breathing with apneic periods of many minutes duration (Smith et al., 1983).

Water flow through the mouth and over the gills is laminar. The Reynolds number for flow through the mouth is around 100 to 300, calculated from the equation $N_R = 2a\bar{U}\rho/\eta$, where $N_R$ = Reynolds number, $\bar{U}$ is mean water velocity in cm sec$^{-1}$, $\rho$ is density in g cm$^{-3}$, and $\eta$ is viscosity in poise (dynes sec cm$^{-2}$). The mouth is considered as a tube of radius $a$; for calculations of Reynolds numbers for flow between lamellae the "$a$" is replaced by "*l*," the length of individual lamellae. $N_R$ for flow between lamellae oscillates around unity.

Holeton and Jones (1975) measured changes in water velocity in the buccal cavity of carp and observed that the oscillations in velocity were reduced close to the gills compared with that at the mouth. Flow over the lamellae, however, is probably still pulsatile. Holeton and Jones also observed a considerable time lag between peak differential pressure and flow velocity (Fig. 1), concluding that gill resistance cannot be determined simply

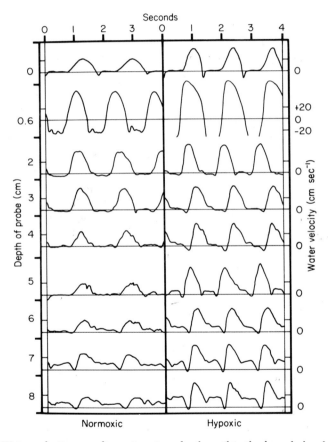

**Fig. 1.** Water velocity recordings at various depths within the buccal chamber of a carp taken during normoxia and mild hypoxia. The recordings are not simultaneous but have been selected so that they correspond with each other temporally. Positive deflections indicate water movement in a posterior direction. Velocity calibration is the same in all cases. The zero-depth recordings were taken immediately outside the mouth. Note the conspicuous backflow of water at the 0.6-cm depth, which is caused by the velocity probe interfering with closure of the buccal valves. (From Holeton and Jones, 1975.)

from measurements of mean differential pressure and mean volume flow because inertial effects cannot be neglected.

The epithelial surface of the lamellae is covered with small ridges of less than a micron in height. These ridges, however, are covered by mucus secreted by cells on the leading edge of the lamellae and gill rakers, so the water–gill interface is smooth. The mucus forms a thin, stable layer next to the gill epithelium. Diffusion coefficients for $Na^+$ and $Cl^-$ are the same for mucus and saline solutions (Marshall, 1978), and oxygen diffuses only a little more slowly through mucus than through water (Ultsch and Gros, 1979). Thus, the mucus can be considered as equivalent to a layer of water next to the epithelium but with a higher viscosity. A thick layer of mucus can build up on the gills under some conditions, for example if fish are exposed to acidic conditions. Boundary layers of water will exist adjacent to the lamellar surface, and it has been estimated that these boundary layers may constitute 25–50% of the total gill diffusion resistance to oxygen (Hills, 1972; Piiper and Scheid, 1977; Scheid and Piiper, 1976).

These studies, however, assume steady-state conditions, which probably only exist in ram-ventilating fish, swimming at a constant speed. A thick mucus layer will enlarge the boundary layer and increase both the diffusion resistance to oxygen transfer and the resistance to water flow through the gills (see Ultsch and Gros, 1979, for discussion), both factors impairing gas exchange. The flow oscillations with each breathing cycle will disturb these boundary layers and promote back-mixing. As flow velocity changes, the boundary layer will be further exacerbated by any flutter of the lamellae caused by blood or water flow. Thus, the magnitude of the boundary layer is difficult to determine; however, it can be no larger than half the width of the gap between lamellae—that is, around 10 to 12 $\mu$m—and is probably much less than this value. Compartmental analysis of the importance of various components of the blood–water barrier to diffusion, however, is incomplete. For instance, the role of the diffusing capacity of erythrocytes has never been considered in fish, because estimates of either the rate of $O_2$ uptake by blood or the capillary blood volume in the gills are not available. Thus, there are no estimates of the relative importance of the gill epithelial diffusing capacity, the plasma diffusing capacity, and the erythrocyte diffusing capacity along the lines already described for mammals by Weibel (1973).

Oxygen diffuses more slowly through tissues than through water, the ratio of the Krogh permeation coefficients being somewhere between 2 and 10 depending on the tissue in question. The Krogh permeation coefficient for fish gills has not been measured, but the value for eel skin of $4.34 \times 10^{-6}$ nmol sec cm$^{-1}$ mm Hg$^{-1}$ (Kirsch and Nonnotte, 1977) is an order of magnitude smaller than that for water at 13 to 14°C of $4 \times 10^{-5}$ nmol sec$^{-1}$ cm$^{-1}$ mm Hg$^{-1}$ (extrapolated from data in Grote, 1967). The eel skin is not specialized for gas transfer; if one compares Krogh permeation coefficients for lung tissue and water the ratio is around 2, oxygen diffusing more rapidly through water (Grote, 1967).

The gill epithelium is a complex tissue and consists of several cell layers with tight junctions between epithelial cells. If we assume the Krogh oxygen permeation coefficient for gill tissue is somewhere between that for the lung and that for eel skin, but approaches that observed for the lung, then a value of 3 seems reasonable for the ratio of permeation coefficients for oxygen in water compared with gill tissue. As oxygen diffuses three times more slowly through gill tissue as water, the boundary layer must be three times as thick as the epithelium if the same $P_{O_2}$ gradient is to exist across each barrier. The gill epithelium is 5 μm thick in trout, so the water boundary layer would have to be 15 μm. This is clearly impossible, as the slit is only 25 μm wide and would have to be greater than 30 μm wide to accommodate two 15-μm boundary layers on each lamellar surface. If the boundary layer is the same thickness as the gill epithelium, then the $O_2$ gradient across the gill epithelium will be three times that across the boundary layer of water, whereas if the boundary layer is 3 μm the $P_{O_2}$ drop across the gill epithelium will be five times that across the water boundary layer. Using this latter example, if we assume that the mean $P_{O_2}$ difference between water and blood across the gill represents the $O_2$ drop across the water boundary layer and the "in series" gill epithelium, then as this is about 45 mm Hg in the trout (see Jones and Randall, 1978), the $P_{O_2}$ drop across the boundary layer will be 7 mm Hg and that across the gill epithelium, 38 mm Hg. This mean $P_{O_2}$ difference between blood and water represents the maximum possible $P_{O_2}$ gradient across the boundary layer and gill epithelium. The drop in $P_{O_2}$ across these two diffusion resistances will in fact be less, because the mean difference between blood and water also reflects ventilation:perfusion inequalities, oxygen gradients in blood, as well as oxygen consumption by gill tissue. Thus, the $P_{O_2}$ gradient across the boundary layer is unlikely to be more than a few millimeters of Hg, that is, $P_{O_2}$ gradients in the lamellar water at right angles to the flow are probably negligible, and the gill epithelium represents the major resistance to oxygen transfer.

The proportion of the total available oxygen removed from water passing over the gills (percentage utilization) varies between species, being 30–45% in trout (Jones and Randall, 1978), 71% in tuna (Stevens, 1972), and 70–80% in carp (Lomholt and Johansen, 1979). There is little change in oxygen utilization with exercise in trout (Kiceniuk and Jones, 1977) or with hypoxia in carp (Lomholt and Johansen, 1979), even though flow increases, reducing both the resistance time for water at the gills and the magnitude of the boundary layer.

The excretion of $CO_2$ into water as it passes over the gills results in only small changes in $P_{CO_2}$ in water, usually less than a few millimeters of Hg. This is because not only is $CO_2$ more soluble in water than oxygen, it also reacts with water to form carbonic acid, which dissociates into bicarbonate and $H^+$ ions. The result is that $P_{O_2}$ may change by as much as 100 mm Hg, whereas $P_{CO_2}$ changes by only a few millimeters of Hg as water passes over the gills. Blood $P_{CO_2}$ levels are only a few millimeters of Hg above ambient, because $CO_2$ permeation coefficients across fish gills are high and only small $P_{CO_2}$ differences exist across the epithelium between water and blood. Thus, blood $P_{CO_2}$ values are usually of the order of 2 to 3 mm Hg in fish. Blood $P_{CO_2}$ levels are increased if water $P_{CO_2}$ levels are raised (Janssen and Randall, 1975) or if gill diffusing capacity is small, as in many air-breathing fish (Randall et al., 1981). Ventilation of the gills affects blood $CO_2$ levels; for example, when gill water flow increases during hypoxia in trout, blood $P_{CO_2}$ levels fall and the fish exhibits a respiratory alkalosis (Eddy, 1974; Thomas and Hughes, 1982). If gill water flow is reduced, as for example during hyperoxia, blood $P_{CO_2}$ levels increase and the fish suffers a respiratory acidosis (Randall and Jones, 1973; Dejours, 1973; Wood and Jackson, 1980).

The effects of $CO_2$ excretion on water pH are complex and poorly described. The addition of $CO_2$ to water will reduce pH, the extent of the reduction depending on the buffering capacity of the water. The uncatalyzed $CO_2$ hydration reaction is slow, of the order of several seconds or minutes. Water is resident at the gills for only several hundred milliseconds; thus, if the reaction occurs at the uncatalyzed rate the excretion of $CO_2$ will not cause any significant pH change while the water is in contact with the epithelium, but will occur downstream of the gills. Thus, bicarbonate formation in water will not affect molecular $CO_2$ gradients across the epithelium, because bicarbonate is formed downstream of the gills.

This assumes, however, that the $CO_2$ hydration in water at the gills is uncatalyzed. There is the possibility, however, that the gill surface contains carbonic anhydrase activity, perhaps excreted within the mucus. The gill epithelium and erythrocytes contain high levels of the enzyme carbonic anhydrase (Haswell et al., 1980). Dimberg et al. (1981) showed that there was more carbonic anhydrase in epithelial cells of seawater-exposed than freshwater salmon smolt gills. If they incubated the gills for only a short time the stain appeared in only superficial regions of the epithelial cells, and Dimberg et al. (1981) concluded that apical regions of the epithelial cells contained the highest concentrations of the enzyme. Lacy (1982) localized carbonic anhydrase in the cytoplasm of mucous, vesicular, and chloride cells but not in pavement cells of the opercular epithelium of Fundulus. There was some carbonic anhydrase in the extracellular spaces between vesicular and supportive cells. Thus, it seems possible that the apical surface of the epithelial cells and the mucus could contain high levels of carbonic anhydrase activity; however, although the mucous cells contain high levels of the enzyme (Lacy, 1982), it is cytoplasmic and the enzyme is absent from the mucous glands, indicating little or no activity in the mucus. Thus, it is still uncertain whether carbonic anhydrase is available to catalyze $CO_2$ hydration reactions in water, much as the endothelial carbonic anhydrase catalyzes $HCO_3^-$ dehydration in plasma in the mammalian lung.

The pH changes in water flowing over the gills are further complicated by ammonia and proton movements. The gill epithelium is permeable to protons (McWilliams and Potts, 1978), and these will be transferred across the gills depending on the size and direction of the proton gradient (Van den Thillart *et al.*, 1983). Ammonia is excreted both as ammonia gas and as ammonium ions. Ammonia excreted into the water will bind a proton to form ammonium ions and will raise pH; depending on the pH, ammonium ions excreted into the water may release a proton and lower pH. These reactions are rapid and will occur while the water is in contact with the gill epithelium. Cameron and Heisler (1983) concluded that, in the rainbow trout, most ammonia is "simply cleared . . . as a respiratory gas." They also concluded that, if external ammonia was raised, then active extrusion mechanisms, presumably $Na^+/NH_4^+$ exchange, were activated. Thus, the pH of water flowing over the gills will be affected by ammonia and proton excretion, the change in pH being determined by the buffering capacity of the water and the magnitude of the ammonia and proton fluxes. As water flows away from the gills, the pH will continue to change as the uncatalyzed $CO_2$ hydration reaction proceeds to equilibrium.

Thus, considering pH changes in water as it flows over the gills of fish in water of neutral pH, one would expect an initial increase due to $NH_3$ excretion followed by a fall in pH as water flows away from the gills as a result of $CO_2$ hydration. If the fish was acidotic one might predict an initial fall in pH if proton excretion exceeded $NH_3$ excretion. The extent of the pH changes will depend on the excretion rates in proportion to the buffering capacity of the water.

## III. GILL BLOOD FLOW

### A. Introduction

Blood flows through the lamellae in a countercurrent arrangement to the water flow. The ventilation:perfusion ratio in fish is between 10 and 20. In resting fish, for example the eel (Smith *et al.*, 1983), breathing may stop for several minutes, and during this period of apnea the ventilation:perfusion

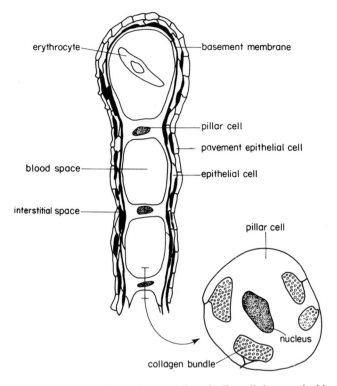

**Fig. 2.** Sections through a teleost secondary lamella and pillar cell showing the blood–water barrier. The drawing is approximately to scale for a trout, *Salmo gairdneri*. (From Randall, 1982b.)

ratio will be zero. The cardiac output is directed through the gills and then into the body circulation, and, unlike mammals, the fish respiratory circulation is subjected to much higher pressures than the systemic circuit (Johansen, 1972). Pressures within the secondary lamellae oscillate with each heartbeat, having a mean pressure of about 20 to 40 mm Hg, with an oscillation of about 4 mm Hg (Randall, 1982a).

Blood flows through the lamellae around pillar cells that hold the epithelial layers together. The pillar cells have collagen columns embedded in them, and these collagen strands extend around the blood space over the pillar cell flanges, wrapping the blood space (Fig. 2). The pillar cells also contain actin and myosin threads; however, these are unlikely to be able to generate enough force to shorten the pillar cells but more likely may play some role in organizing the collagen along lines of stress to maximize their effectiveness in reducing expansion of the lamellar blood sheet (see Booth, 1979b). The thickness of the blood sheet is about 9 to 12 μm and is very

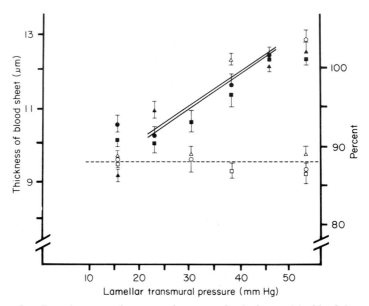

**Fig. 3.** The effect of transmural pressure changes on the thickness of the blood sheet (closed symbols) and the vascular space:tissue ratio (open symbols) in the secondary lamellae of the gill of the lingcod (*Ophiodon elongatus*) (from Farrell *et al.*, 1980). The vascular space:tissue ratio is the area occupied by blood expressed as a percentage of the total area observed in a side view of secondary lamellae. No change in this ratio indicates no change in the height and length of secondary lamellae.

dependent on blood pressure. The lamellae show no increase in length or height as blood pressure rises; that is, the vascular space:tissue ratio remains constant (Fig. 3), but the width of the blood sheet increases with pressure (Farrell *et al.*, 1980). In this respect blood flow through the lamellae and the lung alveoli has similar characteristics. Pressures in the lamellae oscillate by about 4 mm Hg with each heartbeat; this means that, assuming no difference in the dynamic and static responses to blood pressure, the blood sheet thickness will increase by about 0.5 μm and the lamellar blood volume by about 5% with each systole. One possible role of the myosin filaments found within pillar cells may be to contract in phase with the pressure pulse to reduce changes in lamellar volume and the thickness of the blood sheet with each contraction of the heart (Smith and Chamley-Campbell, 1981).

The thickness of the lamellar blood sheet will be related to transmural pressure, that is, the difference between blood and gill water pressure. Water pressure oscillates with each breath, and blood pressure with each heartbeat. In resting fish buccal pressure oscillations are small, and there is no coordination between breathing and heart rate. During hypoxia, however, when buccal and blood pressure oscillations increase in magnitude, synchrony occurs between heart and breathing rates, thus reducing the oscillations in lamellar transmural pressure (Randall, 1982a). Gill water pressures in tuna may be elevated to reduce transmural pressures across the gill epithelium in the face of high arterial pressures. Tuna have a membrane with holes punched through it stretching across the opercular cavity, restricting the exhalant water flow (B. J. Gannon, personal communication). The function of this membrane is unknown, but one possibility is to elevate gill water pressure during ram ventilation, to thin the gill blood sheet. Tuna have a very thin gill epithelium (Hughes and Morgan, 1973) and may generate a high blood pressure to maintain high blood flow in systemic capillaries. The elevated gill water pressure will reduce the gill epithelial transmural pressure and reduce the requirements for a stiff, thick respiratory epithelium. At the same time the tuna can maintain a high input pressure to the systemic circuit. Unfortunately the transmural pressure across the gill epithelium and the lamellar compliance has not been measured in tuna.

Only 60% of all secondary lamellae are perfused at rest (Booth, 1978), there being preferential perfusion of more basal lamellae. Increases in blood pressure cause lamellar recruitment, and this probably occurs during hypoxia and exercise (Randall, 1982b). Total functional lamellar volume exceeds cardiac output, so more than one heartbeat is required to move blood through lamellae (Jones and Randall, 1978). The lamellar blood transit time is about 3 sec at rest and about 1 sec during exercise (Randall, 1982b; Hughes *et al.*, 1981). The time for blood oxygenation within the gills was shown to be approximately 1 sec in carp and eel (Hughes and Koyama, 1974), that is, within the transit time of blood through the secondary lamellae. Hills *et al.* (1982) concluded that a diffusional rather than chemical reaction resistance was the main barrier to oxygen transfer in the gills.

## B. Blood Flow and Pressure

### 1. GENERAL CONSIDERATIONS

Within the branchial circulation, blood flow is presumed to be laminar, having a characteristic parabolic velocity profile across any vessel. For blood flow to be classified as laminar or "disturbed" laminar in smooth vessels, the value of the empirically derived Reynold's number, $N_R$, must be no greater than 1000 (Attinger, 1968). This value is derived from the relationship:

$$N_R = 2\dot{Q}/\pi r K \tag{1}$$

where $\dot{Q}$ is the flow rate (ml sec$^{-1}$), $r$ is the inside radius of the vessel (cm), and $K$ is the kinematic viscosity of the blood. The value for $K$ is an expression that takes into account the viscosity and density of the fluid, where

$$K = \rho/\eta \tag{2}$$

when $\eta$ is the viscosity (dyne sec cm$^{-2}$) and $\rho$ is the density. It is obvious from Eq. (1) that the larger the viscosity, the lower the incidence of turbulent flow. The presence of red blood cells in the blood—and hence of an increase in viscosity with increasing hematocrit—therefore, is assurance that flows will nearly always remain laminar but slightly flattened in profile. Blood velocities in fish seldom are high enough to create turbulence in any case.

The pumping action of the heart generates pressures that are dissipated when the blood is forced to flow through vessels, and these pressures decrease as the blood passes through the gills. The relationship between pressure and laminar flow in rigid tubes can be described by Poiseuille's law, where the flow rate

$$\dot{Q} = \Delta P \Pi r^4/8L\eta_r \tag{3}$$

and where $\Delta P$(dyne cm$^{-2}$ and 1 cm H$_2$O $= 980.64$ dyne cm$^{-2}$) is the pressure drop along a tube of length $L$ (cm), $r$ is the internal radius of the tube (cm), and $\eta_r$ is the effective viscosity of the blood within that tube. It can be seen from Eq. (3) that a very small change in the radius of a tube can have profound effects on the flow velocity. This effective viscosity can be derived from

$$\eta_r = \eta_\infty/(1 + d/r^2) \tag{4}$$

where $\eta_\infty$ is the viscosity of the blood in a tube of infinite radius, $d$ is the diameter of the red blood cell, and $r$ is the internal radius of the tube.

It must be pointed out that the Poiseuillian equation is applicable only to Newtonian fluids flowing steadily in straight, rigid tubes. Within normally encountered physiological limitations, blood does appear to behave as a Newtonian fluid (see Cokelet, 1980), despite the obvious two-phase nature of this viscoelastic fluid. However, blood flow and pressure are pulsatile, and the blood vessels are neither straight nor rigid. Therefore, oscillations in pressure and flow are not necessarily in phase and therefore cannot be described adequately by the Poiseuillian relationship. In an attempt to describe the degree of asynchrony between the pressure and flow pulse, and hence the validity of values derived from Poiseuille's equation, a nondimensional constant, $\alpha$, is used:

$$\alpha = r\frac{\sqrt{2\Pi nf\rho}}{\eta_r} \tag{5}$$

$$= \sqrt{\omega\eta_r/\rho}$$

where $r$ is the vessel radius, $n$ is the order of the harmonic component, $\rho$ and $\eta_r$ are the density and viscosity of the blood in that vessel, $f$ is the frequency of pulsations, and $\omega$ is the angular velocity, $2\pi f$. If this value lies below 0.8, then the Poiseuillian equation adequately describes the relationship between flow and pressure (McDonald, 1960). The pressure pulse is a wave complex of several harmonics, the higher frequencies traveling at higher velocities within the vessels. However, the distances traveled by these waves are too small to allow their summation, and therefore only the first harmonic generally is considered. Since the value for $\alpha$ within the gill and associated vasculature satisfies the preceding criterion, Poiseuille's equation is valid for the description of pressure–flow relationships in fish gills. The degree of pressure–flow asynchrony in a tube also can be described by the impedance phase:

$$\Phi = \text{-arctangent}(2\Pi f T) \tag{6}$$

where $T = t/\ln(P_0/P_t)$ and $t$ is the duration of the diastolic pulse (sec), $P_0$ is the peak diastolic pressure (dyne $cm^{-2}$), $P_t$ is the end diastolic pressure (dyne $cm^{-2}$), and $f$ is the frequency of the heartbeat (Hz).

Fluid flowing in a tube, or as we now assume, any branchial blood vessel, will have some resistance to that flow. This resistance can be derived from a rearrangement of the Poiseuillian equation:

$$R = \Delta P/\dot{Q} = 8L\eta_r/\Pi r^4. \tag{7}$$

Poiseuillian resistance generally is expressed in peripheral resistance units (where 1 PRU = 1330 dyne sec $cm^{-5}$). The preceding expression is convenient, since the internal radii of branchial blood vessels are not readily measurable, especially in vivo under dynamic conditions. This relationship is equivalent to ohmic resistance in an electrical analogy, where

$$R(\text{ohms}) = \frac{\text{potential difference between two points, } V \text{ (volts)}}{\text{current flow, } I \text{ (amperes)}} \tag{8}$$

If, however, a secondary lamella is treated as two sheets of epithelia that are separated by pillar cells, then sheet flow analyses of the pressure–flow relationships have been shown to be applicable (Fung and Sobin, 1969; Glazier et al., 1969; Sobin et al., 1979; Scheid and Piiper, 1976; Farrell, 1979; Farrell et al., 1980). Therefore, the flow of blood through a lamellar sheet can be estimated by

$$Q_{\text{lam}} = 1/C(h_a^4 - h_v^4) \tag{9}$$

where $h_a$ and $h_v$ are the sheet thicknesses at the arterial and venous ends of the lamella, which are pressure dependent, so that

$$\bar{h} = h_0 + (\text{a constant} \times \text{lamellar pressure}) \qquad (10)$$

when $h_0$ is the approximate height of the pillar cell posts, and where the compliance of this sheet,

$$C_{\text{sheet}} = 4\eta_r k f \bar{L}^2 \alpha / SA \qquad (11)$$

In this equation, $k$ and $f$ are constants, $\bar{L}$ is the average length of a lamellar blood "channel," $\alpha$ is the compliance coefficient, $S$ is the vascular space:tissue ratio (VSTR; see Fung and Sobin, 1972), and $A$ is the lamellar area.

Since blood vessels are not straight, rigid tubes but contain varying amounts of elastic elements, they are able to distend and thereby change volume (radially) as flow and pressure oscillate with each beat of the heart. The ratio of the change in internal volume to a change in applied pressure is the compliance or capacitance of that vessel. The equation for compliance of the lamellar sheet already has been described in Eq. (11). However, the compliance of the other vessels can be described by

$$C_{\text{tube}} = 2r^3(1 - \sigma^2)L/Eh \qquad (12)$$

where $\sigma$ is the Poisson ratio. The value for this usually is taken to be between 0.4 and 0.48. When used it generally is given the value of 0.48, since there is only about an 8% difference between $C$ values calculated for the same vessel for $\sigma = 0.4$ and 0.48 (see Wainwright $et\ al.$, 1976). The Poisson ratio is a measure of the relative ability of any material to resist dilation (change in volume) and shearing (change in shape). $Eh$ is the elastance term (dyne $cm^{-1}$) and can be calculated from

$$Eh = (P_2 r_2) - (P_1 r_1)/[(r_2 - r_0)/r_0] - [(r_1 - r_2)/r_0] \qquad (13)$$

where $r_0$, $r_1$, and $r_2$ are the internal radii of the vessel when relaxed, and with applied pressures $P_1$ and $P_2$ (dyne $cm^{-2}$). The individual terms within the elastance are $E$ (Young's modulus, which is not constant but varies with many factors such as vessel tonus and elastin:collagen ratio) and $h$, the vessel wall thickness, which also can change with pulsatile pressure. Therefore, $Eh$ is not an easily calculable value, and so the capacitance of a tube can be estimated more readily from

$$C = 2\Pi r L(\Delta r / \Delta P) \qquad (14)$$

assuming that the change in vessel radius, $\Delta r$ (cm) of length $L$ (cm) can be measured for any change in pressure, $\Delta P$.

One final aspect of fluid flow in tubes that should be considered is the "start-up" resistance to flow in a vessel due to the volume of fluid contained within it. This is the inertance or inductance and can be calculated from

$$L = 9\rho Z / 4\Pi^2 r^2 \qquad (15)$$

where ρ is the density of the blood and we assume that the flow will be laminar in profile through radius r and length Z (Rideout and Katra, 1969).

Given the preceding relationships and the limited information on pressure and flow in the gill circulation, it may be possible to model the gill circulation in some way.

## 2. A GILL MODEL

There have been several attempts to analyze or construct electrically equivalent models to describe blood flow through the branchial circulation of fishes (Langille and Jones, 1976; Farrell et al., 1979; Farrell, 1980; Olson and Kent, 1980). In the following simplified model, we also have constructed an electric circuit analog that incorporates the elements of resistance, capacitance, and inductance properties of the branchial vasculature in a series-parallel network (Fig. 4). The model takes into account only the physical properties of the vessels with respect to passive properties, as affected by resting hemodynamics and flow. No consideration of the contribution of the venolymphatic system is included in the model, nor of any influences on the possible flow patterns by the intervention of neuronal or humoral agents.

The numbers associated with each parallel unit represent the number of individual components that comprise that unit in parallel. From the model it can be seen that each of these units is in series with both the previous and the successive parallel unit in a complete circuit. In order to resolve the values for the individual units and thus the whole circuit, the preceding equations for estimating R, L, and C (in fluid electric units) were used in conjunction with the following assumptions being made concerning vessel wall dynamics throughout the pulse cycle. All data were taken from extensive measurements of the branchial morphometrics and in vivo pressures and flows from a marine teleost, Ophiodon elongatus, made by Farrell (1979). Although the model is based solely on one species, the validity of the interpretations from it will vary only in a quantitative fashion and is thought to be representative of all fishes. The assumptions are as follows:

1. For af.BA, it was assumed that a maximum increase of 50% in the internal radius of this vessel could occur through the pulse pressure cycle.

2. For af.FA, using histological data from vessel cross sections and calculating the stress modulus for the given vessel dimensions, the conclusion was that these vessels would not change radius enough to allow the detection of any more than a 2% change, and therefore the static resistance measurement would apply over the pulse cycle (see Table II).

3. The lamellae were considered as single units and sheet flow analysis applied, rather than a system of an average of 20 parallel "Poiseuillian" tubes per lamella.

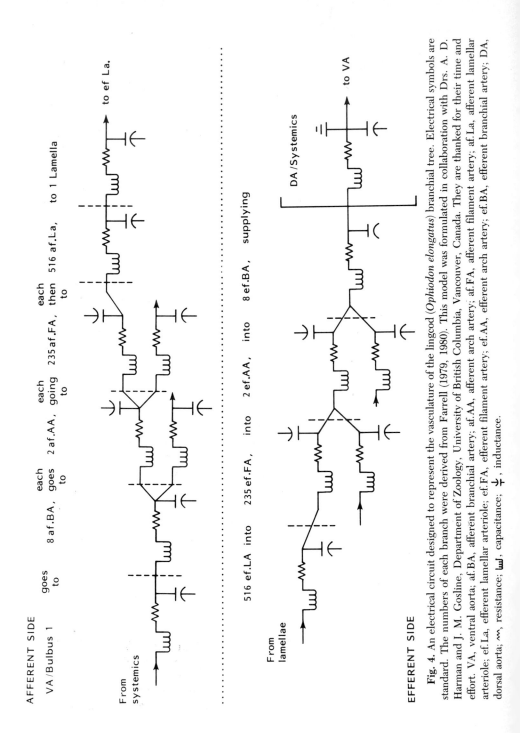

AFFERENT SIDE

VA/Bulbus 1    goes    8 af.BA,    each    2 af.AA,    each    235 af.FA,    each    516 af.La,    to 1 Lamella
                to               goes               going                then               to
                                  to                  to

From
systemics

. . . . . . . . . . . . . . . . . . . . . . . . . . . . . . . . . . . . . . . . . . . . . . . . . . . . . . . . . . . . . . . . . . . . . . . . . . . . . . . .

516 ef.LA    into    235 ef.FA,    into    2 ef.AA,    into    8 ef.BA,    supplying

From
lamellae

DA/Systemics

to VA

EFFERENT SIDE

**Fig. 4.** An electrical circuit designed to represent the vasculature of the lingcod (*Ophiodon elongatus*) branchial tree. Electrical symbols are standard. The numbers of each branch were derived from Farrell (1979, 1980). This model was formulated in collaboration with Drs. A. D. Harman and J. M. Gosline, Department of Zoology, University of British Columbia, Vancouver, Canada. They are thanked for their time and effort. VA, ventral aorta; af.BA, afferent branchial artery; af.AA, afferent arch artery; af.FA, afferent filament artery; af.La, afferent lamellar arteriole; ef.La, efferent lamellar arteriole; ef.FA, efferent filament artery; ef.AA, efferent arch artery; ef.BA, efferent branchial artery; DA, dorsal aorta; ∿∿, resistance; ⊔⊔, capacitance; ⊥, inductance.

**Table II**

Single Component Values Calculated for $L$, $R$, and $C$ (Fluid Units) in the Branchial
Vasculature of the Lingcod, *Ophiodon elongatus*[a]

| Vessel | Inertance ($L$, Henry) | Resistance ($R$, ohms) | Capacitance ($C$, ufarads) |
|---|---|---|---|
| VA/Bulbus | 6.11 (dynamic) | $2.39 \times 10^2$ (static) | 7.1 (dynamic) |
|  | 17.9 (static) | $2.85 \times 10^2$ (pulse) | 30 (pulse) |
|  |  | 41.9 (diastolic) |  |
|  |  | 21.8 (systolic) |  |
| af.BA | 27.7 (dynamic) | $2.58 \times 10^3$ (static) | 5.7 |
|  |  | $8.15 \times 10^3$ (diastolic) |  |
|  | 76.2 (static) | $2.02 \times 10^3$ (systolic) |  |
| af.AA | $1.46 \times 10^3$ | $6.36 \times 10^5$ (static) | $1.82 \times 10^{-1}$ (mean) |
| af.FA | $3.43 \times 10^3$ | $1.83 \times 10^7$ (static) | $1.22 \times 10^{-3}$ (mean) |
| af.La | $1.07 \times 10^4$ | $5.53 \times 10^9$ (static) | $3.81 \times 10^{-7}$ (mean) |
| lamella | $4.76 \times 10^3$ | $9.17 \times 10^8$ (Poiss.) | $5.21 \times 10^{-5}$ (mean) |
|  | (Poiseuillian) | $4.50 \times 10^8$ (sheet) |  |
| ef.La | $1.09 \times 10^3$ | $2.92 \times 10^8$ (static) | $2.09 \times 10^{-8}$ (mean) |
| ef.FA | $3.96 \times 10^1$ | $2.34 \times 10^7$ (static) | $7.81 \times 10^{-3}$ (mean) |
| ef.AA | $1.46 \times 10^3$ | $6.36 \times 10^5$ (static) | $1.82 \times 10^{-1}$ (mean) |
| ef.BA | $1.91 \times 10^2$ | $6.37 \times 10^3$ (static) |  |
|  |  | $1.09 \times 10^4$ (diastolic) | 5.7 |
|  |  | $3.97 \times 10^3$ (systolic) |  |
| DA/Systemics | $7.36 \times 10^2$ | $1.79 \times 10^5$ | Variable |

For VA/Bulbus capacitance: 7.1 (dynamic) / 30 (pulse) = 18.6

[a]Given: static = mean pressure measured in that vessel (or assumed) for calculations; mean = average value calculated for vessel through the pressure pulse; dynamic = with pressure pulse applied to vessel, over mean pressure in it; systolic = value calculated for radii of vessels at systolic pressure; diastolic = value calculated for radii of vessels at diastolic pressure; sheet = calculation using sheet flow assumptions. Poiseuillian (Poiss.) = calculations using Poiseuillian equations for fluid flow in cylindric tubes.

4. For ef.BA, we again assumed that the internal tube radius could increase by a maximum of 50% throughout the pulse cycle.

5. The dorsal aortic (DA) pressure was reduced to 39.6 cm $H_2O$ from a ventral aortic (VA) pressure of 52 cm $H_2O$ across the gill vasculature. The dorsal aortic pulse pressure was reduced to 6 cm $H_2O$ from an input pulse pressure of 12.4 cm $H_2O$. Therefore, assuming that the ef.BA is of the same construction as the af.BA, it can be assumed that the pulse wave could only change the internal radius by 25% through the pulse cycle, since this pulse has been reduced by 50% of the input VA pulse.

6. The DA values in Table II have been lumped into the systemic unit. The data from Farrell (1979) indicate that the systemic resistance in this fish is approximately 2.3 times the gill resistance.

7. It was assumed that the af.AA had no compliance inasmuch as they

possess enough tonus to resist deformation over the pulse pressure range. This is a reasonable assumption, since the maximum pulse pressure seen by these vessels at rest is 13 cm $H_2O$ ($1.22 \times 10^4$ dyne cm$^{-2}$), but the elastic modulus, $E$, of fully contracted smooth muscle is in the order of $10^6$ dyne cm$^{-2}$. This assumption applies equally well for the af.FA, af.La, ef.La, and ef.AA. Farrell (1979) measured the internal radius changes of these vessels at a pulse pressure of 50 cm $H_2O$ (af.FA and ef.FA) and found no change in $r$. Moreover, since the maximum pulse pressure measured at rest was approximately 13 cm $H_2O$, it is quite safe to assume no changes in compliance due to vessel dilation. Therefore, only the length of the vessels would contribute to compliance.

By using the foregoing assumptions and applying the previous equations to the morphological and physiological data from Farrell (1979, 1980), the following values for the electrically equivalent characteristics of the lingcod branchial vasculature were obtained (Table II). Reference to this table and its implications to the elucidation of control mechanisms for blood flow in fish gills will be dealt with in the following section.

## 3. Effects of Changes in Input–Output Dynamics on Blood Flow Distribution

The branchial vasculature represents 20–40% of the total vascular resistance to blood flow in fishes. As is shown in the model (Fig. 4), the ventral aorta divides into arch vessels. The bulbus, ventral aorta and these arch arteries are quite compliant and thus act to damp the large pressure oscillations resulting from the contraction of the heart. These afferent arch arteries supply the filament arteries, which then divide to supply blood to the lamellae. From Table II it can be seen that the afferent and efferent lamellar arterioles constitute the major sites of resistance to blood flow in the gills. We are assuming that, for the sake of simplicity, cardiac output will be representative of lamellar blood flow, because we are considering fishes that have no or few afferent filamental connections to the central sinus (i.e., bypass shunts). In addition, the input pressures to all lamellae are equal, since filamental arteries are large, and although tapering distally (see Fig. 5),

---

Fig. 5. The geometry of the afferent filament artery from vascular corrosion casts. (A) The diameters of the base of vessels versus their length ($n = 17$ filaments), to show that shorter filaments have narrower afferent filament arteries at their base. (B) The afferent filament artery tapers along its length in a nonlinear fashion. Over the proximal 20–30% of its length there is little or no taper. (C) The change in resistance of the afferent filament artery with respect to vessel length as a result of tapering. Average resistance values were calculated from $l/r^4$ and a base radius ($r$) of 1. The average taper was obtained from (A), and the line was fitted by eye. In a nontapering vessel the resistance would be constant with respect to length, that is, $l/r^4 = 1$. (From Farrell, 1980.)

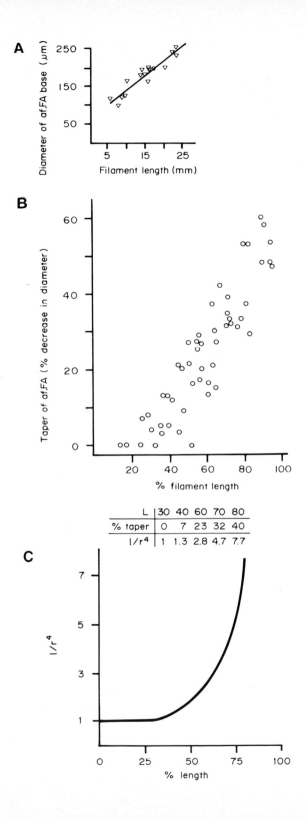

would have reduced flow toward the tip as blood passes into successive lamellae. The diameter of lamellar arterioles also decreases with distance along the filament (Farrell, 1980), and therefore the distal lamellae would have a higher resistance to flow than the more basal lamellae. Because of this anatomic arrangement and because there also appears to be more smooth muscle around more distal lamellar arterioles, basal lamellae would tend to be preferentially perfused. Once out of the lamellae, the blood is collected in the dorsal aorta, the main conduit for blood flow distribution to the rest of the systemic circulation. The dorsal aorta must be less compliant than the ventral aorta because if this were not so, there would be a rapid acceleration of blood through the gills with each beat of the heart when the DA was full of blood, resulting in an increase in the pulsatile flow of blood through the gills. The dorsal aorta lies, for the most part, tightly within the hemal arch of the vertebral column, and in many fishes also has a strong ligament inside it to ensure stiffness (De Kock and Symmons, 1959). This stiffness in the dorsal aorta also will lead to a closer synchrony between pressure and flow pulsatility, as has been demonstrated by a hydraulic model (Langille and Jones, 1976) and in vivo (Jones *et al.*, 1974).

The relationship between input and output pressure and flow and lamellar resistance to blood flow has been investigated using an isolated holobranch preparation from lingcod (*Ophiodon elongatus*) (Farrell *et al.*, 1979). In this preparation, using a constant pulsatile flow of saline, there was a tendency to perfuse only those lamellae on the basal two-thirds of each filament, and similar observations have been made for the trout in vivo (Booth, 1978). An increase in outflow pressure was not matched by an equal increase in inflow pressure, and thus the pressure drop and resistance to flow through the gills decreased with increasing transmural pressure (Fig. 6), demonstrating that the gill vessels are compliant. If the input pressure and flow were increased in these isolated lingcod holobranchs, the result was lamellar recruitment. Using a constant flow rate, a decrease in pulse pressure with an increased pulse frequency was associated with a decrease in the number of lamellae perfused, whereas an increase in pulse pressure with a decrease in pulse frequency caused an increase in the number of lamellae perfused. Farrell *et al.* (1979) concluded that, since the distal lamellar arterioles are narrower they had a higher critical opening pressure than more proximal arterioles, and therefore lamellar recruitment could be explained in terms of an increased input pressure opening previously closed lamellar arterioles. Once patent, these lamellae would remain so at lower pressures because the critical opening pressure undoubtedly exceeds the closing pressure by several torr. Thus, the pressure increase need not be maintained to cause lamellar recruitment. An increase in peak pulse pressure (even if mean pressure is unchanged) would be sufficient to cause lamellar recruitment.

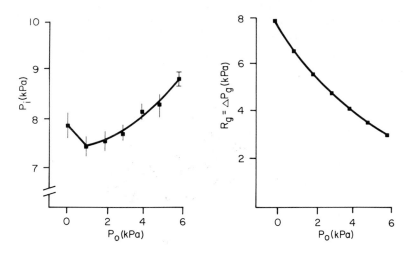

**Fig. 6.** The effect of increasing outflow pressure ($P_o$) on input pressure ($P_i$) and gill resistance ($P_g$) which at constant $\dot{Q}$ is equivalent to the resistance to flow in the gills. (From Farrell et al., 1979.)

The resistance changes, however, were related to regions other than the lamellae, because gill resistance could not be correlated to the number of lamellae perfused in any distinct manner (Fig. 7). We do not imply that the lamellar volume remains constant with pressure, only that any such changes either are balanced by the noted lamellar recruitment or are relatively unimportant in determining the gill resistance to flow.

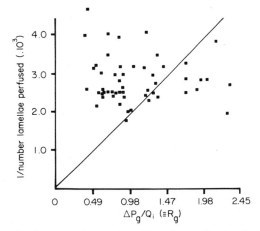

**Fig. 7.** The relationship between the number of lamellae perfused and gill resistance. The straight line indicates the expected relationship for a simple ohmic resistance with a variable number of resistors in parallel. (From Farrell et al., 1979.)

Changes in pressure also affect blood distribution within the secondary lamellae. It already has been pointed out that blood flow through the lamellae of fish can be described as sheet flow as in the lungs of human and dog. In lingcod, the vascular space:tissue ratio of the gill lamellae is not altered, but the thickness of the blood sheet is increased with the transmural pressure (Fig. 3), indicating little or no change in height or length of the lamellae. They may become more rigid and show therefore less tendency to bend in the water flow at higher pressures. Neither do pillar cell posts change length or diameter with increases in thickness of the vascular sheet; rather the flanges bulge outward, causing some thinning of these flanges, thus reducing the space between the pillar and epithelial cells (Fig. 8). Whether the dynamic responses are the same as the static responses of the gills to changes in pressure is not known at this time. Nonetheless, although lingcod gill has a thicker respiratory epithelium and a wider vascular sheet than mammals, and although less compliant than that of humans, this vascular sheet does show about the same compliance as observed in the blood sheet of dog lungs.

Since a portion of each lamella is buried in the body of the filament, the exposed regions would be more compliant when exposed to pressure changes in the physiological range. A pressure increase, therefore, will expand these central, exposed regions more than basal regions of the lamellae.

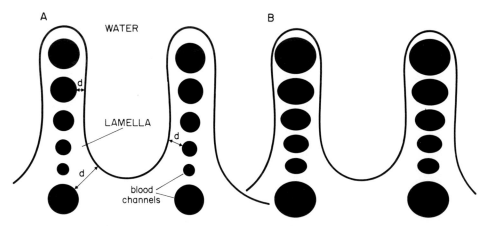

**Fig. 8.** A schematic diagram depicting lamellar channels as seen in a cross section through a lamella (see Fig. 2). Range of water–blood diffusion distances ($d$) is indicated. (A) A possible representation of a resting state. (B) Probable changes in vascular dimensions when $\Delta P_{lam}$ and flow are elevated. Note that in (B) diffusion distances have changed to accommodate increased vascular volume, and a greater proportion of increased volume is located in lamellar subregions with lower diffusion distances. This schematic diagram was based on data presented in Figs. 6 and 7. (From Farrell *et al.*, 1980.)

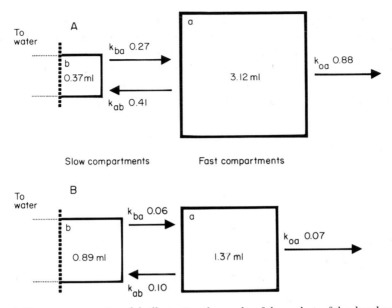

**Fig. 9.** Two-compartment models illustrating the results of the analysis of the dorsal aortic ethanol washout curves during (A) pulsatile constant flow perfusion and (B) nonpulsatile constant pressure perfusion. During pulsatile perfusion EtOH is lost more rapidly from the gills ($k_{oa}$), and exchanges between the fast and slow compartments ($k_{ab}$, $k_{ba}$) are greater. Also during pulsatile perfusion, the fast (vascular) compartment is relatively larger than the slow compartment, when compared with the compartment sizes during nonpulsatile perfusion. (From Davie and Daxboeck, 1982.)

As flow is proportional to $h^4$ [see Eq. (11)], this expansion results in a marked increase in flow through these central regions of the gill, whereas at lower physiological pressures, basal and marginal flow will predominate. Thus, a rise in transmural pressure, although increasing basal flow, also causes a greater increase in flow in regions of the lamellae exposed to water flow (Fig. 8). Since basal channels are separated from the environment by a larger diffusion barrier than marginal channels, the shift toward marginal flow will increase the gas-diffusing capacity of the gills.

Changes in pressure and flow clearly alter perfusion of the secondary lamellae. In addition, Davie and Daxboeck (1982) have shown that pulsatility also affects the permeability of the epithelial barrier in a way similar to that reported by Isaia *et al.* (1978a) and Haywood *et al.* (1977) for epinephrine. Their study showed that ethanol was cleared much more rapidly using pulsatile compared with nonpulsatile flow, and subsequent compartmental analysis (Fig. 9) indicated that this was consistent with a fourfold increase in the permeability of the basal (blood) barrier of the gill epithelium. Since gill permeability to ethanol and water is similar, it is probable that pulsatility also will increase epithelial water permeability and perhaps oxygen permeability. We already have pointed out that not all lamellae are perfused in resting fish. These unperfused lamellae will have a reduced pulse pressure compared with perfused lamellae, and be less permeable to water. Recruitment of lamellae will be associated, therefore, with an increase in water flux across the gills, which is known to occur in swimming fish (Wood and Randall, 1973; Jones and Randall, 1978).

The efferent and, to a lesser extent, the afferent filament arteries give off capillaries that carry blood into a central sinus, which in turn drains blood back to the heart (Vogel *et al.*, 1976; Dunel and Laurent, 1977; Vogel, 1978). The capillaries from the filament arteries to the central sinus are small, are relatively few in number, and often have long and tortuous pathways; therefore, they have a high resistance to blood flow. Thus, there is a considerable blood pressure drop from the filament arteries to the central sinus. Farrell (1979) was able to record pressures within the filament that may have been central sinus blood pressures, and found they were pulsatile and of the order of only a few millimeters of Hg. The pulsatility was probably transmitted to the central sinus through tissues rather than through the connecting blood capillaries, because the oscillations were large in relation to the mean pressure and were not always correlated with blood pressure fluctuations in the dorsal and ventral aortas.

Extracellular fluid can drain from the filament and lamellae into the extensively valved low-pressure central filament sinus. The sinus and allied vessels are closely associated with the high-pressure lamellar blood circuit. Changes in volume in the lamellar circulation may cause flow in the central sinus by mechanical interaction between neighboring vessels. Pulsations of the pillar cell walls and other vessels may also aid in moving extracellular fluid into the central sinus. Certainly, in the isolated, saline-perfused trout head, an increase in pulse rather than mean pressure, at constant input flow, caused a marked increase in central sinus outflow (Table III). It seems probable that the effect of pulsatile lamellar flow is to empty the extravascular spaces and to cause flow in the sinus vessels, preventing any fluid backup or rise in pressure in the system.

A portion of the cardiac output may bypass the lamellae and enter the low-pressure recurrent system via the afferent filament artery to central sinus capillaries. The extent of this flow will depend on the number of afferent connections, which are small in salmonid fish but large in eels. Hughes *et al.* (1982) concluded that about 30% of the cardiac output could bypass the lamellae via this route in eels. Adrenaline sharply reduced the magnitude of this lamellar bypass in eels. Johansen and Pettersson (1981) calculated that 20% of the cardiac output passes through the gill venous network of the cod. Thus, the central sinus appears to have a high blood flow.

**Table III**

Summary of Dorsal Aortic and Anterior Venous Outflow from the Isolated Trout Head (*Salmo gairdneri*) Preparation during Perfusion by Three Different Input Regimes[a]

| Variables[b] | Perfusion regime[c] | | |
|---|---|---|---|
| | Constant pulsatile flow (n = 16) | Constant pressure (n = 12) | Constant nonpulsatile flow (n = 8) |
| $\dot{Q}_{DA}$ (ml min$^{-1}$) | 1.54 ± 0.13 | 1.43 ± 0.22 | 1.34 ± 0.18 |
| $\dot{Q}_{AV}$ (ml min$^{-1}$) | 1.39 ± 0.09[d] | 1.09 ± 0.11 | 1.01 ± 0.11 |
| $R_g$ (cm H$_2$O ml$^{-1}$ min$^{-1}$ 100 g$^{-1}$) | 14.41 ± 1.36 | 18.11 ± 2.85[e] | 23.53 ± 5.22 |
| $R_{AV}$ (cm H$_2$O ml$^{-1}$ min$^{-1}$ 100 g$^{-1}$) | 11.67 ± 0.98 | 14.97 ± 1.83 | 17.83 ± 2.60 |
| $\dot{Q}_{VA}$ (ml min$^{-1}$) | 4.79 ± 0.19 | 4.41 ± 0.29 | 4.62 ± 0.13 |
| Saline loss (%) | 38.71 ± 3.08 | 42.26 ± 3.75 | 48.86 ± 4.33 |

[a]From Daxboeck and Davie (1982).

[b]$\dot{Q}_{DA}$ = dorsal aorta flow; $\dot{Q}_{AV}$ = anterior venous–venolymphatic combined flow; $\dot{Q}_{VA}$ = ventral aortic or input flow.

[c]$n$, number of observations, taken from eight fish.

[d]This is only value that was significantly different because of changes in the perfusion regime used.

[e]Note that $R_g$ (Wood, 1974) at same dorsal aortic pressure and flow rate was 18.8 cm H$_2$O ml$^{-1}$ min$^{-1}$ 100 g$^{-1}$.

The blood in the central sinus contains fewer erythrocytes but more white cells than the rest of the circulation (Skidmore and Tovell, 1972; Booth, 1978; Soivio *et al.*, 1981). This could be a result of plasma skimming at the entrance to capillaries leading from the filament artery to the central sinus (Vogel *et al.*, 1976) as well as dilution of blood with extracellular fluid. Plasma skimming at the level of the afferent artery will raise lamellar hematocrit (Soivio and Hughes, 1978), whereas skimming at the level of the efferent filament artery will raise systemic hematocrit.

The functions of the central venous circulation are not clear. It appears to be a high-flow, low-hematocrit, low-pressure system. It clearly has a nutritive role, supplying the tissues of the gill filament. The flow, however, appears to be much greater than required to meet the metabolic requirements of these tissues, which, in the cod, receive 20% of the cardiac output but utilize only 1% of the oxygen uptake of the fish (Johansen and Pettersson, 1981).

The central sinus could act as a blood reservoir (Vogel *et al.*, 1973). Girard and Payan (1976) suggested that as much as 5% of the total blood volume may be moved from the central sinus into the general circulation by stimulation of α-adrenergic receptors that cut off flow to and empty the central sinus. These findings were not substantiated by either Holbert *et al.* (1979) or Booth (1979a).

The central sinus probably serves as a low-pressure supply to chloride cells. The chloride cells in the gills of saltwater fish are coupled to accessory cells with large paracellular channels between the cells. The fact that these leaky channels are served by a low-pressure system presumably reduces filtration across this region of the gill epithelium.

The function of the large venous flow, in particular that from the afferent to central sinus flow, is less clear, especially if it is 30% of cardiac output as reported by Hughes *et al.* (1982). Why would an animal recirculate 30% of the cardiac output through this pathway? It can have nothing to do with gas transfer because the lamellae are bypassed. The gill venous circulation could be involved in regulating the blood composition, removing or adding substances to the plasma as it passes through the system. The high flow could also constitute a means of rapidly flushing the central sinus and, in the presence of plasma skimming, be a means of adjusting lamellar and systemic erythrocyte numbers. Another possibility is that the venous shunt is used to adjust pressure and flow in the respiratory portion of the gill circulation. None of these suggestions, however, has been investigated experimentally.

## IV. BLOOD

There is considerable variability in hemoglobin structure and function between fish species, and many fish have multiple hemoglobins. These hemoglobins have variable pH and temperature sensitivities, generally assumed but seldom demonstrated to be of physiological significance (see Weber, 1982; Riggs, 1979). Teleost hemoglobin often shows a marked Bohr and Root shift (Riggs, 1979), whereas the Root shift is absent and the Bohr shift weak in elasmobranchs (Weber *et al.*, 1983). Hemoglobin–oxygen reaction velocities and Bohr and Root shifts are generally rapid and, even at the low temperatures at which some fish exist, are probably not rate limiting for $O_2$ and $CO_2$ transfer. This aspect of gas transfer, however, has received little attention in fish. Maren and Swenson (1980) showed that the rates of the Bohr shift were similar in a wide range of vertebrates including dogfish and goosefish. Forster and Steen (1969) showed that the reaction velocity of eel hemoglobin Root and Bohr shift at 15°C was similar to that observed for mammal hemoglobin at 37°C.

Fish hemoglobin is packaged in nucleated erythrocytes, which are flat, platelike structures. The number of circulating erythrocytes varies with the physiological state of the fish (see Weber, 1982). During exercise, for example, circulating red blood cell numbers may increase because of release of erythrocytes from the spleen (Yamamoto *et al.*, 1980) or a reduction in plasma volume (Wood and Randall, 1973) due either to a reduction in total plasma volume or to plasma skimming (Nikinmaa *et al.*, 1981). The volume percentage of erythrocytes (hematocrit) may also increase because of changes in volume of individual cells. Catecholamines and $CO_2$ cause red blood cells to swell (Weber, 1982; Nikinmaa, 1982) and will cause increases in hematocrit in the absence of any change in red blood cell numbers.

The hemoglobin oxygen affinity is quite variable between fish, and within fish it is affected by pH, nucleotide triphosphate levels, $CO_2$, and temperature. Urea has been shown to increase hemoglobin–oxygen affinity in the dogfish (Weber *et al.*, 1983). The erythrocytic membrane is permeable to $H^+$ (Forster and Steen, 1969), bicarbonate (Cameron, 1978; Obaid *et al.*, 1979; Heming and Randall, 1982), and chloride (Haswell *et al.*, 1978), as well as oxygen and carbon dioxide. There appears to be a SITS-sensitive (*i.e.*, inhibited by the drug SITS) $HCO_3^-$ / $Cl^-$ exchange mechanism in the fish erythrocytic membrane (Perry *et al.*, 1982). Erythrocytic pH is about 0.3 to 0.4 units below that of plasma (Fig. 10) and $H^+$ is usually passively distributed across the cell membrane according to the membrane potential. Adrenaline causes rainbow trout red blood cells to swell via β-adrenergic stimulation, which in vitro leads to a rise in intracellular pH and a fall in the gradient across the erythrocyte membrane (Nikinmaa, 1982). Clearly such a rise in intracellular pH will increase hemoglobin–oxygen affinity. In addition, if plasma pH is reduced because of a respiratory or metabolic acidosis, then the release of catecholamines into the blood and their subsequent action on erythrocytes will ameliorate the effects of a fall in plasma pH on erythrocyte pH. The end result will be a more stable red blood cell intracellular pH in the face of oscillations in plasma pH. Protons cross the erythrocytic membrane rather slowly, the half-time for equilibration being of the order of 9 sec in eel red blood cells (Forster and Steen, 1969), much longer than the transit time for blood through the gills. Thus, most of the physiologically important rapid changes in erythrocytic pH and, therefore, hemoglobin–oxygen affinity, are related to movements of carbon dioxide between plasma and the red blood cell, and the subsequent $CO_2$ hydration–dehydration reaction.

Nucleotide triphosphates (NTP), in particular adenosine triphosphate (ATP) and guanosine triphosphate (GTP), also affect fish hemoglobin–oxygen affinity, increases in NTP levels within the erythrocyte decreasing hemoglobin–oxygen affinity (Weber, 1982). Changes in erythrocyte NTP levels are involved in long-term modulation of hemoglobin–oxygen affinity, whereas pH changes are involved in rapid adjustments in oxygen binding. For example, NTP levels in fish erythrocytes have been shown to increase during long-term exposure to hypoxia (Weber and Lykkeboe, 1978; Soivio *et al.*, 1980), but the ratio of NTP:hemoglobin does not change during exercise (Jensen *et al.*, 1983).

An increase in temperature decreases fish hemoglobin–oxygen affinity;

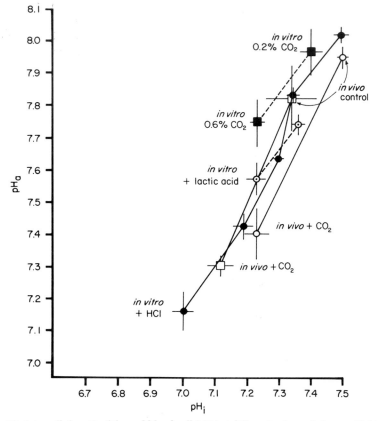

**Fig. 10.** Intracellular pH of the red blood cell ($pH_i$) at different values of plasma pH ($pH_a$) in rainbow trout blood both in vivo and in vitro; plasma pH was varied by adding lactic acid, HCl, or $CO_2$ to plasma.

an exception is found in the tuna, where the hemoglobin is pH sensitive but temperature insensitive (Weber, 1982). Houston and Smeda (1979) measured erythrocyte ion levels and found that $K^+$ and $Mg^{2+}$ levels were higher in cells from cold-acclimated rainbow trout and carp, whereas $Cl^-$ and $Ca^{2+}$ levels were higher in warm-acclimated fish. The effect of temperature acclimation in the carp was to reduce the magnitude of the temperature-induced shift in hemoglobin–oxygen affinity (Albers *et al.*, 1983), presumably by altering the ionic composition and pH of the erythrocytic intracellular environment. Dobson and Baldwin (1982), however, found increased NTP levels in the red blood cells of the Australian blackfish acclimated to higher temperatures, which acted synergistically with the increase in temperature to reduce hemoglobin–oxygen affinity. In this instance acclimation amplified rather than ameliorated the change in the system associated with temperature change.

Carbon dioxide will bind with terminal $NH_2$ groups to form carbamino compounds. These groups on the $\alpha$ chain of the globin may be acetylated and account for the reduced $CO_2$ binding of fish hemoglobins. The $CO_2$ content of erythrocytes is low, however, and as a result carbamino $CO_2$ may constitute a considerable portion of intracellular $CO_2$ content (Farmer, 1979). Carbamino formation decreases hemoglobin–oxygen affinity. There is some competition for the common binding site on the $\beta$ chain of hemoglobin for $CO_2$ and NTP, with GTP competing more effectively than ATP for the site (Weber and Johansen, 1979). Thus, changing levels of NTP will alter the action of carbamino formation on hemoglobin–oxygen affinity.

Thus, there is an impressive body of literature on the properties of hemoglobin, and some information on the characteristics of the erythrocyte intracellular environment in which hemoglobin operates; there is, however, little information on the physiological significance of many of these observed properties.

## V. OXYGEN TRANSFER

Increases in oxygen uptake across the gills are associated with an increase in both blood and water flow. The ratio of water flow to blood flow increases from rest to exercise (Kiceniuk and Jones, 1977). In addition, there may be changes in the blood distribution within the gills resulting in a more even distribution of blood flow within each of the lamellae, recruitment of previously poorly perfused lamellae, and a thinning of the gill water–blood barrier (Randall, 1982b). If the rise in oxygen transfer is associated with an increase in circulating catecholamines, these will alter blood distribution in the gills and increase the permeability of the epithelial membrane to oxygen (see Chapter 1, Volume XB, this series), both effects will increase the diffusing capacity of the gills to oxygen.

During exercise there is a reduction in venous blood oxygen content and an increase in cardiac output as well as a rise in the number of circulating erythrocytes (Jones and Randall, 1978; Yamamoto *et al.*, 1980). All of these factors increase the capacity of the fish to transport oxygen from the gills to the tissues. The increase in cardiac output results in a decrease in blood residence time in the gills. The transit time is not linearly related to cardiac output, because increased cardiac output is usually associated with a rise in ventral aortic pressure, which causes lamellar recruitment and expansion of the blood sheet in each lamella. This increase in gill blood volume will reduce the rate of decrease in transit time with increasing cardiac output.

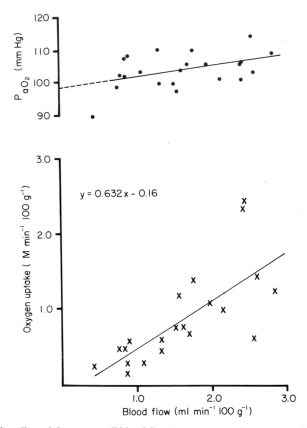

**Fig. 11.** The effect of changes in gill blood flow in oxygen transfer across the blood-perfused, spontaneously ventilating trout. (From Daxboeck et al., 1982.) $P_{a_{O_2}}$, arterial blood oxygen tension. (From Randall, 1982b.)

An increase in water flow results in a reduction in water residence time from 200 to 300 msec at rest to 20 to 50 msec at high rates of exercise. The magnitude of the water boundary layer will be reduced with increasing velocity; the time for axial diffusion and back-mixing of oxygen, however, will be reduced as well. Oxygen utilization from water does not change with increases in $\dot{V}_g$ (Kiceniuk and Jones, 1977; Lomholt and Johansen, 1979).

In a series of experiments on blood-perfused fish (Daxboeck *et al.*, 1982), oxygen uptake was found to be proportional to blood flow (Fig. 11). Manipulations of input pressure, found to promote lamellar recruitment in other preparations (Farrell *et al.*, 1979), did not affect oxygen transfer or arterial blood $P_{O_2}$ at any flow rate, indicating that the system was not diffusion limited but was entirely dependent on the rate of perfusion and gill ventilation. The range of transit times was similar in the blood-perfused fish to that observed in vivo over a wide range of exercise levels. In the blood-perfused fish, oxygen transfer was clearly perfusion limited; the oxygen transfer rate, however, spanned a narrow range from rest to routine activity because blood hematocrit was low. In intact fish much higher rates of oxygen uptake are observed at similar flows (see Jones and Randall, 1978), and under these conditions, increases in gill diffusing capacity may be significant in achieving the high rates of oxygen transfer associated with exercise.

The gill oxygen-diffusing capacity probably increases during hypoxia (Fisher et al., 1969) and exercise (Randall et al., 1967). The factors that contribute to this increase are manifold. An increase in blood pressure promotes lamellar recruitment and a more even blood distribution within the lamellae. Increases in circulating catecholamines have been implicated in lamellar recruitment (Booth, 1978; Holbert et al., 1979) as well as increases in membrane permeability (Isaia et al., 1978b). An increase in the blood pressure, pulse, and synchronization of heart and breathing movements may cause the movement of fluid out of extracellular spaces in the lamellae into the central sinus (Daxboeck and Davie, 1982) and an increase in epithelial permeability (Davie and Daxboeck, 1982). A rise in blood pressure, which accompanies exercise and hypoxia, may also thin the gill epithelium (Soivio and Tuurala, 1981) with an increase in the thickness of the blood sheet (Fig. 8), making the lamellae more rigid and erect. The relative importance, if any, of each of these mechanisms in augmenting gill diffusing capacity is not clear.

## VI. CARBON DIOXIDE TRANSFER

The passage of venous blood through the gills results in a 10–20% reduction in blood total $CO_2$ levels, largely due to a 20% reduction in plasma bicarbonate. The transit time for blood in the gills is between 1 and 3 sec, and this is too rapid for any appreciable formation of $CO_2$ from bicarbonate at the uncatalyzed rate (Forster and Steen, 1969).

Hoffert and Fromm (1973) found that acetazolamide produced a reduction in $CO_2$ excretion and an acidotic state in trout. Thus, bicarbonate excretion across fish gills presumably involves the catalyzed dehydration of plasma bicarbonate. The enzyme, carbonic anhydrase (ca), is found in high concentration in both the gill epithelium and the erythrocyte but not the plasma. The gill epithelium is known to be quite permeable to protons (McWilliams and Potts, 1978), and so bicarbonate could diffuse into the gill epithelium along with protons to form $CO_2$, which then diffuses into the water. Alternatively, plasma bicarbonate could enter the erythrocyte and be

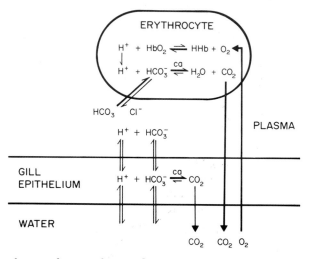

**Fig. 12.** A schematic diagram of $CO_2$ and $H^+$ movement between plasma, erythrocyte, the gill epithelium, and water. ca, carbonic anhydrase.

dehydrated to $CO_2$, which then diffuses into the medium across the gill epithelium (Fig. 12).

A chloride shift ($HCO_3^-/Cl^-$ exchange across the erythrocytic membrane) has been demonstrated in the red snapper and rainbow trout (Cameron, 1978), and the dogfish (Obaid et al., 1979). Haswell et al. (1978) concluded that the trout erythrocyte might be $HCO_3^-$ impermeable, but Heming and Randall (1982) later showed that technical problems with the experimental protocol had led to this conclusion. Wood et al. (1982) observed a respiratory acidosis during anemia in trout, and Perry et al. (1982) showed that the addition of the disulfonic stilbene derivative SITS, known to block anion movement in mammalian erythrocytes (Cabantchik and Rothstein, 1972), reduced $CO_2$ excretion in a blood-perfused trout preparation (Table IVC). In addition, $CO_2$ excretion is proportional to hematocrit (Table IVB) in this same preparation (Perry et al., 1982). All of these studies lead to the conclusion that plasma bicarbonate is excreted by erythrocytic dehydration catalyzed by carbonic anhydrase and that the $CO_2$ so formed then diffuses into the water across the gill epithelium. The experiments also indicate that the epithelium does not play an important role in $HCO_3^-$ dehydration, a conclusion supported by the fact that saline-perfused gills show little $CO_2$ excretion even when saline bicarbonate levels are elevated to very high levels (Table IVA). Thus, gill epithelial carbonic anhydrase does not appear to play a significant role in plasma bicarbonate dehydration.

Carbon dioxide excretion in the experiments of Perry et al. (1982) was perfusion limited in the blood-perfused trout preparation. Carbon dioxide

## Table IV

Factors Affecting $CO_2$ Excretion across the Gills in the Blood-Perfused Trout Preparation (Perry *et al.*, 1982) at 7°C[a]

**A.** Carbon dioxide excretion ($\dot{M}_{CO_2}$) increases with input bicarbonate only in the presence of erythrocytes.
(1) Blood-perfused fish: trout—*Salmo gairdneri*[a] (hematocrit ≃ 10%; $T$ = 7°C)

| Input blood $C_{CO_2}$ (mM) | $\dot{M}_{CO_2}$ ($\mu M$ 100 g$^{-1}$ min$^{-1}$) | RQ | Blood pH gill | |
|---|---|---|---|---|
| | | | Input | Outlet |
| 9.86 | 1.63 | 1.4 | 7.74 | 7.64 |
| 24.76 | 6.50 | 7.7 | 7.96 | 7.93 |

$\dot{Q}$ = 1.59 ± 0.07 ml min$^{-1}$ 100 g$^{-1}$

(2) Saline-prefused fish: coho salmon—*Oncorhynchus kisutch* ($T$ = 11°C)

| Input saline $C_{CO_2}$ (mM) | $\dot{M}_{CO_2}$ ($\mu M$ 100 g$^{-1}$ min$^{-1}$) | RQ | Saline pH gill | |
|---|---|---|---|---|
| | | | Input | Outlet |
| 11.86 | 0.16 | 2.05 | 7.93 | 7.98 |
| 33.15 | −0.17 (net uptake) | −2.02 | 8.44 | 8.34 |

$\dot{Q}$ = 2.15 ml 100 g$^{-1}$ min$^{-1}$

**B.** Carbon dioxide excretion ($\dot{M}_{CO_2}$) is proportional to hematocrit ($T$ = 7°C).

| Hematocrit (%) | $\dot{M}_{CO_2}$ ($\mu M$ 100 g$^{-1}$ min$^{-1}$) | RQ | Blood pH gill | |
|---|---|---|---|---|
| | | | Input | Outlet |
| 4.3 | 1.34 | 2.9 | 7.81 | 7.72 |
| 11.3 | 2.62 | 2.0 | 7.78 | 7.75 |
| 20.2 | 3.87 | 1.6 | 7.75 | 7.75 |

$\dot{Q}$ = 1.63 ± 0.05 ml 100 g$^{-1}$ min$^{-1}$

**C.** Carbon dioxide excretion ($\dot{M}_{CO_2}$) is reduced by SITS ($T$ = 7°C).

| | $\dot{M}_{CO_2}$ ($\mu M$ 100 g$^{-1}$ min$^{-1}$) | RQ | Blood pH | |
|---|---|---|---|---|
| | | | Input | Outlet |
| Normal | 2.57 | 2.1 | 7.90 | 7.79 |
| 10$^{-4}$ SITS | 0.80 | 1.1 | 7.83 | 7.71 |

$\dot{Q}$ = 1.59 ± 0.04 ml min$^{-1}$ 100 g$^{-1}$

[a] $\dot{Q}$ = perfusate flow; $\dot{M}_{CO_2}$ = $CO_2$ excretion; $C_{CO_2}$ = $CO_2$ content; RQ = respiratory quotient.

excretion ($\dot{M}_{CO_2}$) was related to blood flow ($\dot{Q}$) through the gills in the following way:

$$\dot{M}_{CO_2} \; (\mu M) = 1.043 \; \dot{Q} \; (ml) - 0.13$$

The conditions in these experiments were as follows: blood flowing into the gills was of pH 7.76, $CO_2$ content = 10.3 m$M$, hematocrit = 10.3%, and temperature 7°C (Perry *et al.*, 1982). Under these same conditions oxygen transfer was also perfusion limited, so perhaps it is not surprising that the system is perfusion limited for $CO_2$. Obviously, bicarbonate entry into the erythrocyte could be a rate-limiting step in $CO_2$ excretion, but this seems unlikely in the intact fish at low levels of $CO_2$ excretion, because when bicarbonate levels were raised in the blood-perfused preparation, there was a marked increase in $CO_2$ excretion resulting in the very high respiratory quotient of 7.7. Interestingly, the blood plasma pH was not changed during passage through the gills, even though there were large changes in plasma bicarbonate (Table IVA). Oxygenation of hemoglobin will produce protons for bicarbonate dehydration but not sufficient to maintain an RQ of 7.7. Other possible sources of protons are gill metabolism, influx from the environment, or buffering by the contents of the erythrocyte. It seems unlikely that the gill will produce the required $H^+$, so intracellular buffering within the red blood cell or influx from the water seem the most probable sources. The gill epithelium is considered to be very permeable to protons (McWilliams and Potts, 1978), and so the water could be a source of protons for bicarbonate dehydration within the animal. Penetration of $H^+$ into red blood cells (Forster and Steen, 1969), however, is somewhat slower than required to reach equilibrium during blood transit through the gills. The erythrocytes are the site of $HCO_3^-$ dehydration, and it would seem that, although the gills are permeable to $H^+$, there may be limitations on the number of $H^+$ ions that can move from water into the red blood cell as it passes through the gills. Thus, there is no adequate quantitative answer to explain the result reported in Table IVA at this time.

## VII. THE INTERACTION BETWEEN $CO_2$ AND $H^+$ EXCRETION

If the gill epithelium is very permeable to protons, then a reduction in blood pH will result in an increased net efflux of acid into the water if water pH remains high. Conversely, a reduction in water pH will result in a net uptake of acid. Burst activity in coho salmon in seawater leads to a large production of lactic acid; blood pH falls, and as a result proton excretion is

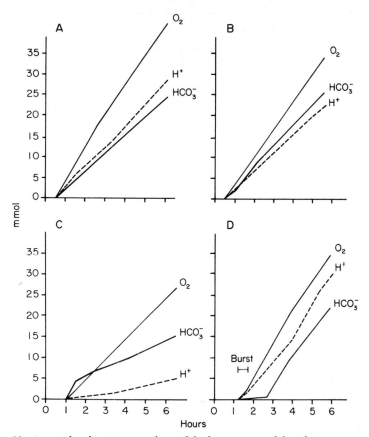

**Fig. 13.** Accumulated oxygen uptake and hydrogen ion and bicarbonate excretion into seawater containing a swimming coho salmon, *O. kisutch* (600 g weight), at 13°C in a closed respirometer under four different conditions. $U_{crit}$ is the critical swimming speed, which can be defined as the maximum prolonged swimming speed. (A) Seawater pH 7.95, normal saltwater control, 80% $U_{crit}$. (B) Seawater pH 7.95, low-bicarbonate seawater, 80% $U_{crit}$. (C) Seawater pH 7.10, normal seawater, 50% $U_{crit}$. (D) Seawater pH 7.95, normal seawater, 50% $U_{crit}$ after a burst swim at 120% $U_{crit}$ to exhaustion. (From Van den Thillart *et al.*, 1983.)

elevated in fish following a burst swim (Fig. 13D). If seawater pH is reduced, then $H^+$ excretion by the fish is also reduced or even reversed. In coho salmon swimming aerobically at 80% critical velocity in seawater at pH 7.95, $H^+$ and $HCO_3^-$ levels increased in seawater by approximately the same amount, and the simplest assumption is that they were formed from $CO_2$ excreted across the gills and that $H^+$ and $HCO_3^-$ flux is negligible (Fig. 13A). If seawater $HCO_3^-$ levels are changed, altering the $HCO_3^-$ gradient across the gills, there is little effect on $CO_2$ excretion (Fig. 13B); Van den

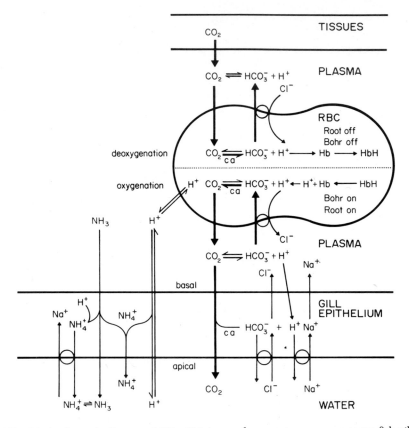

**Fig. 14.** A schematic diagram of $CO_2$, $H^+$ ion, and ammonia movement across fish gills.

Thillart *et al.* (1983) concluded that there could be a large and separate flux of protons across the gills. Thus, depending on the circumstances, there may be a flux of protons either into or out of the fish that is quite separate from $CO_2$ movement. There is some interaction between the two processes, however, because a reduction of water pH not only decreased $H^+$ excretion but also bicarbonate excretion in coho salmon exercising in seawater. The exact nature of this interaction is not clear (Van den Thillart *et al.*, 1983).

The outer membrane of the gill epithelium of both saltwater and freshwater fish is thought to contain mechanisms for the exchange of $Na^+/H^+$ and $HCO_3^-/Cl^-$ (see Chapter 8, Volume XB, this series), in which NaCl is taken up in exchange for $H^+$ and $HCO_3^-$ (Fig. 14). These exchange processes operate at a rate such that the $CO_2$ excreted via these pathways accounts for about 10% of the total $CO_2$ excretion of the resting fish (Cameron, 1976). Inhibition of these processes results in acid–base changes in the animal. SITS inhibition of $Cl^-/HCO_3^-$ results in a rise in blood pH, whereas amiloride inhibition of $Na^+/H^+$ exchange results in a fall in blood pH. As these mechanisms appear to be on the external surface of the gill epithelium, amiloride will cause a reduction in epithelial pH, whereas SITS will cause a rise in epithelial pH. Thus, SITS will affect $H^+$ concentrations in the region of the $Na^+/H^+$ exchange site, and amiloride will affect $HCO_3^-$ availability at the $HCO_3^-/Cl^-$ exchange location. Hence, it is not surprising that SITS affects $Na^+/H^+$ as well as $HCO_3^-/Cl^-$ exchange, and amiloride, $HCO_3^-/Cl^-$ exchange as well as $Na^+/H^+$ exchange (Perry and Randall, 1981). The action on the secondary exchange site in each case is thought to be mediated by changes in gill epithelial cell pH rather than direct binding to the exchange site. This theory is based on work with these drugs in other tissues indicating that amiloride binds to $Na^+/H^+$ exchange sites and not $HCO_3^-/Cl^-$, whereas the reverse is true for SITS.

These exchange sites produce a net uptake of NaCl, and if modulated differentially, these systems will result in a change in epithelial pH which, in turn, results in a change in plasma pH. One might expect these systems to be modulated, for instance, during hypercapnia, increasing $H^+$ excretion but decreasing $HCO_3^-$ excretion to correct blood pH. Perry et al. (1981) could find no evidence of the modulation of NaCl uptake during imposed hypercapnia in the trout, even though NaCl flux could be markedly reduced by inhibition with amiloride and SITS. Cameron (1976), however, showed that the ratio of $Na^+$ to $Cl^-$ uptake was increased during hypercapnic acidosis, the response appropriate for increased $H^+$ excretion and $HCO_3^-$ retention by the fish. Wood and Randall (1973) showed that $Na^+$ influx remained remarkably stable during exercise, indicating a stable $Na^+/H^+$ exchange process in the intact animal. We think it is still not clear what role, if any, $Na^+/H^+$ and $HCO_3^-/Cl^-$ exchange mechanisms on the gills play in acid–base regulation in the fish.

There are excellent quantitative descriptions of changes in net flux of NaCl and $HCO_3^-$ and $H^+$ plus $NH_4^+$ between the fish and the surrounding water under a variety of conditions (see Chapter 6, this volume). What is not clear is the extent to which these fluxes are coupled directly to a carrier-mediated exchange site in the gill membrane or indirectly through separate passive fluxes across the gills to maintain electrical neutrality of plasma and transepithelial potential across the gill epithelium.

## VIII. CONTROL OF OXYGEN AND CARBON DIOXIDE TRANSFER

Gas transfer across the gills is modulated by altering water and blood flow, the distribution of blood within the gills, the properties of the gill epithelium, and the number and characteristics of circulating erythrocytes.

Changes in water flow and, to a somewhat lesser extent, blood flow have been described in a number of fish under a variety of conditions; the changes in the properties and numbers of circulating erythrocytes have also been described for several species, but almost nothing is known of changes in the characteristics of the gill epithelium between species or in a single species under a variety of conditions.

## A. Control of Water Flow

Water flow over the gills of many fish increases with hypoxia, hypoxemia caused by either anemia (Cameron and Davis, 1970; Wood *et al.*, 1979) or carbon monoxide (Holeton, 1971), hypercapnia (Janssen and Randall, 1975), a rise in temperature, or exercise (Randall and Cameron, 1973; Jones and Randall, 1978). Hyperoxia causes a reduction in gill water flow (Randall and Jones, 1973; Dejours, 1973). Those fish that increase gill water flow in response to aquatic hypoxia are able to maintain oxygen uptake over a wide range of water $P_{O_2}$ levels. Other fish do not show this increase in ventilation with aquatic hypoxia. The first group, the oxygen regulators or nonconformers, have been extensively studied and include the goldfish, *Carassius auratus* (Prosser *et al.*, 1957); trout, *Salmo gairdneri* (Holeton and Randall, 1967); bluegill, *Lepomis macrochirus* (Marvin and Heath, 1968; Spitzer *et al.*, 1969); dogfish, *Scyliorhinus canicula* (Hughes and Umezawa, 1968); mullet, *Mugil cephalus* (Cech and Wohlschlag, 1973); flounders, *Pseudopleuronectes americanus, Platichthys flesus*, and *Platichthys stellatus* (Watters and Smith, 1973; Cech *et al.*, 1977; Kerstens *et al.*, 1979); carp, *Cyprinus carpio* (Itazawa and Takeda, 1978; Lomholt and Johansen, 1979); tench, *Tinca tinca* (Eddy, 1974); catfish, *Ictalurus punctatus* (Burggren and Cameron, 1980); lingcod, *Ophiodon elongatus* (Farrell and Daxboeck, 1981); and black prickleback, *Xiphister atropurpureus* (Daxboeck and Heming, 1982). In the other group of fish, termed oxygen conformers, oxygen uptake falls with oxygen availability. This group includes the toadfish, *Opsanus tau* (Hall, 1929); the dragonet, *Callionymus lyra* (Hughes and Umezawa, 1968); the catfish, *Ictalurus nebulosus* (Marvin and Heath, 1968); the sturgeon, *Acipenser transmontanus* (Burggren and Randall, 1978); the dogfish, *Scyliorhinus stellaris* (Piiper *et al.*, 1970); the red grouper, *Epinephelus akaara;* and the black sea bream, *Mylio macrocephalus* (Wu and Woo, 1983). In many of these species it has been shown that breathing frequency and gill water flow decrease with oxygen levels in the water. Thus, these animals respond to hypoxia by decreasing gill water flow and presumably energy expended in ventilating the gills. The division between oxygen regulators and conformers is not absolute. At low oxygen levels oxygen regulators eventually become oxygen conformers.

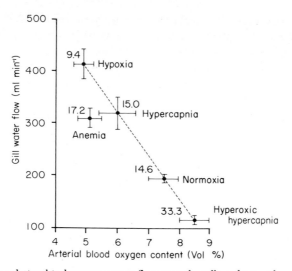

**Fig. 15.** The relationship between water flow over the gills and arterial oxygen content of dorsal aortic blood from rainbow trout, *Salmo gairdneri*. The lines indicate ±SE. The number by each point is the corresponding arterial oxygen tension in kPa. Arterial oxygen content was changed by bubbling $N_2$, $CO_2$, and/or $O_2$ into the water containing the fish. (Data from Smith and Jones, 1982.)

The control of ventilation has only been studied in oxygen regulators. It appears that these fish increase gill ventilation to maintain oxygen content in arterial blood efferent to the gills (Smith and Jones, 1982), thus hypoxia tends to reduce, whereas hyperoxia increases, the oxygen content of the blood, and each has the reverse effect on gill water flow (Fig. 15). Hypercapnia also increases gill ventilation via changes in arterial oxygen content due to the Root shift. At high $CO_2$ levels, however, Smith and Jones (1982) found that the increase in ventilation was not simply related to changes in arterial blood oxygen content. Thus, there may be a direct effect of $CO_2$ on respiration as well as an effect via pH-induced changes in oxygen content.

The location of the oxygen content receptors is not known, except that they are efferent to the gills. The receptors may be sensitive to flow as well as to oxygen content, because anemia, which is associated with large increases in blood flow, results in a slightly reduced ventilatory response to that predicted from studies of hypoxia (Fig. 15) (Randall, 1982b).

Hyperoxia results in a reduction in gill ventilation in rainbow trout (Randall and Jones, 1973; Wood and Jackson, 1980), white sucker (Wilkes *et al.*, 1981), and tench (Dejours, 1973). The overall effect is a rise in both arterial blood $P_{O_2}$ and $P_{CO_2}$, the latter being associated with a fall in blood pH. Artificial increases in gill ventilation during hyperoxia cause a reduction in $P_{ACO_2}$, but not to normoxic levels, so the rise in $P_{ACO_2}$ is not simply a reflection of reduced gill ventilation during hyperoxia (Wood and Jackson, 1980). Wilkes *et al.* (1981) reported a marked reduction in gill diffusing capacity, which would also contribute to the elevation in $P_{ACO_2}$ during hyperoxia. These results, along with those of Smith and Jones (1982), show that gill ventilation is adjusted to the oxygen requirements of the fish even at the expense of $P_{CO_2}$ and pH levels within the body, and that oxygen is the major respiratory gas involved in the control of breathing in fish.

Peyraud-Waitzenegger (1979) showed that catecholamine infusion caused a rise in $P_{AO_2}$ and an increase in gill ventilation, whereas, following β-receptor blockade, catecholamine infusion caused a reduction in gill ventilation associated with a transient increase followed by a decrease in $P_{AO_2}$. Thus, it was concluded that the rise in $P_{AO_2}$ following catecholamine infusion in the intact fish was the result of the elevated gill ventilation.

The effects of catecholamines on gas transfer are manifold. Catecholamines alter (*a*) cardiac output and blood distribution in the gills, (*b*) the permeability of the gill epithelium and the movements of ions across the gills, (*c*) red blood cell pH, and (*d*) indirectly, hemoglobin–oxygen affinity. Finally, catecholamines have metabolic actions and actions on many other tissues in addition to the gills. To dissect out direct and indirect actions of catecholamines will prove difficult and arduous, and to understand the responses in the intact animal will be most difficult because of the multitude of actions and mode of release of this versatile group of hormones.

The systems involved in initiating and regulating increases in gill ventilation during exercise are not known. The switch from rhythmic to ram ventilation may involve receptors in the gill cavity that detect water velocity (Randall, 1982b), but there is little evidence to support this or any other contention.

## B. Control of Blood Flow

There is little change in blood flow during hypoxia in oxygen regulators, but there is a marked bradycardia that is offset by an increase in stroke volume to maintain cardiac output. The slowing of the heart is due to increased activity in vagal cholinergic fibers, which have an inhibitory effect on heart rate (Randall, 1982b). The afferent arm of this reflex appears to be from oxygen receptors in the first gill arch in the region of the efferent arch vessel (Daxboeck and Holeton, 1978; Smith and Jones, 1978). Synchrony between heartbeat and a specific phase of the breathing cycle also develops during hypoxia (see Randall, 1982a). Interestingly, hyperoxia, as well as hypoxia, is associated with a marked bradycardia, and it seems the same receptors are involved, that is, those on the first gill arch in the region of the efferent arch

**A**

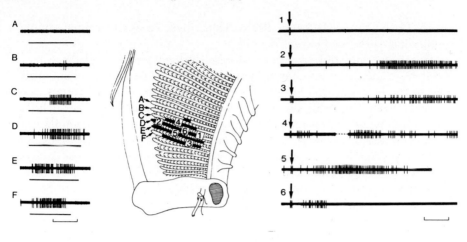

**B**

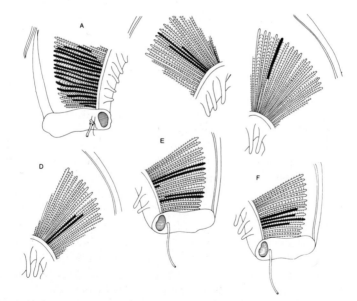

**Fig. 16.** (A) Discharge patterns of a branchial receptor. The receptive field is marked in black. Oscilloscope traces **A–F** (left) record discharge caused by lightly stroking filaments **A–F**; black line records duration of the stimulus. Traces **1–6** on the right record discharge evoked by PDG-soaked filter paper islands 1–6. Arrow marks moment of application: 5 sec have been removed from the break in trace 4 to show the recurring bursts of discharge. The brief response to the mechanical stimulus of the filter paper is well shown in traces **1–6**. The filter paper islands have been enlarged ×2 in the illustration, and the spikes have been retouched. Time calibration is 5 sec. (B) Six receptive fields located in different regions of a gill: in **E** the field is broken into two parts separated by three unresponsive filaments. (From Poole and Satchell, 1979. We thank Professor Satchell for supplying the original negatives.)

vessel (Daxboeck and Holeton, 1978). Thus, these receptors are not simply responding to reduced oxygen levels.

Heart rate and stroke volume increase during exercise, and these may result from increased adrenergic activity and decreased vagal cholinergic tone to the heart, as well as an increase in circulating catecholamines (see Randall, 1982b, for further discussion). Gill vascular resistance changes little with either hypoxia or exercise. There may be some control of blood flow at the level of the branchial arch. Cameron *et al.* (1971) closed off various gill slits of the stingray to water flow but found that arterial blood was always saturated. They concluded that nonventilated gill arches received little blood flow, possibly because of hypoxic vasoconstriction of the arch blood vessels. If this was the case one would expect an increase in gill vascular resistance during hypoxia; this was not measured in the stingray, but in other fish hypoxia caused little change in gill resistance to blood flow.

Many air-breathing fish can preferentially direct oxygenated blood through some gill arches and deoxygenated blood through others (Johansen *et al.*, 1968; Randall *et al.*, 1981). Separate bloodstreams have been shown to exist in the air-breathing teleost fish, *Channa argus,* which has two ventral aortas (Ishimatsu and Itazawa, 1983) for directing oxygenated blood to the posterior arches and the body, and the deoxygenated blood to the anterior gill arches and the air-breathing organ. Smith and Gannon (1978) showed in *Hoplerythrinus,* another air-breathing fish, that some control may be exerted on blood distribution between gill arches such that during periods of air breathing more blood flows to the posterior arches, which, in this fish, direct blood to the air-breathing organ.

Respiratory nociceptors have been located in dogfish gills (Satchell, 1978; Poole and Satchell, 1979) that have similarities to the type J receptors of mammals. The receptors are located in the gill filaments and have fairly large receptive fields, sometimes encompassing several filaments (Fig. 16). These receptors discharge in response to phenyldiguanide (PDG), 5-hydroxytryptamine (5-HT), and mechanical stimulation, as do J receptors (Poole and Satchell, 1979). The receptors appear to be located between blood vessels and the gill epithelium, because they respond to PDG in either the water or blood. They may function to protect the gills against edema, as do J receptors, because they respond to the edema-inducing drug, alloxan.

Stimulation of the nociceptors in the dogfish gill caused apnea, bradycardia, hypotension, and inhibited swimming (Satchell, 1974). These changes result in a reduction in blood pressure, which, as suggested by Satchell (1974), will reduce edema in the gills. This is a protective mechanism, because fluid accumulation in the gills will increase diffusion distances between blood and water and will impair gas transfer.

## REFERENCES

Albers, C., Manz, R., Muster, D., and Hughes, G. M. (1983). Affect of acclimation temperature on oxygen transport in the blood of the carp *Cyprinus carpio*. *Respir. Physiol.* **52**, 165–179.

Altman, P. L., and Dittmer, D. S., eds. (1971). "Respiration and Circulation." Fed. Am. Soc. Exp. Biol., Bethesda, Maryland.

Attinger, E. O. (1968). Analysis of pulsatile blood flow. *In* "Adv. Biomed. Eng. Med. Phys." (S. N. Levine, ed.), Vol. 1, pp. 1–59. Wiley (Interscience), New York.

Booth, J. H. (1978). The distribution of blood flow in the gills of fish: application of a new technique to rainbow trout (*Salmo gairdneri*). *J. Exp. Biol.* **73**, 119–129.

Booth, J. H. (1979a). The effects of oxygen supply, epinephrine and acetylcholine on the distribution of blood flow in trout gills. *J. Exp. Biol.* **83**, 31–39.

Booth, J. H. (1979b). Circulation in trout gills: The relationship between branchial perfusion and the width of the lamellar blood space. *Can. J. Zool.* **57**, 2193–2185.

Brett, J. R., and Zala, C. A. (1975). Daily pattern of nitrogen excretion and oxygen consumption of sockeye salmon (*Oncorhynchus nerka*) under controlled conditions. *J. Fish. Res. Board Can.* **32**, 2479–2486.

Burggren, W. W. (1978). Gill ventilation in the sturgeon, *Acipenser transmontanus*: Unusual adaptations for bottom dwelling. *Respir. Physiol.* **34**, 153–170.

Burggren, W. W., and Cameron, J. N. (1980). Anaerobic metabolism, gas exchange, and acid-base balance during hypoxic exposure in the channel catfish, *Ictalurus punctatus*. *J. Exp. Zool.* **213**, 405–416.

Burggren, W. W., and Randall, D. J. (1978). Oxygen uptake and transport during hypoxic exposure in the sturgeon, *Acipenser transmontanus*. *Respir. Physiol.* **34**, 171–184.

Cabantchik, Z., and Rothstein, A. (1972). The nature of the membrane sites controlling anion permeability of human red blood cells as determined by studies with disulfonic stilbene derivatives. *J. Membr. Biol.* **10**, 311.

Cameron, J. N. (1976). Branchial ion uptake in aortic grayling: Resting values and effects of acid-base disturbance. *J. Exp. Biol.* **56**, 711–725.

Cameron, J. N. (1978). Chloride shift in fish blood (1). *J. Exp. Zool.* **206**, 289–295.

Cameron, J. N., and Davis, J. C. (1970). Gas exchange in rainbow trout (*Salmo gairdneri*) with varying blood oxygen capacity. *J. Fish. Res. Bd. Can.* **27**, 1069–1085.

Cameron, J. N., and Heisler, N. (1983). Studies of ammonia in rainbow trout: Physico-chemical parameters, acid-base behaviour and respiratory clearance. *J. Exp. Biol.* **105**, 107–125.

Cameron, J. N., Randall, D. J., and Davis, J. C. (1971). Regulation of the ventilation-perfusion ratio in the gills of *Dasyatis sabina* and *Squalus suckleyi*. *Comp. Biochem. Physiol. A* **39A**, 505–519.

Cech, J. J., Jr., and Wohlschlag, D. E. (1973). Respiratory responses of the striped mullet, *Mugil cephalus* to hypoxic conditions. *J. Fish Biol.* **5**, 421–428.

Cech, J. J., Jr., Rowell, D. M., and Glasgow, J. S. (1977). Cardiovascular responses of the winter flounder *Pseudopleuronectes americanus* to hypoxia. *Comp. Biochem. Physiol. A* **57A**, 123–215.

Cokelet, G. R. (1980). Rheology and hemodynamics. *Annu. Rev. Physiol.* **42**, 311–324.

Cowey, C. B., and Sargent, J. R. (1979). Nutrition. *In* "Fish Physiology" (W. S. Hoar, D. J. Randall, and J. R. Brett, eds.), Vol. 8, pp. 1–69. Academic Press, New York.

Davie, P., and Daxboeck, C. (1982). Effect of pulse pressure on fluid exchange between blood and tissues in trout gills. *Can. J. Zool.* **60**, 1000–1006.

Daxboeck, C., and Davie, P. (1982). Effect of pulsatile perfusion on flow distribution within an isolated saline-perfused trout head preparation. *Can. J. Zool.* **60**, 994–999.

Daxboeck, C., and Heming, T. A. (1982). Bimodal respiration in the intertidal fish *Xiphister astropurpureus* (Kittlitz). *Mar. Behav. Physiol.* **9**, 23–34.

Daxboeck, C., and Holeton, G. F. (1978). Oxygen receptors in the rainbow trout, *Salmo gairdneri. Can. J. Zool.* **56**, 1254–1259.

Daxboeck, C., Davie, P. S., Perry, S. F., and Randall, D. J. (1982). The spontaneously ventilating blood perfused trout preparation: Oxygen transfer across the gills. *J. Exp. Biol.* **101**, 35–45.

Dejours, P. (1973). Problems of control of breathing in fishes. *In* "Comparative Physiology" (L. Bolis, K. Schmidt-Nielsen, and S. H. P. Maddrell, eds.), pp. 117–133. North-Holland Publ., Amsterdam.

De Kock, L. L., and Symmons, S. (1959). A ligament in the dorsal aorta of certain fishes. *Nature (London)* **187**, 194–195.

Dimberg, K., Hoglund, L. B., Knutsson, P. G., and Riderstrale, Y. (1981). Histochemical localization of carbonic anhydrase in gill lamella from young salmon (*Salmo salar* L.) adapted to fresh and salt water. *Acta Physiol. Scand.* **112**, 218–220.

Dobson, G. P., and Baldwin, J. (1982). Regulation of blood oxygen affinity in the Australian blackfish *Gadopsis marmoratus.* II. Thermal acclimation. *J. Exp. Biol.* **99**, 245–254.

Dunel, S., and Laurent, P. (1977). La vascularization branchiale chez l'Anguille: Action de l'acetylcholine et de l'adrenaline sur la repartition d'une resine polymerisable dans les différents compartiments vasculaires. *C. R. Hebd. Seances Acad. Sci.* **284**, 2011–2014.

Eddy, F. B. (1974). Blood gases of the tench (*Tinca tinca*) in well aerated and oxygen-deficient water. *J. Exp. Biol.* **60**, 71–83.

Farmer, M. (1979). The transition from water to air breathing: Effects of $CO_2$ on hemoglobin function. *Comp. Biochem. Physiol. A* **62A**, 109–114.

Farrell, A. P. (1979). Gill blood flow in teleosts. Ph.D. Thesis, University of British Columbia, Vancouver.

Farrell, A. P. (1980). Gill morphometrics, vessel dimensions and vascular resistance in lingcod, *Ophiodon elongatus. Can. J. Zool.* **58**, 807–818.

Farrell, A. P., and Daxboeck, C. (1981). Oxygen uptake in the lingcod *Ophiodon elongatus* during progressive hypoxia. *Can. J. Zool.* **59**, 1272–1275.

Farrell, A. P., Daxboeck, C., and Randall, D. J. (1979). The effect of input pressure and flow on the pattern and resistance to flow in the isolated perfused gill of a teleost fish. *J. Comp. Physiol.* **133**, 233–240.

Farrell, A. P., Sobin, S. S., Randall, D. J., and Crosby, S. (1980). Intralamellar blood flow patterns in fish gills. *Am. J. Physiol.* **239**, R428–R436.

Fisher, T. R., Coburn, R. F., and Forster, R. E. (1969). Carbon monoxide diffusing capacity in the bullhead catfish. *J. Appl. Physiol.* **26**, 161–169.

Forster, R. E., and Steen, J. B. (1969). The rate of the "Root shift" in eel red cells and eel haemoglobin solutions. *J. Physiol. (London)* **204**, 259–282.

Freadman, M. A. (1979). Swimming energetics of striped bass (*Morone saxatilis*) and bluefish (*Pomatonus saltratrix*): Gill ventilation and swimming metabolism. *J. Exp. Biol.* **83**, 217–230.

Freadman, M. A. (1981). Swimming energetics of striped bass (*Morone saxatilis*) and bluefish (*Pomatomus saltatrix*): Hydrodynamic correlates of locomotion and gill ventilation. *J. Exp. Biol.* **90**, 253–265.

Fung, Y. C., and Sobin, S. S. (1969). Theory of sheet flow in the lung alveoli. *J. Appl. Physiol.* **26**, 472–488.

Fung, Y. C., and Sobin, S. S. (1972). Elasticity of the pulmonary alveolar sheet. *Circ. Res.* **30**, 451–469.

Girard, J.-P., and Payan, P. (1976). Effect of epinephrine on vascular space of gills and head of rainbow trout. *Am. J. Physiol.* **230**, 1555–1560.

Glazier, J. B., Hughes, J. M. B., Maloney, J. E., and West, J. B. (1969). Measurements of capillary dimensions and blood volume in rapidly frozen lungs. *J. Appl. Physiol.* **26**, 65–76.

Grote, J. (1967). Die Sauerstoffdiffusionskonstanten im Lungengewebe und Wasser und ihre Temperaturabhängigkeit. *Pflüegers Arch. Gesamte Physiol. Menschen Tiere* **295**, 245–254.

Hall, F. G. (1929). The influence of varying oxygen tensions upon the rate of oxygen consumption in marine fishes. *Am. J. Physiol.* **88**, 212–218.

Haswell, M. S., Zeider, R., and Kim, H. D. (1978). Chloride transport in red cells of a teleost, *Tilapia mossambica. Comp. Biochem. Physiol. A* **61A**, 217–220.

Haswell, M. S., Randall, D. J., and Perry, S. F. (1980). Fish gill carbonic anhydrase: Acid-base regulation or salt transport? *Am. J. Physiol.* **238**, R240–R245.

Haywood, G. P., Isaia, J., and Maetz, J. (1977). Epinephrine effects on branchial water and urea flux in rainbow trout. *Am. J. Physiol.* **232**, R110–R115.

Heming, T. A., and Randall, D. J. (1982). Fish erythrocytes are bicarbonate permeable: Problems with determining carbonic anhydrase activity using the modified boat technique. *J. Exp. Zool.* **219**, 125–128.

Hills, B. A. (1972). Diffusion and convection in lungs and gills. *Respir. Physiol.* **14**, 105–114.

Hills, B. A., Hughes, G. M., and Koyama, T. (1982). Oxygenation and deoxygenation kinetics of red cells in isolated lamellae of fish gills. *J. Exp. Biol.* **98**, 269–275.

Hoffert, J. R., and Fromm, P. O. (1973). Effect of acetazolamide on some hematological parameters and ocular oxygen concentrations in rainbow trout. *Comp. Biochem. Physiol. A* **45A**, 371–378.

Holbert, P. W., Boland, E. J., and Olsen, K. R. (1979). The effect of epinephrine and acetylcholine on the distribution of red cells within the gills of the channel catfish (*Ictalarus punctatus*). *J. Exp. Biol.* **79**, 135–146.

Holeton, G. F. (1971). Oxygen uptake and transport by the rainbow trout during exposure to carbon monoxide. *J. Exp. Biol.* **54**, 239–254.

Holeton, G. F., and Jones, D. R. (1975). Water flow dynamics in the respiratory tract of the carp (*Cyprinus carpio* L.). *J. Exp. Biol.* **63**, 537–549.

Holeton, G. F., and Randall, D. J. (1967). The effect of hypoxia upon the partial pressure of gases in the blood and water afferent and efferent to the gills of rainbow trout. *J. Exp. Biol.* **46**, 317–327.

Houston, A. H., and Smeda, J. S. (1979). Thermoacclimatory changes in the ionic microenvironment of hemoglobin in the stenothermal rainbow trout (*Salmo gairdneri*) and eurythermal carp (*Cyprinus carpio*). *J. Exp. Biol.* **80**, 317–340.

Hughes, G. M. (1960a). The mechanism of gill ventilation in the dogfish and skate. *J. Exp. Biol.* **37**, 11–27.

Hughes, G. M. (1960b). A comparative study of gill ventilation in marine teleosts. *J. Exp. Biol.* **37**(1), 28–45.

Hughes, G. M. (1966). The dimensions of fish gills in relation to their function. *J. Exp. Biol.* **45**, 177–195.

Hughes, G. M., and Ballintijn, C. M. (1965). The muscular basis of the respiratory pumps in the dogfish (*Scyliorhinus canicula*). *J. Exp. Biol.* **43**, 363–383.

Hughes, G. M., and Koyama, T. (1974). Gas exchange of single red blood cells within secondary lamellae of fish gills. *J. Physiol. (London)* **24b**, 82–83.

Hughes, G. M., and Morgan, M. (1973). The structure of fish gills in relation to their respiratory function. *Biol. Rev. Cambridge Philos. Soc.* **48**, 419–475.

Hughes, G. M., and Shelton, G. (1958). The mechanisms of gill ventilation in three freshwater teleosts. *J. Exp. Biol.* **35**, 807–823.

Hughes, G. M., and Umezawa, S. I. (1968). On respiration in the dragonet *Callionymus lyra* L. *J. Exp. Biol.* **49**, 565–582.

Hughes, G. M., Horimoto, M., Kikuchi, Y., Kakiuchi, Y., and Koyama, T. (1981). Blood flow velocity in microvessels of the gill filaments of the goldfish (*Carassius auratus* L.). *J. Exp. Biol.* **90**, 327–331.

Hughes, G. M., Peyraud, C., Peyraud-Waitzenegger, M., and Soulier, P. (1982). Physiological evidence for the occurrence of pathways shunting blood away from the secondary lamellae of eel gills. *J. Exp. Biol.* **98**, 277–288.

Isaia, J., Girard, J. P., and Payan, P. (1978a). Kinetic study of gill epithelial permeability to water diffusion in the freshwater trout, *Salmo gairdneri*: Effect of adrenaline. *J. Membr. Biol.* **41**, 337–347.

Isaia, J., Maetz, J., and Haywood, G. P. (1978b). Effects of epinephrine on branchial non-electrolyte permeability in rainbow trout. *J. Exp. Biol.* **74**, 227–237.

Ishimatsu, A., and Itazawa, Y. (1983). Difference in blood oxygen levels in the outflow vessels of the heart of an air-breathing fish *Channa argus*. Do separate blood streams exist in a teleostean heart? *J. Comp. Physiol.* **149**, 435–440.

Itazawa, Y., and Takeda, T. (1978). Gas exchange in the carp gills in normoxic and hypoxic conditions. *Respir. Physiol.* **35**, 263–269.

Janssen, R. G., and Randall, D. J. (1975). The effects of changes in pH and $P_{CO_2}$ in blood and water on breathing in rainbow trout, *Salmo gairdneri*. *Respir. Physiol.* **25**, 235–245.

Jensen, F. B., Nikinmaa, M., and Weber, R. E. (1983). Effects of exercise stress on acid-base balance and respiratory function in blood of the teleost *Tinca tinca*. *Respir. Physiol.* **51**, 291–301.

Johansen, K. (1972). Heart and circulation in gill, skin and lung breathing. *Resp. Physiol.* **14**, 193–210.

Johansen, K., and Pettersson, K. (1981). Gill $O_2$ consumption in a teleost fish, *Gadus morhua*. *Respir. Physiol.* **44**, 277–284.

Johansen, K., Lenfant, C., and Hanson, D. (1968). Cardiovascular dynamics in the lungfishes. *Z. Vergl. Physiol.* **59**, 157–186.

Jones, D. R., and Randall, D. J. (1978). The respiratory and circulatory systems during exercise. *In* "Fish Physiology" (W. S. Hoar and D. J. Randall, eds.), Vol. 7, pp. 425–501. Academic Press, New York.

Jones, D. R., Langille, B. L., Randall, D. J., and Shelton, G. (1974). Blood flow in the dorsal and ventral aortae of the cod (*Gadus morhua*). *Am. J. Physiol.* **266**, 90–95.

Kerstens, A., Lomholt, J. P., and Johansen, K. (1979). The ventilation, extraction and uptake of oxygen in undisturbed flounders, *Platichthys flesus*: Responses to hypoxia acclimation. *J. Exp. Biol.* **83**, 169–179.

Kiceniuk, J. W., and Jones, D. R. (1977). The oxygen transport system in trout (*Salmo gairdneri*) during sustained exercise. *J. Exp. Biol.* **69**, 247–260.

Kirsch, R., and Nonnotte, G. (1977). Cutaneous respiration in three freshwater teleosts. *Respir. Physiol.* **29**, 339–354.

Lacy, E. R. (1983). Histochemical and biochemical studies of carbonic anhydrase activity in the opercular epithelium of the euryhaline teleost, *Fundulus heteroclitus*. *Am. J. Anat.* **166**, 19–39.

Langille, B. L., and Jones, D. R. (1976). Examination of elastic nonuniformity in the arterial system using a hydraulic model. *J. Biomech.* **9**, 755–761.

Laurent, P., and Dunel, S. (1980). Morphology of gill epithelia in fish. *Am. J. Physiol.* **238**, R147–R159.

Lomholt, J. P., and Johansen, K. (1979). Hypoxia acclimation in carp—how it affects $O_2$ uptake, ventilation and $O_2$ extraction from water. *Physiol. Zool.* **52**, 38–49.

McDonald, D. A. (1960). "Blood Flow in Arteries." W. & J. McKay & Co., Ltd., Chatham.

McWilliams, P. G., and Potts, W. T. W. (1978). The effects of pH and calcium concentrations on gill potentials in the brown trout, *Salmo trutta. J. Comp. Physiol.* **126**, 277–286.

Maren, T. H., and Swenson, E. R. (1980). A comparative study of the kinetics of the Bohr effect in vertebrates. *J. Physiol. (London)* **303**, 535–547.

Marshall, W. S. (1978). On the involvement of mucous secretion in teleost osmoregulation. *Can. J. Zool.* **56**, 1088–1091.

Marvin, D. E., and Heath, A. G. (1968). Cardiac and respiratory responses to gradual hypoxia in three ecologically distinct species of fresh-water fish. *Comp. Biochem. Physiol.* **27**, 349–355.

Nikinmaa, M. (1982). Effects of adrenaline on red cell volume and concentration gradients of protons across the red cell membrane in the rainbow trout, *Salmo gairdneri. Mol. Physiol.* **2**, 287–297.

Nikinmaa, M., Soivio, A., and Railo, E. (1981). Blood volume of *Salmo gairdneri*: Influence of ambient temperature. *Comp. Biochem. Physiol. A* **69A**, 767–769.

Obaid, A. L., Critz, A. M., and Crandall, E. D. (1979). Kinetics of bicarbonate/chloride exchange in dogfish erythrocytes. *Am. J. Physiol.* **237**, 132–138.

Olson, K. R., and Kent, B. (1980). The microvasculature of the elasmobranch gill. *Cell Tissue Res.* **209**, 49–63.

Perry, S. F., and Randall, D. J. (1981). Effects of Amiloride and SITS on branchial ion fluxes in rainbow trout, *Salmo gairdneri. J. Exp. Zool.* **215**, 225–228.

Perry, S. F., Haswell, M. S., Randall, D. J., and Farrell, A. P. (1981). Branchial ionic uptake and acid-base regulation in the rainbow trout (*Salmo gairdneri*). *J. Exp. Biol.* **92**, 289–303.

Perry, S. F., Davie, P. S., Daxboeck, C., and Randall, D. J. (1982). A comparison of $CO_2$ excretion in a spontaneously ventilating, blood-perfused trout preparation and saline-perfused gill preparations: Contribution of the branchial epithelium and red blood cell. *J. Exp. Biol.* **101**, 47–60.

Peyraud-Waitzenegger, M. (1979). Simultaneous modification of ventilation and arterial $P_{O_2}$ by catecholamines in the eel, *Anguilla anguilla* L.: Participation of α and β effects. *J. Comp. Physiol.* **129**, 343–354.

Piiper, J., and Scheid, P. (1977). Comparative physiology of respiration: Functional analysis of gas exchange organs in vertebrates. *Int. Rev. Physiol.* **14**, 219–253.

Piiper, J., Baumgarten, D., and Meyers, M. (1970). Effects of hypoxia upon respiration and circulation in the dogfish *Scyliorhinus stellaris. Comp. Biochem. Physiol.* **36**, 513–520.

Poole, C. A., and Satchell, G. H. (1979). Nociceptors in the gills of the dogfish *Squalus acanthias. J. Comp. Physiol.* **130**, 1–7.

Prosser, C. L., Barr, L. M., Pranc, R. O., and Laver, C. Y. (1957). Acclimation of goldfish to low concentrations of oxygen. *Physiol. Zool.* **30**, 137–141.

Randall, D. J. (1972). Respiration. *In* "The Biology of Lampreys" (M. W. Hardisty and I. C. Potter, eds.), Vol. 2, pp. 287–306. Academic Press, New York.

Randall, D. J. (1982a). Blood flow through fish gills. *Symp. Soc. Exp. Biol.* pp. 173–191.

Randall, D. J. (1982b). The control of respiration and circulation in fish during hypoxia and exercise. *J. Exp. Biol.* **100**, 275–288.

Randall, D. J., and Cameron, J. N. (1973). Respiratory control of arterial pH as temperature changes in rainbow trout *Salmo gairdneri. Am. J. Physiol.* **225**, 997–1002.

Randall, D. J., and Daxboeck, C. (1982). Circulatory changes during exercise in fish. *Can. J. Zool.* **60**, 1135–1140.

Randall, D. J., Holeton, G. F., and Stevens, E. D. (1967). The exchange of oxygen and carbon dioxide across the gills of rainbow trout. *J. Exp. Biol.* **46,** 339–348.

Randall, D. J., and Jones, D. R. (1973). The effect of deafferentation of the pseudobranch on the respiratory response to hypoxia and hyperoxia in the trout (*Salmo gairdneri*). *Respir. Physiol.* **17,** 291–301.

Randall, D. J., Heisler, N., and Drees, F. (1976). Ventilatory response to hypercapnia in the larger spotted dogfish *Scyliorhinus stellaris. Am. J. Physiol.* **230,** 590–594.

Randall, D. J., Burggren, W. W., Farrell, A. P., and Haswell, M. S. (1981). "The Evolution of Air Breathing in Vertebrates." Cambridge Univ. Press, London and New York.

Rideout, V. C., and Katra, J. A. (1969). Computer simulation study of the pulmonary circulation. *Simulation* **12,** 239–245.

Riggs, A. (1979). Studies of the hemoglobins of Amazonian fishes: An overview. *Comp. Biochem. Physiol. A* **62A,** 257–271.

Sardet, C. (1980). Freeze fracture of the gill epithelium of euryhaline teleost fish. *Am. J. Physiol.* **23B,** R207–R212.

Satchell, G. H. (1974). The J reflex in the fish. *Proc. Int. Union Physiol. Sci.* **11.** 26th Int. Congr. Physiol. Sci., New Dehli.

Satchell, G. H. (1978). The J reflex in fish. *In* "Respiratory Adaptations, Capillary Exchange, and Reflex Mechanisms" (A. S. Paintol, P. Gill Kumar, eds.), pp. 422–441.

Saunders, R. L. (1961). The irrigation of the gills in fishes. I. Studies of the mechanism of branchial irrigation. *Can. J. Zool.* **39,** 637–653.

Saunders, R. L. (1962). The irrigation of the gills in fishes. II. Efficiency of oxygen uptake in relation to respiratory flow activity and concentrations of oxygen and carbon dioxide. *Can. J. Zool.* **40,** 817–862.

Scheid, P., and Piiper, J. (1971). Theoretical analysis of respiratory gas equilibration in water passing through fish gills. *Respir. Physiol.* **13,** 305–318.

Scheid, P., and Piiper, J. (1976). Quantitative functional analysis of branchial gas transfer: Theory and application to *Scyliorhinus stellaris* (Elasmobranchii). *In* "Respiration of Amphibious Vertebrates" (G. M. Hughes, ed.), pp. 17–38. Academic Press, New York.

Shelton, G. (1970). Regulation of breathing. *In* "Fish Physiology" (W. S. Hoar and D. J. Randall, eds.), Vol. 4, pp. 293–359. Academic Press, New York.

Skidmore, J. F., and Tovell, P. W. A. (1972). Toxic effects of zinc sulphate on the gills of rainbow trout. *Water Res.* **6,** 217–230.

Smith, D. G., and Chamley-Campbell, J. (1981). Localization of smooth muscle myosin in branchial pillar cells of snapper (*Chrysophys auratus*) by immunofluorescence histochemistry. *J. Exp. Zool.* **215,** 121–124.

Smith, D. G., and Gannon, B. J. (1978). Selective control of branchial arch perfusion in an air-breathing Amazonian fish *Hoplerythrinus unitaeniatus. Can. J. Zool.* **56,** 959–964.

Smith, D. G., Duiker, W., and Cooke, I. R. C. (1983). Sustained branchial apnea in the Australian short-finned eel, *Anguilla australis. J. Exp. Zool.* **226,** 37–43.

Smith, F. M., and Jones, D. R. (1978). Localization of receptors causing hypoxic bradycardia in trout (*Salmo gairdneri*). *Can. J. Zool.* **56,** 1260–1265.

Smith, F., and Jones, D. R. (1982). The effect of changes in blood oxygen-carrying capacity on ventilation volume in the rainbow trout—"*Salmo gairdneri.*" *J. Exp. Biol.* **97,** 325–334.

Sobin, S. S., Fung, Y. C., Tremer, H. M., and Lindall, R. G. (1979). Distensibility of human pulmonary capillary blood vessels in the intra-alveolar septa. *Microvasc. Res.* **17,** 5–87.

Soivio, A., and Hughes, G. M. (1978). Circulatory changes in secondary lamellae of *Salmo gairdneri* gills in hypoxia and anaesthesia. *Ann. Zool. Fenn.* **15,** 221–225.

Soivio, A., and Tuurala, H. (1981). Structural and circulatory responses to hypoxia in the secondary lamellae of Salmo gairdneri gills at two temperatures. *J. Comp. Physiol.* **145**, 37–43.

Soivio, A., Nikinmaa, M., and Westman, K. (1980). The blood oxygen binding properties of hypoxic *Salmo gairdneri*. *J. Comp. Physiol.* **136**, 83–87.

Soivio, A., Nikinmaa, M., Nyholm, K., and Westman, K. (1981). The role of gills in the responses of *Salmo gairdneri* during moderate hypoxia. *Comp. Biochem. Physiol. A* **70A**, 133–139.

Spitzer, K. W., Marvin, D. E., and Heath, A. G. (1969). The effect of temperature on the respiratory and cardiac response of the bluegill sunfish to hypoxia. *Comp. Biochem. Physiol.* **30**, 83–90.

Steffensen, J. F., Lomholt, J. P., and Johansen, K. (1982). Gill ventilation and $O_2$ extraction during graded hypoxia in two ecologically distinct species of flatfish, the flounder (*Platichthys flesus*) and the plaice (*Pleuronectes platessa*). *Environ. Biol. Fishes* **7**, 157–163.

Stevens, E. D. (1972). Some aspects of gas exchange in tuna. *J. Exp. Biol.* **56**, 809–823.

Thomas, S., and Hughes, G. M. (1982). A study of the effects of hypoxia on acid-base status of rainbow trout blood using an extracroporeal blood circulation. *Respir. Physiol.* **40**, 371–382.

Ultsch, G. R., and Gros, G. (1979). Mucus as a diffusion barrier to oxygen: possible role in $O_2$ uptake at low pII in carp (*Cyprinus carpio*) gills. *Comp. Biochem. Physiol.* **62A**, 685–689.

Van den Thillart, G., Randall, D. J., and Lin Hoa-ren. (1983). $CO_2$ and $H^+$ excretion by swimming coho salmon, *Oncorhynchus kisutch*. *J. Exp. Biol.* **107**, 169–180.

Vogel, W. O. P. (1978). Arteriovenous anastamoses in the afferent region of trout gill filaments (*Salmo gairdneri*). *Zoomorphologie* **90**, 205–212.

Vogel, W., Vogel, V., and Kremers, H. (1973). New aspects of the intrafilamental vascular system in gills of the euryhaline teleost *Tilapia mossambica*. *Z. Zellforsch. Mikrosk. Anat.* **144**, 573–583.

Vogel, W., Vogel, V., and Pfautsch, M. (1976). Arterio-venous anastamoses in rainbow trout gill filaments. *Cell Tissue Res.* **167**, 373–385.

Wainwright, S. A., Biggs, W. D., Currey, J. D., and Gosline, J. M. (1976). "Mechanical Design in Organisms." Wiley, New York.

Watters, K. W., Jr., and Smith, L. S. (1973). Respiratory dynamics of the starry flounder *Platichthys stellatus* in response to low oxygen and high temperature. *Mar. Biol.* **19**, 133–148.

Weber, R. E. (1982). Intraspecific adaptation of hemoglobin junction in fish to oxygen availability. *In* "Exogenous and Endogenous Influences on Metabolic and Neural Control" (A. D. F. Addink and N. Sprank, eds.), pp. 87–102. Pergamon, Oxford.

Weber, R. E., and Johansen, K. (1979). Oxygenation linked binding of carbon dioxide and allosteric phosphate cofactors by lungfish hemoglobin. *In* "Animals and Environmental Fitness" (R. Gilles, ed.), Vol. 2, pp. 49–50. Pergamon, Oxford.

Weber, R. E., and Lykkeboe, G. (1978). Respiratory adaptations in carp blood. Influences of hypoxia, red cell organic phosphates, divalent cations and $CO_2$. *J. Comp. Physiol.* **128**, 127–137.

Weber, R. E., Wells, R. M. G., and Rossetti, J. E. (1983). Allosteric interactions governing oxygen equilibric in the haemoglobin system of the spiny dogfish *Squalas acanthias*. *J. Exp. Biol.* **103**, 109–120.

Weibel, E. R. (1973). Morphological basis of alveolar-capillary gas exchange. *Physiol. Rev.* **53**, 419–495.

Wilkes, P. R. H., Walker, R. L., McDonald, D. G., and Wood, C. M. (1981). Respiratory, ventilatory, acid-base and ionoregulatory physiology of the white sucker, *Catostomus commersoni*. The influence of hyperoxia. *J. Exp. Biol.* **60**, 241–265.

Wood, C. M. (1974). A critical examination of the physical and adrenergic factors affecting blood flow through the gills of the rainbow trout. *J. Exp. Biol.* **60**, 241–265.

Wood, C. M., and Jackson, E. B. (1980). Blood acid-base regulation during environmental hyperoxia in the rainbow trout *Salmo gairdneri. Respir. Physiol.* **42**, 351–372.

Wood, C. M., and Randall, D. J. (1973). The influence of swimming activity on sodium balance in the rainbow trout (*Salmo gairdneri*). *J. Comp. Physiol.* **82**, 207–233.

Wood, C. M., McMahon, B. R., and McDonald, D. G. (1979). Respiratory gas exchange in the resting starry flounder *Platichthys stellatus:* a comparison with other teleosts. *J. Exp. Biol.* **78**, 167–179.

Wood, C. M., MacDonald, D. G., and McMahon, B. R. (1982). The influence of experimental anaemia on blood acid-base regulation *in vivo* and *in vitro* in the starry flounder (*Platichthys stellatus*) and the rainbow trout (*Salmo gairdneri*). *J. Exp. Biol.* **96**, 221–237.

Wu, R. S. S., and Woo, N. Y. S. (1983). Respiratory responses and tolerance to hypoxia in two marine teleosts *Epinephelus akaara* (Temminck and Schlegel) and *Mylio macrocephalus* (Basilewsky). Unpublished manuscript.

Yamamoto, K.-I., Itazawa, Y., and Kobayashi, H. (1980). Supply of erythrocytes into the circulating blood from the spleen of exercised fish. *Comp. Biochem. Physiol. A* **65A**, 5–11.

# 6

# ACID–BASE REGULATION IN FISHES*

*NORBERT HEISLER*

Abteilung Physiologie
Max-Planck-Institut für Experimentelle Medizin
Göttingen, Federal Republic of Germany

## I. INTRODUCTION

Maintenance of a constant pH in the body fluids at a given temperature is one of the important tasks of the regulatory systems for homeostasis in animals. Since most enzyme systems catalyzing metabolic reactions possess

*Dedicated to my friend Ernst Meißner on the occasion of his seventy-fifth birthday.

FISH PHYSIOLOGY, VOL. XA

pH optima in their activity distribution, changes in pH are expected to result in reduced metabolic performance. During normal steady-state conditions there is continuous production of surplus $H^+$ or $OH^-$ ions, which are eliminated from the body fluids by the excretory organs of the animals at the same rate as they are produced, such that pH in the body compartments is kept constant within narrow limits. During temperature change and stress conditions, however, the capacity of the excretory organs is usually not large enough to prevent transient acid–base disturbances, which are then limited by buffering of surplus $H^+$ or $OH^-$ ions in the body fluids and, at least in higher vertebrates, by respiratory compensation of nonrespiratory disturbances.

In fish, the buffer values of blood and intracellular compartments are generally much smaller than in higher vertebrates. The product of the relative volume and the hematocrit of the blood is smaller by a factor of 5 to 6, and also the nonbicarbonate buffer values of skeletal muscle tissues (cf. Albers, 1970; Heisler and Neumann, 1980; Cameron and Kormanik, 1982a; Heisler, 1984) are only 50–70% of those in mammalian muscle (Heisler and Piiper, 1971, 1972). Also, the buffer value of the bicarbonate–$CO_2$ buffer system, which is active only against $H^+$ ions originating from dissociation of nonvolatile acids, is smaller in fish by a factor of three to eight, depending on species and body compartment type (for review, see Heisler, 1980).

Fish are also handicapped in the respiratory compensation of nonrespiratory acid–base disturbances because of physical limitations (Rahn, 1966b) or the energetic problems with long-term hyperventilation of the viscous gas exchange medium, water. Accordingly, gill ventilation in fish is little affected during acid–base disturbances (e.g., Babak, 1907; Van Dam, 1938; Saunders, 1962; Peyraud and Serfaty, 1964; Dejours, 1973) or is increased only transiently (Janssen and Randall, 1975; Randall et al., 1976) to an extent much smaller than observed in air-breathing vertebrates.

Because the buffering mechanisms and adjustments of gill ventilation are limited, excretory mechanisms are of much greater importance in fish than in air-breathing vertebrates not only for steady-state pH adjustment, but also for the transient acid–base regulation during stress conditions. The basic principles of acid–base regulation in fish have been carefully delineated by Albers (1970) in Vol. IV of this series. The present chapter is intended to describe the experimental techniques for studies of acid–base relevant ion transfer between fish and environmental water, to review the observed regulatory patterns in the extracellular and intracellular compartments of fish, and to analyze the contributions of transepithelial and transmembrane ion transfer for the acid–base regulation during various stress conditions.

## II. TECHNIQUES FOR MEASUREMENT OF ACID–BASE ION TRANSFER

### A. General Considerations

#### 1. ACID–BASE RELEVANT SUBSTANCES

Any description of the acid–base status must focus on the pH value as a parameter of large significance for many biological functions. The pH in a body compartment can, according to its definition, be affected by addition or removal of $H^+$ ions from the compartment fluid. This can result from changes in the concentration of the volatile anhydride of carbonic acid, $CO_2$, or that of nonvolatile $H^+$-dissociating substances (acids) and substances capable of transferring $H^+$ ions into the nondissociated state (bases), induced by changed metabolic production. A number of other possibilities—such as changes in the affinity of buffer bases to $H^+$ ions by variation in ionic strength, or changes in binding of other electrolytes to buffer bases induced by large variations in the activity of these ions (cf. Jackson and Heisler, 1982)—are rarely encountered by fish.

Also, transepithelial and transmembrane transfer of acid–base relevant ions may have considerable influence on the compartmental acid–base status. The acid–base relevant ions may be divided into two groups:

1. The absolutely acid–base relevant ions are those that, when transferred, affect the acid–base status of both compartments involved under all possible conditions ($H^+$, $OH^-$, and $HCO_3^-$ ions).
2. The potentially acid–base relevant ions are those that—only under certain conditions—affect the acid–base status in either the compartment of origin or the compartment to which they are transferred. This group includes various buffer ions that change their dissociation when transferred between fluids with different pH values, as well as ammonium ions ($NH_4^+$).

Ammonia ($NH_3$) is the main nitrogenous waste product in ammoniotelic fish and is—at physiological pH values—almost completely ionized according to its high p$K$ value (e.g., Emerson *et al.*, 1975; Bower and Bidwell, 1978; Cameron and Heisler, 1983). The ionization results in equimolar removal of $H^+$ ions from the fluid. Transfer into other compartments is equivalent to transfer of $H^+$ ions, but because the ionization of $NH_4^+$ in biological fluids is rarely ever changed significantly (p$K$ ~9.6), the acid–base status of the receiving compartment is not affected.

The analysis of acid–base relevant transfer processes requires accordingly a determination of the net transfer of $H^+$, $OH^-$, and $HCO_3^-$ via the effect of these ions on the acid–base equilibrium in the respective compartment, and a separate determination of the transfer of potentially acid–base relevant substances. The effect of $H^+$, $OH^-$, and $HCO_3^-$ transfer can be determined from the changes in pH, the buffering characteristics, and the volumes of the respective compartments, whereas the $H^+$-equivalent contribution of potentially acid–base relevant transfer processes has to be individually evaluated on the basis of physicochemical parameters of the respective substances and involved compartments. In fish, the transfer of absolutely acid–base relevant ions ($H^+$, $OH^+$, $HCO_3^-$) and transfer of $NH_3/NH_4^+$ is predominant, and the effect of other transfer processes on acid–base regulation is usually negligible.

## 2. BUFFERING CHARACTERISTICS AND VOLUME OF BODY COMPARTMENTS

*a. Extracellular Space and Blood.* The volume of the extracellular space is about 20 to 25% of the total body water (e.g., Heisler *et al.*, 1976a; Heisler, 1978, 1984; Cameron, 1980), which in turn represents about 68 to 80% of the body weight (e.g., Heisler, 1978; Cameron, 1980; Holeton *et al.*, 1983). The extracellular space of water-breathing fish contains much less bicarbonate (4–13 m$M$; see Heisler, 1980, and Section III,B) than in species of higher classes of vertebrates, and only very small concentrations of nonbicarbonate buffers. Accordingly, the total buffer value of the extracellular space is rather small and is not much higher for the combined system of extracellular space and blood, since the blood volume is small (3–5% of the body weight). The nonbicarbonate buffer value is also generally low (6–10 mEq (pH liter)$^{-1}$; see Albers, 1970) because of the relatively low hematocrit of fish blood (15–25%). Only in some air-breathing fish species are the extracellular bicarbonate concentration and blood nonbicarbonate buffer value exceptionally high, being comparable to or even higher than in mammals (e.g., DeLaney *et al.*, 1974, 1977; Heisler, 1982a). As a result of the generally low buffer capacity (= buffer value × compartment volume), acid–base relevant ion transfer of significant quantity has a relatively large effect on the extracellular acid–base status.

*b. Intracellular Body Compartments.* The intracellular space, which represents about 75 to 80% of the body fluids, is a rather heterogeneous body compartment (see Heisler, 1978). The largest proportion is represented by the intracellular volume of muscle tissues (50–75% of the intracellular body fluids, calculated from data of Stevens, 1968; Cameron, 1974; Heisler, 1978; Neumann *et al.*, 1983), of which 2–20% is intracellular fluid of red muscle and 80–98% of white muscle tissue. Buffer values of intracellular

**Table I**

Buffering Properties of Intracellular Compartments of Fish[a]

| Species | Tissue | $\beta_{im}$ | $\beta_{ph}$ | $\beta_{im}/\beta_{ph}$ | $\beta_{tot}$ | Reference |
|---|---|---|---|---|---|---|
| *Scyliorhinus stellaris* | White muscle | 38.9 | 10.6 | 3.7 | 49.5 | Heisler and Neumann (1980) |
| | Red muscle | 20.5 | 17.9 | 1.1 | 38.4 | Heisler and Neumann (1980) |
| | Heart muscle | 27.1 | 10.4 | 2.2 | 37.5 | Heisler and Neumann (1980) |
| *Ictalurus punctatus* | White muscle | — | — | — | 35.0 | Cameron and Kormanik (1982a) |
| *Synbranchus marmoratus* | White muscle | 36.1 | 9.2 | 3.9 | 45.3 | Heisler (1982a, 1984) |
| | Heart muscle | 24.9 | 9.2 | 2.7 | 34.1 | Heisler (1982a, 1984) |

[a] Buffer values given in mEq (pH liter cell water)$^{-1}$. Total buffer value, $\beta_{tot}$; imidazole-like buffer value, $\beta_{im}$; phosphate-like buffer value, $\beta_{ph}$.

muscle compartments range from 37 to 49 mEq (pH kg cell water)$^{-1}$ for white muscle, from 34 to 37 mEq (pH kg cell water)$^{-1}$ for heart muscle, and 38 mEq (pH kg cell water)$^{-1}$ for red muscle (Table I). Although these values are lower than in mammalian muscle tissue (Heisler and Piiper, 1971, 1972), the intracellular muscle compartments of fish are still much better buffered than the extracellular space. More than 90% of the total buffer capacity of fish has to be attributed to the intracellular body compartment.

At constant temperature the buffer values in various types of muscle tissue are different by only a factor of 1.2 to 1.3. At variable temperature, however, the buffering characteristics of various muscle types differ considerably as a result of largely different contributions to the overall buffer value of "phosphate-like" and "imidazole-like" buffers with their different temperature coefficients (Heisler and Neumann, 1980; Heisler, 1984; Table I). These differences have considerable effect for the determination of transmembrane acid–base relevant ion transfer after changes of temperature (Section IV,A,2).

3. COMPOSITION OF NATURAL WATERS

The effect of excretion or uptake of various substances by the fish on the acid–base status of the environmental water is very much dependent on the composition of this water. Natural waters inhabited by fish may contain a large number of different ionized and nonionized substances at quite variable concentrations. The osmolarity may be much higher than that of seawater (>600 mOsm) in reservoirs with only part-time connection to the open sea and a high rate of water evaporation, or in inland saltwater lakes. In contrast, freshwater osmolarity of lower than 50–100 μOsm is found in areas of solute-poor soil, for example in some parts of the Amazon basin.

Most of the solutes in natural waters are strong electrolytes and have an important role as counterions for electroneutral ion exchange mechanisms and for the osmoregulation, but do not directly interact with acid–base relevant ions secreted or taken up by fish. The concentration of buffer substances in seawater is exceedingly low as compared to the other solutes (Harvey, 1974), but may in fresh water represent a sizable proportion of the total osmolarity.

The most important environmental buffer is the carbon dioxide–carbonic acid–bicarbonate–carbonate system. The concentration of bicarbonate, the largest component of this system, is rather constant in various oceans and is—depending on salinity—in the range of 2 to 2.5 m$M$. According to the temperature- and photosynthesis-induced variations of $P_{CO_2}$ from about 0.15 to 0.3 mm Hg, pH in the surface layers of the seawater is usually in the range of 7.8 to 8.5. In greater depths of 200 to 500 m, $P_{CO_2}$ may rise to values much higher than those near the surface (5–10 mm Hg; Harvey, 1974) as a result of anaerobic metabolism of bacteria and organic debris, combined with thermostratification.

In freshwater fish habitats the bicarbonate concentration is much more variable than in seawater and may range from concentrations of several times that in seawater to less than 1/100 of the seawater concentration ($<10$ $\mu M$) in some areas of the Amazon basin (e.g., Sioli, 1954, 1955, 1957; Heisler, 1982a). In recent years similarly low bicarbonate concentrations have been found also in many lakes of the northeastern United States and Canada as well as of southern Norway and Sweden as a result of acid precipitation mainly originating from oxidation of sulfurous contaminations in fossil fuels used in energy production.

Also, $P_{CO_2}$ in natural fresh waters is much more variable than in seawater, ranging from values of less than the atmospheric partial pressure ($\sim 0.26$ mm Hg $\cong 0.035\%$) due to photosynthesis, up to values as high as 60 mm Hg (cf. Heisler et al., 1982). Such high water $P_{CO_2}$ values are usually due to hindrance of gas exchange between water and air by dense water surface vegetation or by thermostratification. This results in accumulation of $CO_2$ produced by aerobic metabolism and $CO_2$ liberated from water bicarbonate by $H^+$ ions as the metabolic end product of various anaerobic pathways of microorganisms.

As a result of largely variable bicarbonate concentrations and $P_{CO_2}$ values, pH in fresh water covers a much larger range—as compared to seawater—from about 8.5 to 9 down to pH 4 (and even to lower values, as a result of acid precipitation, which, however, is then not compatible with sustained fish life). Probably the highest pH ($\sim 10$) in a fish-inhabited body of water is found in the hot Lake Magadi, in Kenya (temperature 25–40°C), the water of which contains about 87 m$M$ carbonate and bicarbonate, and 210 m$M$ Na$^+$ (F. B. Eddy, personal communication).

Nonbicarbonate buffers are usually much less concentrated than bicarbonate. In seawater of buffers with p$K$ values in the range of $\pm 1.5$ pH Units of the water pH, only borate (p$K$ $\sim 8.75$, 15°C, salinity 36‰) is available in appreciable concentration ($\sim 0.5$ m$M$) (Harvey, 1974). Fresh water may also contain considerable concentrations of phosphate as a result of rain washout from artificially fertilized soils.

## B. Experimental Approach

The transfer of acid–base relevant substances cannot be monitored directly to date but has to be deduced from the effect of the transferred species on the acid–base status of one or more of the involved fluid compartments.

This is generally performed by determination of pH, $P_{CO_2}$, and bicarbonate concentration, and of the nonbicarbonate buffer value of the compartment fluid. The amount of surplus $H^+$ ions can also be determined by direct titration. This method, however, is not applicable for biological fluids and has also some drawbacks with application to environmental water, which will be discussed later (Section II,C).

## 1. Transepithelial Transfer Processes

*a. Fick Principle: Blood.* The excretion or resorption of acid–base relevant ions to or from the environmental water should be accountable by application of the Fick principle from measurement of blood flow ($\dot{Q}$) through the exchange organ (e.g., the gills) and the arteriovenous difference in bicarbonate equivalents in the blood ($\Delta[HCO_3^-]_{a-v}$):

$$\Delta HCO_3^-{}_{e \to w} = \dot{Q}[\Delta HCO_3^-]_{a-v} \qquad (1)$$

The difference in bicarbonate equivalents in the blood is determined after standardization of $P_{CO_2}$ and oxygen saturation by equilibration of both venous and arterial blood samples with the same gas mixture from the difference in total $CO_2$ content ($C_{CO_2}$), and the $H^+$ ions bound to nonbicarbonate buffers.

$$[\Delta HCO_3^-]_{a-v} = \Delta C_{CO_2\,a-v} + \Delta pH \cdot \beta_{NB\,bl} \qquad (2)$$

where $\beta_{NB\,bl}$ is the nonbicarbonate buffer value of the blood.

This procedure eliminates the influence of the respiratory $CO_2$ gas exchange, and the Bohr effect on the acid–base status of the blood. Simple determination of total $CO_2$ differences in nonstandardized blood (e.g., Cross *et al.*, 1969) is therefore inappropriate.

However, even proper application of the method just described would not lead to satisfying accuracy in the determination of acid–base relevant ion transfer. Since the resting $H^+$ ion excretion rate of fish is generally in the range of $-0.8$ to $+0.8$ μmol (min kg fish)$^{-1}$ (Table III), and cardiac output in fish averages about 30 ml (min kg fish)$^{-1}$ (cf. Neumann *et al.*, 1983), the expected bicarbonate-equivalent a–v difference is in the range of $\pm 25$ μ$M$, and thus well below the detection limits of blood acid–base measurement techniques. Although the excretion rates increase during acidic stress situations (see later), cardiac output under such conditions is regularly increased as well (cf. Neumann *et al.*, 1983).

*b. Fick Principle: Water Flow-Through System.* The use of a flow-through respirometer also does not provide the required accuracy. The water flow through the fish chamber has to be kept at least five times as high as the respiratory water flow through the gills in order to avoid anomalously high inspired $P_{CO_2}$ and low $P_{O_2}$. Application of the Fick principle will yield more reliable results than the previously described approaches only when the gill ventilation can be accurately measured and when mixed expired water samples can be taken directly from the gill opercula (e.g., by application of the method of collecting the gill effluent water with rubber funnels glued to the skin surface around the opercula; Piiper and Schumann, 1967). With a gill ventilation rate of about 200 to 300 ml (min kg fish)$^{-1}$ (e.g., Piiper and Schumann, 1967; Randall et al., 1976), the inspired–expired water concentration difference for excreted H$^+$ ions (resting rate ±0.8 µmol (min kg fish)$^{-1}$) will be about ±3 µM. This concentration difference is in the range of the resolution limits of the most accurate analytical techniques (Section II,C,1,c) and thus also cannot be expected to yield reliable data (e.g., Randall et al., 1976).

c. Closed Water Recirculation System. Since the available analytical techniques limit the application of the Fick principle to preliminary and qualitative experiments, most recent studies of transepithelial acid–base ion transfer processes have been performed in closed water recirculation systems (Heisler et al., 1976a). Such systems typically consist of a fish box, an oxygenator and bubble trap system, and a water circulation pump (Fig. 1). They are closed for all nonvolatile substances, but open for molecular $CO_2$. The amount of bicarbonate-equivalent ions transferred between fish and environmental water during a time interval (indices 1 and 2) is determined from the change in bicarbonate-equivalent ions in the water ($\Delta[HCO_3^-]_w$), the volume of the water system ($V_w$), and the volume of the fish ($V_f$):

$$\Delta[HCO_3^-]_w = \Delta C_{CO_2\ 1-2} + \Delta pH \cdot \beta_{NB\ w} \tag{3}$$

$$\Delta HCO_3^-{}_{e \to w} = \Delta[HCO_3^-]_w \cdot \frac{V_w}{V_f} \tag{4}$$

where $\beta_{NB\ w}$ is the nonbicarbonate buffer value of the water and $V_f$, according to the reference system used, is either the total amount of body water, or the weight of the fish.

The main advantage of a closed recirculation system is that ions excreted by the fish accumulate to concentration changes in the water that can be measured accurately with the available methods (see Section II,C). Also, the integration of transferred ion species is not biased by changes in the excretion rate between single measurements and the directional error of the individual determinations in an open system. Care, however, has to be taken to prevent buildup of toxic metabolic end products (e.g., ammonia). This is done by exchange of the recirculating water with fresh water of the same $P_{O_2}$, $P_{CO_2}$, and pH, thermostatted to the experimental temperature.

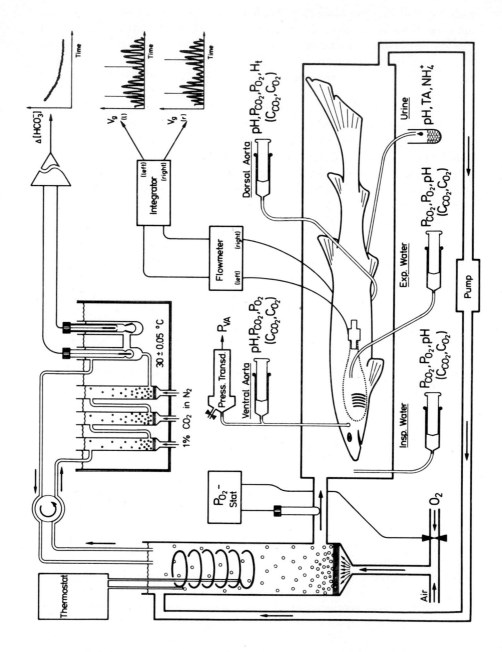

*d. Extrabranchial Excretion.* The relative contribution of gills and other exchange sites to the total transfer of acid–base relevant ions between fish and environmental water can be evaluated when the outputs of the respective sites are separated from the environmental water. This can be accomplished for the gills by application of the method of collecting the gill effluent water by rubber funnels glued to the skin surface around the gill opercula (Piiper and Schumann, 1967), or by the divided-chamber technique (Van Dam, 1938). With this technique the segment of the fish including the gill opercula is separated from the rest of the body by rubber dams such that the gill effluent water can also be collected. However, this method has a relatively low sensitivity (see Section II,B,1,b and c) and is barely tolerated by nonanesthetized fish for longer time periods. The technique of Piiper and Schumann (1967) can be applied with good success only to elasmobranch fish. Therefore, it is desirable to separate instead the output of other sites of excretion.

Urine can be sampled via catheters introduced into the urethra and fixed by a circular ligature (e.g., Heisler *et al.*, 1976a; Heisler, 1978, 1982a; Holeton and Heisler, 1983) or held in position by spherical enlargements on the catheter positioned inside and outside the external urethra sphincter (e.g., Cameron and Kormanik, 1982a). The catheter–urethra fit is made additionally tight by direct application of cyanoacrylate glue (e.g., Heisler *et al.*, 1976a; Heisler, 1978; Holeton and Heisler, 1983) or by suturing a piece of surgical rubber on the surface around the papilla and glueing the rubber to the body surface and the papilla with cyanoacrylate (e.g., Cameron and Kormanik, 1982a). The catheter is filled with water and siphoned with very small negative pressure (1–5 cm $H_2O$). Such arrangements remain functional in elasmobranchs for up to 3 weeks, in teleost fish species not more than 1 week, and in some species not more than 2 days. The collected urine is analyzed for titratable acidity and ammonia concentration.

In elasmobranchs, the fluid released from their unique abdominal pores and rectal glands can be collected by rubber fingerlings glued to the skin surface around the cloaca of female specimens. Skin ion excretion has been studied in elasmobranchs by glueing rubber pockets on the skin surface of

---

**Fig. 1.** Typical closed water recirculation system for the determination of transepithelial acid–base relevant ion transfer processes (here during hyperoxia-induced hypercapnia). The fish is confined in a box that is continuously flushed with recirculated water, oxygenated and thermostatted in a water gas exchange system (left). Water $P_{O_2}$ is regulated by a $P_{O_2}$-stat system (or other parameters like pH, by appropriate stat systems). Water is pumped continuously through a Δ-bicarbonate measurement system (upper middle, see Section II,C,1,c). Gill ventilation is monitored by direct measurement with electromagnetic flowmeters of the gill effluent water collected by rubber pockets glued to the surface of the fish around the gill opercula. For further details see text, Section II,B and C.

sharks, filling with water, and analyzing after time periods of several hours (Heisler *et al.*, 1976a; Heisler, 1978; Holeton and Heisler, 1983; N. Heisler and P. Neumann, unpublished). Separation between branchial and renal plus skin excretion has also been studied in fish pups by rubber condoms placed over the hind end of the whole animal and secured by a thread (Evans, 1982).

## 2. Transmembrane Transfer Processes

The determination of the transfer of acid–base relevant ions between intracellular compartments and the extracellular space requires more analytical efforts than determination of the transepithelial transfer. The Fick principle is usually not applicable because of the same limitations described already (i.e., blood flow measurements, small arteriovenous concentration differences). Accordingly, transmembrane transfer can only be determined by its effect on a specific intracellular compartment, or the transfer can be determined summarily between the "intracellular space" and the extracellular space from the acid–base changes in the extracellular space and the transepithelial transfer.

*a. Effect of Transmembrane Transfer on the Intracellular Acid–Base Status.* Any transfer of acid–base relevant ion species to or from the intracellular compartment affects the intracellular pH ($\Delta pH_i$), which can be determined by a number of methods (e.g., from distribution of 5,5-dimethyl-2,4-oxazolidinedione, DMO; Waddell and Butler, 1959; see also Heisler, 1975; Heisler *et al.*, 1976b). The bicarbonate-equivalent transfer ($\Delta HCO_3^-{}_{i \rightarrow e}$) can be quantified, if the nonbicarbonate buffer value ($\beta_{NB\,i}$) of the intracellular compartment is known (for method, see Heisler and Piiper, 1971; Heisler and Neumann, 1980; Table I):

$$\Delta HCO_3^-{}_{i \rightarrow e} = (\Delta pH_i \cdot \beta_{NB\,i} - \Delta[HCO_3^-]_i) \cdot V_i \qquad (5)$$

where $\Delta[HCO_3^-]_i$ is the change in intracellular bicarbonate concentration and $V_i$ is the volume of intracellular water.

If the transmembrane transfer is to be determined during changes in temperature, the temperature-induced shifts of the nonbicarbonate buffer p$K$ values have also to be taken into account (cf. Heisler and Neumann, 1980).

$$\Delta HCO_3^-{}_{i \rightarrow e} = \{(\Delta pH_i - \Delta pK_{im}) \cdot \beta_{im} + (\Delta pH_i - \Delta pK_{ph}) \cdot \beta_{ph} - \Delta[HCO_3^-]_i\} \cdot V_i \qquad (6)$$

Division of the nonbicarbonate buffers into two groups—imidazole-like buffers (index "im") with a temperature coefficient of $\Delta pK/\Delta t = -0.021$ U/°C, and phosphate-like buffers (index "ph") with $\Delta pK/\Delta t = -0.002$ U/°C—

yields a satisfactory approximation of the buffering characteristics of tissues (Heisler and Neumann, 1978, 1980; Heisler, 1984).

An important prerequisite for this type of approach is that the production of nonvolatile acid–base relevant substances in the cells is equivalent to their elimination, that is, that the intracellular compartment is in a steady state. Any nonsteady-state conditions of the cell metabolism (e.g., after lactic acid production induced by severe exercise) will result in an estimate of the transmembrane transfer that is considerably different from the actual value.

Since during steady-state conditions the basic production of $H^+$ ions is equivalent to its elimination, this approach is not applicable for determination of the steady-state production of acid–base relevant ions.

*b. Quantitative Analysis of Transepithelial Transfer and Changes in the Extracellular Compartment.* The overall acid–base relevant ion transfer between the intracellular and the extracellular compartments can be determined from the amount transferred to the environmental water and the changes of the acid–base status in the extracellular compartment. As long as no acid–base relevant substances are produced or consumed in the extracellular space, the changes in bicarbonate-equivalent ions ($\Delta HCO_3^-{}_e$) in the extracellular space can be described as follows:

$$\Delta HCO_3^-{}_e = (\Delta[HCO_3^-]_e - \Delta pH_e \cdot \beta_{NB\ e}) \cdot V_e + (\Delta[HCO_3^-]_{rc} - \Delta pH_{rc} \cdot \beta_{NB\ rc}) \cdot V_{irc} \tag{7}$$

where indices e and rc denote parameters of the extracellular space or the erythrocytes, respectively, $V$ is the fluid volume, and $\beta_{NB}$ is the nonbicarbonate buffer value. When $P_{CO_2}$ is constant and the total buffer value of blood ($\beta_{tot} = \beta_{NB} + \beta_{Bic}$) is known,

$$\Delta HCO_3^-{}_e = (\Delta[HCO_3^-]_e - \Delta pH_e \cdot \beta_{NB\ e}) \cdot (V_e - V_{pl}) - \Delta pH_e \cdot \beta_{tot} \cdot (V_{pl} + V_{rc}) \tag{8}$$

where $V_{pl}$ is the plasma volume and $\beta_{Bic}$ is the bicarbonate buffer value.

Since the blood volume in fish is usually relatively small (3–5% of the body fluids) as compared to the extracellular space (20–25%), and the hematocrit low, the contribution of the intraerythrocyte nonbicarbonate buffers (hemoglobin, phosphates) can often be neglected, or only roughly approximated without any significant effect on the final result.

The overall amount of bicarbonate-equivalent ions transferred is

$$\Delta HCO_3^-{}_{i \to e} = \Delta HCO_3^-{}_{e \to w} + \Delta HCO_3^-{}_e \tag{9}$$

where $\Delta HCO_3^-{}_{e \to w}$ is the amount transferred between extracellular space and environmental water (Eqs. (3) and (4)).

Acid–base relevant substances are not produced in the extracellular space and, under normal experimental conditions, in the environmental water. Accordingly this approach can only be biased by release or deposition of $CaCO_3$ in the bone structures of the fish, which is, however, a relatively slow process.

## C. Analytical Techniques

Measurement of various parameters is required for all of the foregoing approaches for the determination of transepithelial and transmembrane acid–base relevant ion transfer processes. Quite a number of different methods have been applied. Not all, however, have been proven to be suitable for the intended purpose. This section describes the analytical techniques and evaluates their theoretical and practical accuracy and their applicability.

### 1. METHODS FOR DETERMINATION OF ACID–BASE RELEVANT ION TRANSFER BETWEEN FISH AND ENVIRONMENTAL WATER

Excretion or uptake of $H^+$, $OH^-$, or $HCO_3^-$ ions by fish to or from the environmental water directly affects the acid–base status of the animal and the water (absolutely acid–base relevant ions). Since the body fluids are divided into various compartments with different volumes, buffering characteristics, and acid–base status, transfer of these ions between fish and environment can reliably be determined only on the basis of their effect on the acid–base status of the water. $H^+$, $OH^-$, and $HCO_3^-$ ions are in equilibrium with each other:

$$CO_2 + \boxed{\begin{array}{c} H_2O \\ H_2O \end{array}} \begin{array}{c} \rightleftarrows \ H_2CO_3 \ \rightleftarrows \\ \rightleftarrows \ OH^- \ + \end{array} \boxed{\begin{array}{c} H^+ \\ H^+ \end{array}} + HCO_3^- \ \rightleftarrows 2H^+ + CO_3^{2-} \tag{10}$$

Consequently, addition of $H^+$ ions or removal of $HCO_3^-$ or $OH^-$ to or from a fluid compartment in an equimolar quantity will result in the same changes in the concentrations of the substances in Eq. (10). It therefore is impossible to draw any conclusion about the type of ionic species transferred from the observed concentration changes. The permanent exchange between the contributors of Eq. (10), which is the result of their thermodynamic activity, also negates the application of tracer molecules for an identification of the transferred ion species. The tracer is readily transferred between $H^+$ pool, water pool, and bicarbonate pool ($^3H$) or between bicarbonate–carbonate pool, carbonic acid pool, and the pool of dissolved and volatile molecular $CO_2$ ($^{14}C$). Neither can a conclusive decision be drawn from the net transfer of counterions in opposite direction for maintenance of electroneutrality (see Section V). Since the stoichiometry of the ion exchange mechanisms is not quite clear yet, unidirectional fluxes of $Na^+$ and $Cl^-$ also provide only qualitative information.

Transfer of any of the three different ions ($H^+$, $OH^-$, or $HCO_3^-$) directly affects the bicarbonate concentration of the involved compartments, the smaller the nonbicarbonate buffer value is in the respective fluid. As a result of this primary effect, it is conventional to describe any transfer of absolutely acid–base relevant ions as "bicarbonate equivalent," or simply as "bicarbonate transfer," although other ions are very likely also involved.

*a. Gasometric Methods.* Direct determination of the total $CO_2$ content ($C_{CO_2}$) at the actual pH of the water sample is a theoretically accurate method for the detection of transepithelial "bicarbonate" transfer. This is valid if the nonbicarbonate buffer value of the water is insignificant as compared to the bicarbonate buffer value at constant $P_{CO_2}$($\beta_{Bic} = 2.3[HCO_3^-]$). Such conditions are found in almost all natural fresh waters. Prominent exceptions are the Amazon range, and lakes that are acidified by acid precipitation. When such conditions are mimicked in closed water recirculation systems, relatively small excretion of nonbicarbonate buffer bases like phosphate may introduce a significant error into the measurement. Then the nonbicarbonate buffer value and the change in pH have to be determined several times during the experiment, and the amount of $H^+$ ions buffered by nonbicarbonate buffers has to be taken into account, or a double-titration procedure (see Section II,C,3) has to be performed.

In seawater the natural borate concentration also has to be taken into account, or the water pH has to be standardized to about 7 by equilibration with an elevated $P_{CO_2}$, in order to get out of the range of significant nonbicarbonate buffering.

As long as the water pH is in the range of pH 7, where the nonbicarbonate buffer value of natural waters is about zero and the bicarbonate buffer value relatively high, the changes in water bicarbonate concentration can be taken as representative for surplus $H^+$ or $OH^-$ ions. Then all methods for determination of total $CO_2$ in the water yield sufficiently accurate results within the methodological resolution limits.

The classical gasometric methods (e.g., Van Slyke apparatus, Van Slyke, 1917) can usually not be expected to resolve less than $\pm0.100$ m$M$ in water bicarbonate concentration. Slightly better accuracy is achieved by transfer of bicarbonate into physically dissolved $CO_2$ by acidification and measurement of the change in $P_{CO_2}$ (Cameron, 1971). This method yields a resolution of about $\pm2.5\%$ (equivalent to $\pm0.050$ m$M$ on a background of 2 m$M$ [$HCO_3^-$]), with introduction of an increased sample volume and direct calibration with $HCO_3^-$ standards. The theoretical accuracy of the method of differential conductivity measurement (Maffly, 1968) with an apparatus recently developed (Capnicon III, Cameron Instruments, Inc., Port Aransas, Texas) is much better and only limited to about $\pm 0.5\%$ (equivalent to 0.010 m$M$ on a background of 2 m$M$ [$HCO_3{}^-$]) by the problems involved with volumetric measurements of samples and standards.

  $b.$ *Titration Procedures.* Acid–base relevant ion transfer is often determined by titration of the water to a pH of 4 in order to transfer bicarbonate and carbonate quantitatively into molecular $CO_2$. According to the Henderson–Hasselbalch Equation, less than 1% of the total $CO_2$ is then present as bicarbonate. When the molecular $CO_2$ is eliminated from the sample by equilibration with decarbonated air or pure nitrogen, accuracies of better than $\pm 0.1\%$ are theoretically expected. However, as for the gasometric methods, the accuracy of titration procedures is limited by the volumetric determinations of the sample volume, the volume of the environmental water, and the volume of the added titrant. Furthermore, titrations are biased by the problems associated with production and stability of highly accurate dilute titrants (here the volatile dissolved HCl). Also, reduction of the pH to 4.0 will result in titration of any buffer bases with p$K$ values between about 3 and 9, which covers a large proportion of the biological organic bases.

  Such substances may very well be, and are actually often, unpredictably released by the fish as indicated by the considerable rise of the nonbicarbonate buffer value of the water during experiments especially in the range between pH 6.5 and 4 (N. Heisler, G. F. Holeton, P. Neumann, and H. Weitz, unpublished). This factor can only be eliminated by back-titration to the original pH of the environmental water after careful equilibration with $P_{CO_2} = 0$ (which is a slow process because of the delayed uncatalyzed dehydration of $CO_2$), or by establishment of a nonbicarbonate buffer curve for each analyzed sample. According to the factors just mentioned, the absolute accuracy of titration procedures for the determination of surplus $H^+$ or $OH^-$ ions in the water may very well be drastically reduced as compared to its reproducibility, and cannot be expected to yield a better standard deviation than $\pm 2\%$ (equivalent to 0.040 m$M$ on a background of 2 m$M$ [$HCO_3{}^-$]). Accordingly, titration is less desirable as a method for determination of the water bicarbonate concentration changes than the most accurate gasometric methods, especially the Capnicon method, at least in water with relatively high bicarbonate concentration (e.g., seawater).

  When the effects of low environmental water pH on the acid–base status and the associated acid–base relevant ion transfer processes are studied in a closed water recirculation system, the pH of the water has to be kept constant by titration of the whole recirculation system (Ultsch *et al.*, 1981). As a result of the extremely low bicarbonate concentration (e.g., $< 1$ $\mu M$ at pH 4), the pretitrated water pH will rise within short periods to much higher values as a result of the bicarbonate-equivalent ion release of the fish. At $P_{CO_2}$ of about 1 mm Hg an increase in bicarbonate to only 50 $\mu M$ of water will result in elevation of the pH to about 6. This amount, and the quantity of free $H^+$ ions that have to be titrated between pH 4 and 6 ($\sim$100 $\mu M$), are released by control carp or trout within 6 to 12 hr, if the volume ratio of water to fish is about 10. The release at pH 4, however, is larger by a factor of about two (Ultsch et al., 1981; McDonald and Wood, 1981). With these conditions, the use of a pH-stat titration system is an undispensable prerequisite for reproducible experimental conditions, and provides directly the bicarbonate-equivalent release of the fish.

*c. Potentiometric Methods.* The change in water bicarbonate concentration of the environmental water can also be determined by measurement of pH and $P_{CO_2}$, and calculation based on the Henderson–Hasselbalch equation with application of appropriate values for $pK_1'''$ and $\alpha_{CO_2}$ (see Section II,C,4). The accuracy of this method is almost exclusively determined by the uncertainty of $\pm 4\%$ in the measurement of the extremely low $P_{CO_2}$ values ($<1$ mm Hg) with appropriate electrodes (Heisler et al., 1976b). This problem can be eliminated by standardization of $P_{CO_2}$ before the measurement of pH by equilibration with a gas of constant $P_{CO_2}$. The accuracy achieved by this approach ($\pm 1.5\%$, equivalent to $\pm 0.030$ mM on a background of 2 mM [$HCO_3^-$]) can further be improved by measurement of pH not with the usual blood pH electrode equipment, but with any type of commercially available large-surface, low-temperature glass electrode chain introduced directly into the equilibration device ($\pm 0.015$ mM) (method applied by Heisler et al., 1976a).

When $P_{CO_2}$ and pH are monitored directly in the recirculation system with electrodes selected for extreme stability, and the fluctuations in $P_{CO_2}$ of the water are small, changes in water bicarbonate concentration of about $\pm 0.5\%$ can be detected (method applied by Heisler, 1982a).

The most sensitive and accurate method for this type of measurement to date is the "$\Delta$-bicarbonate measurement system" (Heisler, 1978). Water is pumped continuously by a roller pump from the fish system countercurrent with a gas of constant $P_{CO_2}$ bubbled through the fritted-glass bottoms of three columns connected in series, and is fed over a special pH-sensitive electrode set before being returned to the fish system. The electrode chain consists of a large-surface, spherical pH glass electrode and a double electrolyte bridge sleeve diaphragm reference, which is soaked for a limited time period in 0.1 $N$ HCl, adapted to the ionic strength of the respective water for several months, and then selected for optimal long-term stability. Since all changes in ionic strength of the medium at the electrode chain result in transient instability of the electrode set, calibration should be performed with buffers of the same ionic strength as the water, or, if this is impossible (as in freshwater experiments), the procedure should be kept as short as possible. The equilibration gas, consisting of 0.3 to 1% $CO_2$ in $N_2$, is provided by a gas-mixing pump (Wösthoff, Bochum, Federal Republic of Germany). Equilibration and measurement are performed at a constant temperature, which is usually higher than the temperature in the fish system (30–40 ± 0.02°C), in order to improve the stability by reduction of the glass electrode impedance. By taking these precautions, the drift of the whole system can be kept well below ±0.0005 pH Units 24 hr$^{-1}$ and is usually less than 0.010 pH Units during periods of several months (Heisler, 1978; Claiborne and Heisler, 1984).

Because of its extreme stability, the system is calibrated only once before each experiment and is rechecked after the experiment. This can be performed by additions of known amounts of $NaHCO_3$ to the water of the whole fish recirculation system, or, since the total $CO_2$ content of natural waters is related monoexponentially to pH below pH 7 (see also Section II,C,4), also by only one known addition. Such direct calibration is advantageous in several ways. It eliminates the need for pH buffer calibrations of the electrode with all the instability problems related to the exposure of the electrodes to fluids different from the environmental water. It also eliminates any error in the volume determination of the water recirculation system.

When the recording of the electrode signal is performed via isolation amplifiers and proper grounding of the fish system is conducted, then the noise of the setup is well below the electrode drift, which, since the amount of standard added can be weighed to a least four digits, then determines the accuracy of the system (<0.1%, equivalent to 0.002 m$M$ on a background of 2 m$M$ water bicarbonate).

Since ion-selective electrodes are usually cross-sensitive to changes in the $H^+$ ion concentration (especially glass $Na^+$ electrodes), a similar standardization for $P_{CO_2}$ as performed in the $\Delta$-bicarbonate measurement system is strongly recommended for the measurement of counterion fluxes in experiments with largely changing $P_{CO_2}$ in the recirculation system (e.g., during control periods and subsequent environmental hypercapnia; method applied by Heisler et al., 1982).

## 2. Ammonia Determination

Numerous methods are available for the measurement of ammonia concentrations in natural waters and biological fluids.

The phenolhypochlorite method (Solorzano, 1969) has been widely used for studies in fish physiology and is proven to be very sensitive and reliable when applied to water samples. In biological solutions like plasma this method is often subject to interferences with substances other than ammonia, whereas the recently developed enzymatic methods (Boehringer, Mannheim, Federal Republic of Germany; Sigma Chemicals, St. Louis, Missouri) are absolutely specific and even sensitive enough for many applications in natural waters.

Ammonia electrodes, which consist of a pH electrode in an internal filling solution separated from the sample by a porous hydrophobic membrane, are also rather sensitive ($10^{-7}$ $M$) and specific. The only group of substances interacting with the measurement of ammonia are volatile amines. However, if such amines are actually excreted by the fish (e.g., trimethylammonium, cf. Bilinski, 1960, 1961, 1962), the effect for the acid–base regulation is the same as for ammonia. Therefore, it can only be considered as advantageous for quantitative analyses of the acid–base regulation to determine these substances simultaneously together with ammonia as "ammonia-like" substances. The principle of ammonia electrodes is similar to that of $CO_2$ electrodes: after alkalinization of a sample to pH >12, all ammonium is transferred into physically dissolved ammonia gas, which diffuses through the membrane into the internal filling solution of the electrode and changes pH in a logarithmic function of the $P_{NH_3}$. The changes in pH are sensed by a single-unit glass electrode chain.

Such electrodes have been used for automatic measurements of water ammonia (Claiborne and Heisler, 1984). Controlled by a microprocessor/clock assembly, water samples were drawn from the fish system and mixed with 10 $N$ NaOH in the ammonia electrode chamber. Automatic calibration with $NH_4Cl$ standards introduced by solenoid valves and automatic recalibration of the ion analyzer immediately before every measurement resulted in an accuracy better than 0.001 m$M$ (Claiborne and Heisler, 1984).

Because of the hydrophobic nature of the electrode membrane, the application of ammonia electrodes to biological fluids containing substances with surface tension affecting activity is limited.

## 3. DETERMINATION OF THE EXCRETION BY KIDNEYS, RECTAL GLAND, AND SKIN

Ions that are excreted with the urine or the fluid released from the rectal gland are diluted in a much smaller volume than ions excreted branchially in the environmental water. The urine flow rate is extremely small in elasmobranchs, and the rate of fluid release from the rectal gland is even smaller. Also, in teleost fish species the urine flow rate rarely exceeds 4 ml (hr kg fish weight)$^{-1}$ (for review, see Hunn, 1982). Accordingly, the analysis of excreted ions is much easier than that of those excreted branchially and diluted in the environmental water.

The amount of $H^+$ ions excreted with the urine $(\Delta H^+_u)$ is the sum of ammonia in the urine $(NH_4^+_u)$ plus the titratable acidity $(TA)$ minus the urine bicarbonate:

$$\Delta H^+_u = NH_4^+_u + TA_u - HCO_3^-_u \tag{11}$$

The ammonia concentration is determined by using the same methods as for water samples, the titratable acidity by titration of the urine to the plasma pH of the animal, and the urine bicarbonate either gasometrically (e.g., Van Slyke), from calculation by application of the Henderson–Hasselbalch equation after measurement of pH and $P_{CO_2}$, or by a double-titration procedure. The urine is then titrated to pH <4 with HCl to transfer all bicarbonate to physically dissolved $CO_2$, then back-titrated with NaOH to the original urine pH: the amount of bicarbonate in the urine is then

$$HCO_3^-_u = HCl_{pH_u \to pH4} - NaOH_{pH4 \to pH_u} \tag{12}$$

If the non-ammonia $H^+$ excretion is to be determined summarily, $(TA - HCO_3^-)_u$, this can be performed by the same procedure, but back-titrating to the plasma pH of the animal:

$$(TA - HCO_3^-)_u = NaOH_{pH4 \to pH_{pl}} - HCl_{pH_u \to pH4} \tag{13}$$

The same procedures are applicable for fluid from abdominal pores and rectal gland.

## 4. Constants

The bicarbonate concentration in various compartments of the animal and in the environmental water is an important parameter for model calculations and evaluation of the acid–base status. The most desirable way to determine the bicarbonate concentration is direct determination by one of the large number of available methods (e.g., Van Slyke apparatus, Van Slyke, 1917; the Cameron chamber method, Cameron, 1971; or by differential conductivity measurement, Maffly, 1968, with the "Capnicon III," see Section II,C,1,a).

Direct determination, however, is often impossible because of practical limitations. Then bicarbonate concentration is determined by calculation from the Henderson–Hasselbalch equation:

$$pH = pK_1''' + \log \frac{[HCO_3^-]}{\alpha_{CO_2} \cdot P_{CO_2}} \tag{14}$$

$$[HCO_3^-] = 10^{pH - pK_I'''} \cdot \alpha_{CO_2} \cdot P_{CO_2} \tag{15}$$

The required constants have been determined for various species, fluid types, temperatures, and other parameters. In most cases, however, a complete set of data with respect to temperature, pH, ionic strength, and other border conditions has not been determined, so that application or even extrapolation of the reported values introduces uncertainties into the experimental results.

In theory, the "constants" depend on only a few parameters, as long as the free-water phase, which is exclusively relevant for application of the Henderson–Hasselbalch equation, is considered. The solubility of $CO_2$ ($\alpha_{CO_2}$) varies, then, only with temperature and concentration of dissolved species. Carbamate formation and other side effects like lipid solubility have to be disregarded.

On this basis a formula that is generally valid for pure water, salt solutions, and various body fluids has been developed. It is based on literature data (e.g., Bartels and Wrbitzky, 1960; Van Slyke *et al.*, 1928; Markham and Kube, 1941) and on unpublished measurements of our laboratory on body fluids of a number of animal species (N. Heisler, P. Nissen, K. Winn, F. G. Hagmann, G. Fischer, H. Weitz, P. Neumann, and H. Koch, unpubl.)

$$\alpha_{CO_2} = 0.1008 \; \begin{aligned} &-29.80 \times 10^{-3}[M] + (1.218 \times 10^{-3}[M] - 3.639 \times 10^{-3})t \\ &- (19.57 \times 10^{-6}[M] - 69.59 \times 10^{-6})t^2 \\ &- (71.71 \times 10^{-9}[M] - 559.6 \times 10^{-9})t^3 \; \text{(mmol liter}^{-1} \text{ mm Hg}^{-1}) \end{aligned} \tag{16}$$

where $t$ is temperature (°C) (valid between 0 and 40°C), $M$ is molarity of dissolved species (mol liter$^{-1}$), $\alpha$ refers to the volume of water in a solution; for solubility per liter of solution, appropriate correction is required (i.e., for the volume of proteins and salts).

The $pK_1$ value is similarly independent of other variables except temperature and ionic strength, as long as only bicarbonate and $CO_2$ in the free-water phase are considered. If, however, total $CO_2$ is determined directly by gasometric or titrimetric methods, also $CO_3^{2-}$ and $NaCO_3^-$ (and probably further combinations) are measured together with bicarbonate. If a comparable value is to be calculated from the Henderson–Hasselbalch equation, an "apparent" $pK_1$ value ($pK_{app}$) has to be applied, which is, according to the interaction with other constituents of the solution, slightly variable with pH and at least the sodium concentration. This is in addition to the function of $pK_1$ with temperature and ionic strength. Since complete data sets including the dependence of the apparent $pK$ on temperature, ionic strength, pH, and [Na$^+$] are not available in the literature, a generally applicable formula is given. It was derived on the basis of theoretical considerations, and correction factors according to actual measurements of $pK_1'''_{app}$ in water, salt solutions, and biological fluids of animals from various classes (N. Heisler, P. Nissen, K. Winn, F. G. Hagmann, G. Fischer, P. Neumann, G. F. Holeton, and D. P. Toews, unpublished; Nicol *et al.*, 1983; Jackson and Heisler, 1982, 1983).

$$pK_1'''_{app} = 6.583 - 13.41 \times 10^{-3}t + 228.2 \times 10^{-6}t^2 - 1.516 \times 10^{-6}t^3 - 0.341\, I^{0.323}$$
$$- \log\{1 + 0.00039\,[Pr] + 10^{pH - 10.64 + 0.011t + 0.7371\,I^{0.323}}$$
$$\times (1 + 10^{1.92 - 0.01t - 0.7371\,I^{0.323}} + \log[Na^+] + (-0.494I + 0.651)(1 + 0.0065\,[Pr]))\} \tag{17}$$

where $t$ is temperature (°C, range 0–40°C), $I$ is ionic strength of nonprotein ions $(I = 0.5 \sum_{1}^{n} (CZ^2)$, where $C$ is concentration in mol liter$^{-1}$, and $Z$ is charge of the ion), $[Na^+]$ is sodium concentration (mol liter$^{-1}$), and $[Pr]$ is protein concentration (g liter$^{-1}$).

The formula is valid, as far as can be evaluated, for measurements of pH with low-temperature glass electrodes and double electrolyte bridge reference electrodes that are calibrated with buffers of the same ionic strength as the sample, application of $\alpha_{CO_2}$ values derived from Eq. (16), and gasometric determination of "bicarbonate" (total $CO_2$ less dissolved $CO_2$) with Van Slyke apparatus, Cameron chamber (1971), or "Capnicon" (see earlier).

The conditions of measurements and the design of apparatus used may have large influence especially on measurement of pH. Therefore, any literature data on p$K$ values should be used only with an extreme amount of discretion and considered as a last resort. However, determination of p$K_1'''$ values for the $CO_2$–bicarbonate buffer system is, when correctly performed, a difficult and—because of the pH, $I$, $[Na^+]$, and temperature dependence—also a very time-consuming procedure. Therefore, it is highly advisable to determine the bicarbonate concentration directly, if at all possible.

## III. STEADY-STATE ACID–BASE REGULATION

### A. Disturbing Factors

Fish are rather sensitive experimental animals and can easily be disturbed by apparently minor stimuli. Small changes in light intensity, noise level, hydrostatic pressure on the animal, or the environmental temperature may induce considerable changes in gill ventilation with large changes in arterial $P_{O_2}$ and in the steady-state transepithelial ion transfer rate (N. Heisler, G. F. Holeton, P. Neumann, and H. Weitz, unpublished; Claiborne and Heisler, 1984). Such factors have to be completely eliminated before a normal acid–base status can be determined in fish. But even with all these factors eliminated, confinement in an experimental chamber often results in an elevation of arterial $P_{O_2}$ as compared to values measured in freely swimming, undisturbed fish (Ultsch et al., 1981; Claiborne and Heisler, 1984), and may also induce enforced swimming activity and struggling. Such activity with the associated anaerobic lactic acid production will obviously affect the acid–base status and is clearly an extremely disturbing factor for all measurements regarding the normal acid–base regulation. Since these events are artifacts of the experimental procedure and cannot be completely avoided, they have to be monitored so that the disturbed experiment is not included in the experimental series. This can be accomplished by visual inspection or, in long-term experiments, by measurement of the hydrostatic pressure in the fish box, which fluctuates considerably when struggling occurs (Heisler, 1978).

With this information in mind, we can see that the fish should not be disturbed during blood sampling. This is possible only with indwelling catheters, preferably of polyethylene without steel needle tips (cf. Soivio et al., 1972, 1975; Ultsch et al., 1981; Holeton et al., 1983; Claiborne and Heisler, 1984). Blood sampling by heart or sinus venosus puncture includes capture of the fish with associated struggling. Blood pH is affected to a large extent by such "grab and stab" methods because of lactic acid production and stress-induced transmembrane ionic transfer (e.g., Garey, 1972; cf. Benadé and Heisler, 1978; Holeton and Heisler, 1983). Consequently, the only useful data for normal values are those obtained from fish with chronically implanted blood catheters.

Another factor that is often disregarded is continuous supply of oxygen to the fish during the surgical procedures performed in order to implant blood and urinary catheters, or other devices into the fish. The anesthesia should preferably be induced by immersing the fish into a well-aerated highly concentrated solution of anesthetic (e.g., MS 222 or urethane) in water in order to keep the stress period of struggling and excitation as short as possible. Immediately after loss of reactivity, and before cessation of gill ventilation, the animals should be removed from the highly concentrated anesthesia solution, and anesthesia and also oxygen supply should be maintained during surgery by flushing the gills with a low-concentrated and well-aerated solution of anesthetic. This procedure still allows at least partial gas exchange and also blood gas transport by relatively normal cardiac output. Under such conditions the arterial $P_{O_2}$ can be kept higher than 50 mm Hg and venous $P_{O_2}$ not lower than 15 mm Hg even during operations of several hours duration, indicating a relatively normal oxygen supply to the tissues (Scyliorhinus stellaris; N. Heisler, G. F. Holeton, and D. P. Toews, unpublished). After surgery the fish should be recovered by flushing the gills with fresh water until apparently normal ventilation occurs.

Methods of anesthetizing fish using unstirred and nonaerated anesthetic solutions, without flushing of the gills with aerated solution during surgery, and recovering the fish without force flushing the gills with fresh water results in extremely delayed normalization of the acid–base status and the transepithelial ion fluxes and has to be considered inappropriate (N. Heisler, G. F. Holeton, and P. Neumann, unpublished). This creates spurious data even when performing short surgical procedures like puncture of the dorsal aorta.

Recovery of the fish from anesthesia even with the oxygen supply maintained requires 12–18 hr in marine fish, and 30 hr to 5 days in freshwater fish, before normal acid–base status and transepithelial ion fluxes can be observed (Heisler, 1978; Holeton and Heisler, 1983; Holeton et al., 1983; Claiborne and Heisler, 1984). Taking these factors into account during the experimental procedure is an indispensable prerequisite for the unbiased determination of normal parameters of the acid–base status. Unfortunately, quite a number of studies have neglected the foregoing considerations.

## B. Normal Acid–Base Status

Steady-state arterial plasma pH in fish varies inversely with body temperature. This was first demonstrated by Rahn and Baumgardner (1972), who plotted blood pH measurements obtained from 11 different fish species against their respective environmental temperature. They found that the data points from the various individual fish species and temperatures were grouped within a range of 0.25 pH Units around a constant relative alkalinity and thus confirmed that the overall picture of decreasing pH with rising temperature in ectothermal animals also applied to fish. However, the model of plasma pH regulation in fish toward a constant relative alkalinity proposed by Rahn (1966a, 1967) was in general not confirmed by systematic studies of the acid–base status in individual fish species, which had been acclimated to various temperatures (Fig. 2). The temperature coefficient of plasma pH ($\Delta pH/\Delta t$) was found in the majority of experiments to be in the range of $-0.010$ to $-0.014$, with an average of $-0.012$ U $°C^{-1}$ (Table II). This is significantly different from a constant relative alkalinity ($\Delta pN/\Delta t = -0.019$ U $°C^{-1}$ for the temperature range of 5 to 20°C; $\Delta pN/\Delta t = -0.017$ U $°C^{-1}$, range 20–35°C; Weast, 1975–1976). In only one species, the tropical air breather *Synbranchus marmoratus* (Heisler, 1980), plasma pH changed in parallel ($\Delta pH/\Delta t = -0.017$ U $°C^{-1}$) with a constant relative alkalinity. The uniformity of the temperature coefficients and also of the absolute pH

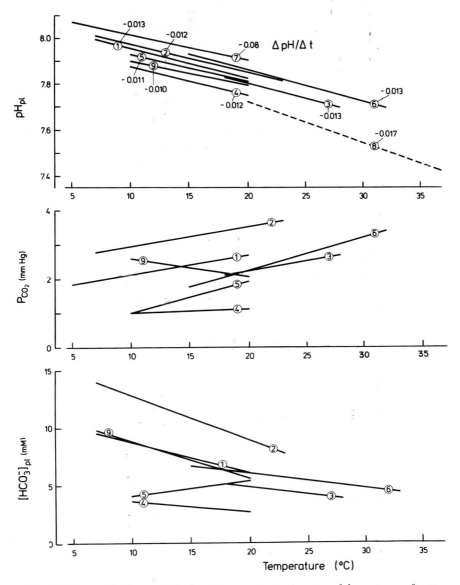

**Fig. 2.** Plasma pH, $P_{CO_2}$, and bicarbonate concentration in various fish species as a function of temperature. The values associated to the regression lines are the temperature coefficients of pH ($\Delta pH/\Delta t$). Species and data sources: (1) *Salmo* (Randall and Cameron, 1973); (2) *Cyprinus* (N. Heisler and P. Neumann, unpublished); (3) *Cynoscion* (Cameron, 1978); (4) and (5) *Scyliorhinus*, juvenile and adults, respectively (Heisler *et al.*, 1980); (6) *Ictalurus* (Cameron and Kormanik, 1982a); (7) *Anguilla* (Walsh and Moon, 1982); (8) *Synbranchus* (Heisler, 1980): (9) *Scyliorhinus*, $\Delta P_{CO_2}/\Delta t$ reversed (Heisler *et al.*, 1980).

**Table II**

Changes of pH with Changes of Temperature ($\Delta pH/\Delta t$) in Extracellular and Intracellular Compartments of Various Fish Species (U $°C^{-1}$)

| Species | Temperature range | Extracellular space | White muscle | Red muscle | Heart muscle | Liver | Reference |
|---|---|---|---|---|---|---|---|
| *Salmo gairdneri* | 7–23 | −0.013 | — | — | — | — | Randall and Cameron (1973) |
| *Scyliorhinus stellaris* | | | | | | | |
| Juvenile | 10–23 | −0.013 | −0.018[a] | −0.033 | −0.010 | — | Heisler *et al.* (1976b) |
| Adult | 10–23 | −0.014 | | | | | |
| *Cynoscion arenarius* | 18–28 | −0.013 | — | — | — | — | Cameron (1978) |
| *Scyliorhinus stellaris* | | | | | | | |
| Juvenile | 10–20 | −0.012 | −0.020[a] | −0.029 | −0.007 | — | Heisler *et al.* (1980) |
| Adult | 10–20 | −0.011 | −0.018[a] | −0.028 | −0.007 | — | Heisler *et al.* (1980) |
| Adult, $\Delta P_{CO_2}/\Delta t$ reversed | 10–20 | −0.010 | −0.018[a] | −0.029 | −0.005 | — | Heisler *et al.* (1980) |
| *Cyprinus carpio* | 7–23 | −0.012 | −0.012 | −0.026 | −0.005 | — | N. Heisler and P. Neumann (unpublished) |
| *Ictalurus punctatus* | 15–31 | −0.013 | −0.014 | −0.018[a] | −0.011 | — | Cameron and Kormanik (1982) |
| *Synbranchus marmoratus* | 20–37 | −0.017[a] | −0.009 | — | −0.003 | — | Heisler (1980) |
| *Anguilla rostrata* | 5–20 | −0.008 | −0.009 | −0.003 | −0.020[a] | −0.018[a] | Walsh and Moon (1982) |

[a]Not significantly different from $\Delta pK_{im}/\Delta t$.

values (pH difference less than 0.15 U at any given temperature) among the seven water-breathing fish species is impressive and suggests a rather tight regulation of plasma pH. Plasma $P_{CO_2}$ is much more variable and ranges at 15°C from 1 mm Hg (*Scyliorhinus*) to about 3.2 mm Hg (*Cyprinus*) (Fig. 2). Also, the relative changes in $P_{CO_2}$ with temperature are quite variable and range from almost no change at all (*Salmo gairdneri*, juvenile *Scyliorhinus*) to a greater change than required for the change in pH at constant bicarbonate concentrations (adult *Scyliorhinus*). Accordingly, the absolute values and the changes of plasma bicarbonate concentration with temperature are rather different among species.

In spite of the fact that the temperature coefficient of the plasma pH in *Synbranchus* is different ($\Delta pH / \Delta t = -0.017$) from the water-breathing species, the absolute pH at 20°C is similar. However, the pattern of $P_{CO_2}$ with values between 20 and 31 mm Hg and of the plasma bicarbonate concentration between 36 and 20 m$M$ (for 20 or 37°C, respectively) appears to be more closely related to the acid–base status observed in amphibians and reptiles than to water-breathing fish (e.g., Heisler *et al.*, 1982; Glass *et al.*, 1984; Toews and Heisler, 1982; Boutilier *et al.*, 1979a,b,c; Hicks and Stiffler, 1980).

Since the $P_{CO_2}$ in fish cannot be freely adjusted as in air-breathing animals because of the physical limitations involved with breathing water as gas exchange medium (Rahn, 1966b), adjustment of the bicarbonate concentration is apparently the predominant acid–base regulatory mechanism in fish. This hypothesis is supported by an experiment on *Scyliorhinus*, where the normally observed pattern of $P_{CO_2}$ with low values at low temperature and higher $P_{CO_2}$ at higher temperature was artificially reversed by exposure of the animals to elevated inspired $P_{CO_2}$ at lower temperature (Heisler *et al.*, 1980). The limited increase in $P_{CO_2}$ was completely compensated by an elevation in plasma bicarbonate, so that the absolute pH and the change in pH with temperature remained completely unaffected (Fig. 2, Table II). Also the intracellular pH in three types of muscle was not significantly affected (Table II).

The changes of intracellular pH with changes in body temperature are not as uniform as for the extracellular pH (Table II). It is apparent as a general pattern at least in four of the five species studied that $\Delta pH / \Delta t$ of red muscle tissue is always the highest, followed by that of white muscle. Heart muscle pH changes least with changes of temperature. This general pattern is reversed only in *Anguilla*, where $\Delta pH_i / \Delta t$ of heart muscle is highest followed by white muscle. Red muscle pH remains essentially unaffected by changes in temperature. In only 4 of the 15 studied intracellular fish compartments (i.e., white muscle of *Scyliorhinus*, red muscle of *Ictalurus*, and heart muscle and liver in *Anguilla*), $\Delta pH / \Delta t$ is not significantly different from a constant relative alkalinity.

These data on the behavior of intracellular pH in fish indicate that the setpoint values of intracellular pH change with temperature in a less uniform way than observed for the extracellular body compartment. This could be interpreted as only loose regulation of the intracellular pH. A much more attractive and likely alternative, however, is that the energy-producing metabolic pathways in various types of tissues possess optimal performance at different pH values. This hypothesis is supported by the repeatability of the measurements and the extremely small effect of external disturbances such as hypercapnia on the regulation of intracellular pH (Heisler *et al.*, 1980, 1982; Heisler, 1980, 1982b).

## C. Imidazole Alphastat Regulation

Based on some experiments performed in air-breathing ectotherms, which resulted in $\Delta pH/\Delta t$ values similar to the changes in p$K$ of histidine imidazole with temperature, Reeves (1972) explained the observed acid–base regulation as adjustment of a constant dissociation of imidazole. His imidazole alphastat hypothesis was that ventilation and thus $P_{CO_2}$ in lower vertebrates is regulated such that the fractional dissociation of peptide-linked histidine imidazole is kept constant. Since the histidine imidazole of hemoglobin and plasma proteins represents by far the most important non-bicarbonate buffer of the extracellular space, this implies that the bicarbonate concentration of the compartment is kept constant, especially when transmembrane and transepithelial acid–base relevant ion transfer is excluded (Reeves and Malan, 1976).

The p$K$ value of biological imidazole compounds changes with temperature $(\Delta pK/\Delta t)$ by $-0.018$ to $-0.024$, depending on ligands and steric arrangement (Edsall and Wyman, 1958). The lower limit of this range is not even approached in the extracellular space of any of the seven water-breathing fish species examined to date (Table II). Also the bicarbonate concentration changes considerably in all studied water-breathing species (Fig. 2) and also in the air-breathing *Synbranchus* (Heisler, 1982a), denying the other criteria of the alphastat hypothesis: constant bicarbonate concentration, and no transmembrane and transepithelial acid–base relevant ion transfer (Reeves, 1972; Reeves and Malan, 1976). Also, the intracellular pH in the majority of the studied tissue species (11 of 15) does not follow alphastat regulation (Table II). A more or less constant dissociation of imidazole can be expected only in white muscle of *Scyliorhinus*, red muscle of *Ictalurus*, and heart and liver of *Anguilla*.

It should be noted in this context that even if $P_{CO_2}$ could be adjusted in fish, in spite of the physical limitations of the breathing medium water, such that constant imidazole dissociation would be achieved in one of the various compartments (e.g., the extracellular space), alphastat regulation could not be expected in the other, intracellular body compartments. Especially in water-breathing fish with their low bicarbonate concentrations, the temperature coefficient of pH ($\Delta pH_i/\Delta t$) is predominantly determined by the ratio of imidazole-like ($\Delta pK/\Delta t \sim -0.021$) over phosphate-like nonbicarbonate buffers ($\Delta pK/\Delta t \sim -0.002$ U $°C^{-1}$) and to only a lesser extent by changes in $P_{CO_2}$ (Glass et al., 1984). Since this ratio is rather variable (1–4, Table I) in different tissue species, constant dissociation of imidazole could only be achieved by transmembrane acid–base relevant ion transfer.

Based on these data and considerations, it has to be concluded that a constant dissociation of imidazole is rarely attained (in only 4 of 21 compartments) and that the alphastat mechanism is only of minor importance for the temperature-dependent acid–base regulation in fish.

## D. Steady-State Excretion

The normal acid–base status of fish just described is the result of a steady state between net metabolic production of acid–base relevant substances and their elimination (see Heisler, 1982b).

The quantitatively most important metabolic end product relevant to the acid–base status is $CO_2$. Depending on the type of metabolic substrates, the amount of $CO_2$ produced is between 0.7 and 1.0 (respiratory quotient) of the oxygen consumption, which, in turn, depends on the environmental temperature (e.g., Ott et al., 1980; Ultsch et al., 1980). The largest proportion of molecular $CO_2$ is eliminated from the fish by diffusion through the gill epithelium, but in small fish and in certain fish species with large relative skin surface (e.g., eels), $CO_2$ is also eliminated to a significant extent by diffusion through the skin (the mechanism of $CO_2$ elimination is dealt with in detail by other chapters of this volume).

Urea and ammonia are the two main nitrogenous metabolic end products in fish. The elasmobranchs are ureotelic, producing more than 70% of their nitrogenous waste as urea. Production of urea is neutral for the acid–base status, and urea is comparatively nontoxic, which makes it possible for these animals to utilize it as a significant fraction of their plasma osmotic activity (~50%) (Smith, 1929a; see also Evans, 1979).

Teleost fish are ammoniotelic, producing predominantly ammonia (60–90%, Smith, 1929b; Wood, 1958; Fromm, 1963) in a relative amount of about 10% of the oxygen consumption (Table III). Since the highly toxic ammonia (e.g., Ball, 1967; Hillaby and Randall, 1979; Tomasso et al., 1980; Arillo et al., 1981; Thurston and Russo, 1981; Schenone et al., 1982) reacts

## Table III

Control Release Rates of Bicarbonate ($\Delta HCO_3^-{}_w$), Ammonium ($\Delta NH_4^+{}_w$), and Net $H^+$ Excretion ($\Delta H^+_{excr}$) in Correlation with the Respetive Oxygen Consumption ($\dot{V}_{O_2}$)[a]

| Species | Series | $\Delta HCO_3^-{}_w$ | $\Delta NH_4^+{}_w$ | $\Delta H^+_{excr}$ | $\dot{V}_{O_2}$ | $\Delta H^+_{excr} \times 100/\dot{V}_{O_2}$ | Reference |
|---|---|---|---|---|---|---|---|
| *Scyliorhinus stellaris* | Hypercapnia | 0.57 | 0.83 | 0.26 | 34.8[b] | 0.74 | Heisler *et al.* (1976a) |
| | Temperature changes | −0.09 | 0.33 | 0.42 | 34.8[b] | 1.20 | Heisler (1978) |
| | Exercise | 0.44 | 0.64 | 0.20 | 34.8[b] | 0.57 | Holeton and Heisler (1983) |
| *Conger conger* | Exercise | 3.39 | 3.75 | 0.36 | — | — | Holeton *et al.* (1984) |
| | Hypercapnia | 2.78 | 3.22 | 0.44 | — | — | Toews *et al.* (1983) |
| *Cyprinus carpio* | Acidic water | 2.89 | 3.75 | 0.86 | 39.1[c] | 2.19 | Ultsch *et al.* (1981) |
| | Hypercapnia | 2.08 | 3.54 | 1.46 | 39.1[c] | 3.74 | Claiborne and Heisler (1984) |
| *Salmo gairdneri* | Exercise | 5.58 | 4.96 | −0.62 | 55.3[c] | −1.12 | Holeton *et al.* (1983) |
| | Acidic water | 3.52 | 6.83 | 3.3 | 55.3[c] | 5.97 | McDonald and Wood (1981) |
| *Symbranchus marmoratus* | Hypercapnia | 4.19 | 4.41 | 0.22 | 52.0 | 0.40 | Heisler (1982a) |
| Human | Normal human | −0.37 | 0.49 | 0.85 | 220 | 0.39 | Pitts (1945) |

[a] Rates given in $\mu$mol (min kg body water)$^{-1}$.
[b] Data from Randall *et al.* (1976).
[c] Data from Ultsch *et al.* (1980).

strongly alkaline (p$K$ ~9.6, see Emerson et al., 1975; Bower and Bidwell, 1978; Cameron and Heisler, 1983), it is to its largest fraction (more than 97% at physiological pH values) immediately ionized after production, which occurs mainly from deamination of α-amino acids (α-AA), mainly in liver, but also in kidneys and gills (Goldstein et al., 1964; Pequin, 1962; Pequin and Serfaty, 1963; Aster et al., 1982; Cameron and Heisler, 1983). Accordingly, it neutralizes a significant fraction of the simultaneous $CO_2$ production, thereby producing bicarbonate and ammonium ions:

$$CO_2 + H_2O \rightarrow H^+ + HCO_3^- \quad \rightarrow HCO_3^-$$
$$+$$
$$\alpha-AA \rightarrow NH_3 \qquad\qquad \rightarrow NH_4^+ \qquad\qquad (18)$$

In this form it is transported to the site of excretion, in water-breathing fish species predominantly the gills (>90%, Smith, 1929b).

If, as has been suggested by a number of investigations, ammonia is eliminated from the organism by ionic exchange of $NH_4^+$ against $Na^+$ (e.g., Maetz, 1973; Kerstetter et al., 1970; Evans, 1977, 1980b; Payan and Maetz, 1973; see also Maetz, 1974; Evans, 1979, 1980a), a significant fraction of metabolically produced $CO_2$ has to be eliminated in the form of $HCO_3^-$, probably in ionic exchange with $Cl^-$ (Krogh, 1939; Maetz and Garcia-Romeu, 1964; De Renzis and Maetz, 1973; De Renzis, 1975; Kerstetter and Kirschner, 1972; Kormanik and Evans, 1979).

If, however, $NH_3$ diffuses out of the organism predominantly in nonionic form through the gill epithelium similar to the mechanism for $CO_2$, as has been suggested by recent studies for environmental water with low ammonia concentration (Cameron and Heisler, 1983), then the ammonia would only support the $CO_2$ elimination by reducing the arteriovenous pH difference by buffering of $CO_2$ during the blood transport.

Regardless of elimination mechanisms, ammonia and bicarbonate are expected to be released to the environmental water always in equimolar amounts. When ammonia is eliminated by ionic exchange, this must during steady state always be accompanied by an equimolar exchange of $HCO_3^-$ (or, with the same effect, $H^+$ in the opposite direction) in order to avoid progressive alkalinization in the fish. If ammonia is eliminated by nonionic diffusion, the alkalinizing effect of the ammonia ionization on the body fluids of the fish is reversed and the $CO_2$ originally bound at the site of ammonia production is set free at the site of elimination:

$$NH_4^+ \rightarrow NH_3 \uparrow + H^+$$
$$+$$
$$HCO_3^- \rightarrow CO_2 \uparrow + H_2O \qquad\qquad (19)$$

If the water is not extremely alkaline, $NH_3$ and $CO_2$ diffusing then into the water will quantitatively recombine to $NH_4^+$ and $HCO_3^-$ (Eq. (18)) and thus produce the same effect in the water as with ionic exchange.

Ammonia and bicarbonate are actually not released in equimolar quantities by the fish (Table III). The small difference has to be attributed to the production or ingestion of surplus nonvolatile $H^+$ ions in metabolism. Fixed acids, dissociating $H^+$ ions, are produced endogenously as end products of sulfur-, phosphorus-, and chlorine-containing organic compounds, and from incomplete oxidation of fatty acids, amino acids, or carbohydrates. These are partially neutralized by fixed bases originated from oxidation of carbohydrate alkali metal salts (which are especially concentrated in herbal diets) or from incomplete oxidation of organic bases. The relative amount of nonvolatile $H^+$ ions excreted (Table III) during steady-state conditions is rather constant with time. It represents usually less than 3.5% of the molar oxygen consumption and is often comparable to that of mammals (Table III).

The active ion exchange mechanisms in freshwater fish, which are likely involved in the steady-state excretion of acid–base relevant ions, are capable of compensating, at least partially, the passive leakage of $Na^+$ and $Cl^-$ to the environment along the electrochemical gradient (cf. Evans, 1979, initially suggested by Krogh, 1939). However, since production and therefore elimination of acid–base relevant ions are strictly correlated with the rate of the respective metabolic pathway, the passive leakage rate depends on various internal and external factors and is usually not matched with the requirements of acid–base regulation. In marine fish, exchanges of endogenous $H^+$, $NH_4^+$, and $HCO_3^-$ against exogeneous $Na^+$ and $Cl^-$ are ionically inappropriate, and in contrast to the conditions in freshwater fish even add to the passive influx of $Na^+$ and $Cl^-$ into the animal along the electrochemical gradient (Evans, 1979). Accordingly, the electroneutral elimination of acid–base relevant ions and maintenance of osmotic equilibrium require mechanisms for acid–base regulation, which can only loosely be interlocked with the mechanisms for osmoregulation (Heisler, 1982b).

## IV. ACID–BASE REGULATION AND IONIC TRANSFER PROCESSES DURING STRESS CONDITIONS

The acid–base status of fish is frequently stressed in addition to the endogenous steady-state load of surplus acid–base relevant ions by various endogenous and environmental factors. The effect of some of these factors on the acid–base status and on ionic transfer processes will be described and discussed in this section. This must naturally be incomplete, since only a few of the factors known to interact with acid—base regulation have been studied, and there are probably many additional, but still unknown important variables. It is also difficult to evaluate an overall pattern of acid—base regulation in fish, since most of the stress factors dealt with have been imposed on only one or very few fish species and many of the reported studies are incomplete with respect to certain parameters.

## A. Temperature Changes

Most fish species are subject to relatively large seasonal temperature changes of their environmental water. The differences between maximal and minimal temperatures may well be 20°C or more. The rate of seasonal temperature changes is slow, and the time available for acclimation is long. Under certain conditions, however, fish may also encounter radical changes in temperature in a short time period or even almost immediately. The temperature of small ponds and seawater reservoirs with only part-time contact with the open sea (e.g., intertidal rock pools) may undergo large diurnal variations of up to 25°C (e.g., Kramer *et al.*, 1978; C. R. Bridges *et al.*, personal communication; Truchot *et al.*, 1980; Morris and Taylor, 1983). According to thermostratification, temperature differences of more than 10 or 15°C may occur in lakes during warm summer periods. Carnivorous fish species usually stay in the deeper and colder water layers during the day and move into the shallow warm water for predatory activities during early morning, late afternoon, and evening. Then they encounter immediate temperature changes of considerable extent.

Thus, changes in environmental temperature are often pronounced and, as outlined already, will significantly affect the acid—base status. The following section delineates the kinetics of changes in pH, $P_{CO_2}$, and $[HCO_3^-]$ on step changes in temperature, and evaluates the role of physicochemical buffering and ion transfer processes involved in the adjustment of the steady state at the new temperature.

### 1. KINETICS OF EXTRACELLULAR ACID—BASE ADJUSTMENT

The kinetics of acid—base adjustment after changes in temperature have been studied in some detail in only two relatively nonrepresentative fish species, the elasmobranch *Scyliorhinus stellaris* (Heisler, 1978) and the tropical air-breathing teleost fish *Synbranchus marmoratus* (Heisler, 1984).

Plasma pH in *Scyliorhinus* undershoots the finally attained steady-state value largely with a step change in temperature from 10 to 20°C. The initial $\Delta pH/\Delta t$ ($\sim -0.039$ U °C$^{-1}$) is more than three times that of the shift finally

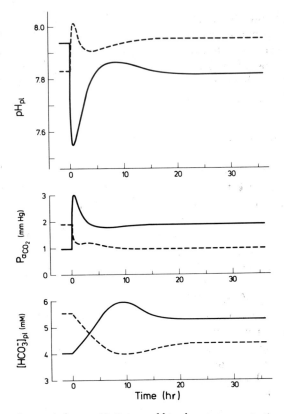

**Fig. 3.** Kinetics of arterial plasma pH, $P_{CO_2}$, and bicarbonate concentration after changes in temperature (solid line, $10 \rightarrow 20°C$; dashed line, $20 \rightarrow 10°C$) in the elasmobranch *Scyliorhinus stellaris*. (Data of Heisler, 1978.)

observed after about 20 hr, when steady-state conditions have been attained $(\Delta pH/\Delta t = -0.012 \text{ U } °C^{-1})$ (Fig. 3, upper panel). This is mainly attributable to an initial, about threefold increase in plasma $P_{CO_2}$ during the first hour. The time course of pH adjustment to the final value is primarily a function of the slow increase in plasma bicarbonate concentration (Fig. 3, lower panel). The time course of adjustment for the reverse temperature change (20 to 10°C) is similar, and the change in plasma bicarbonate resembles more or less a mirror image of the change from 10 to 20°C. Plasma pH, however, overshoots less $(\Delta pH/\Delta t = -0.019 \text{ U } °C^{-1})$, but it still requires 20 hr to attain the new steady state as a result of the slow bicarbonate adjustment.

In air-breathing *Synbranchus* an increase in temperature from 20 to 30°C induces a significant overshoot in $P_{CO_2}$, and undershoot in pH in one of two

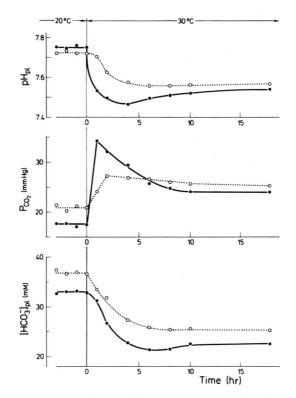

**Fig. 4.** Kinetics of plasma pH, $P_{CO_2}$, and bicarbonate concentration after changes in temperature (20 → 30°C) in two specimens of the air-breathing teleost fish *Synbranchus marmoratus*. (Data of Heisler, 1984.)

studied specimens, before new steady-state values are attained. Plasma bicarbonate is reduced without significant undershoot, and a new steady state is arrived at for all three parameters within 10 hr (Fig. 4).

It is evident that plasma pH in both studied species is not adjusted to the new steady-state value immediately after the change in temperature by regulation of $P_{CO_2}$, but that considerable changes in the extracellular bicarbonate concentration have to be produced. These changes cannot be attributed to nonbicarbonate buffering in the extracellular space. The nonbicarbonate buffer capacity of *Scyliorhinus* is negligible, and it alters bicarbonate in the opposite direction of the observed change. In *Synbranchus* the histidine imidazole, which represents more than 95% of the extracellular nonbicarbonate buffer capacity, is not titrated, since pH changes by about the same amount as p$K$ of the buffers ($\Delta$pH/$\Delta t$ = −0.017 U °C$^{-1}$). Also, physicochemical buffering is almost instantaneous and would not require the observed time period of 10 hr. Accordingly the changes in bicarbonate have to be attributed to transepithelial and/or transmembrane transfer processes of acid–base relevant ions.

## 2. Transepithelial and Transmembrane Ion Transfer

Transepithelial transfer of acid–base relevant ions has been studied in three fish species, utilizing the experimental approach of the closed water recirculation system (Section II,B,1,c). The results are diverse among species: for a 10°C increase in temperature, *Scyliorhinus* releases about 0.4 mmol kg body water$^{-1}$ of bicarbonate-equivalent ions (Heisler, 1978, 1984), and the channel catfish *Ictalurus punctatus*, 0.55 mmol kg body water$^{-1}$ (Cameron and Kormanik, 1982a). In contrast, *Synbranchus* behaves more or less as a closed buffer system, taking up only the insignificant amount of 0.08 mmol kg body water$^{-1}$ (Heisler, 1984) ($\Delta HCO_3^-{}_{e \to w}$; Table IV).

Based on the transepithelial ion transfer (Eq. (4)) and the change of extracellular bicarbonate-equivalent ions (Eq. (8)), the amount of acid–base relevant ions transferred between the extracellular space and the overall intracellular space can be determined (Eq. (9)) (for details, see Heisler, 1984). The overall bicarbonate-equivalent transfer is considerable in *Synbranchus*, where about 2 mmol kg body water$^{-1}$ are transferred from the extracellular to the intracellular compartment when temperature is raised by 10°C. In the other two species, bicarbonate is transferred in the opposite direction (0.77 or 0.36 mmol kg body water$^{-1}$ for *Scyliorhinus* and *Ictalurus*, respectively) (Table IV).

This approach of quantitative analysis of transepithelial transfer and changes in the bicarbonate pool of the extracellular space provides only information about the overall transfer, which is the weighted average of all tissue species. More specific information can be gained by the opposite approach.

The net amount of bicarbonate transferred (i.e., in excess of the steady-state flux) between the intracellular compartment of a specific tissue and the extracellular space is determined on the basis of the changes in the intracellular bicarbonate pool, and the amount of bicarbonate that is produced or has been decomposed by the intracellular nonbicarbonate buffers. Production or decomposition of bicarbonate depends on the intracellular nonbicarbonate buffer value, and on the extent of titration (i.e., the effective change in pH). Since the p$K$ values of the intracellular buffers also change with temperature, the effective change in pH ($\Delta pK - \Delta pH$) is very much dependent on the type of buffer (i.e., imidazole-like or phosphate-like buffer). If the fractional contribution of these buffer groups to the overall buffer value is

**Table IV**

Net Bicarbonate-Equivalent Ion Transfer between Intracellular Space and Extracellular Space ($\Delta HCO_3^-{}_{i\rightarrow e}$), and between Extracellular Space and Environmental Water ($\Delta HCO_3^-{}_{e\rightarrow w}$) with a 10°C Increase in Temperature[a]

| Species | $\Delta HCO_3^-{}_{e\rightarrow w}$ | $\Delta HCO_3^-{}_{i\rightarrow e}$ | Reference |
|---|---|---|---|
| *Scyliorhinus stellaris* | +0.41 | +0.77 | Heisler (1978); Heisler *et al.* (1980) |
| *Ictalurus punctatus* | +0.55 | +0.36 | Cameron and Kormanik (1982a) |
| *Synbranchus marmoratus* | −0.08[a] | −2.01 | Heisler (1984) |

[a]Values given as mmol kg body water$^{-1}$.
[b]Nonsignificant.

## Table V

Production of Bicarbonate by Intracellular Nonbicarbonate Buffers ($\Delta[HCO_3^-]_{NB_i}$) and Net Transfer of Bicarbonate-Equivalent Ions per Unit Volume of Intracellular Tissue Compartments ($\Delta[HCO_3^-]_{i \rightarrow e}$) after a 10°C Increase in Temperature[a,b]

| Species | Tissue | $\Delta[HCO_3^-]_{NB_i}$ | $\Delta[HCO_3^-]_{i \rightarrow e}$ | Reference |
|---|---|---|---|---|
| *Scyliorhinus stellaris* (adult specimens) | White muscle | +0.53 | +0.34 | Heisler and Neumann (1980); Heisler *et al.* (1980) |
| | Red muscle | +7.24 | +7.49 | |
| | Heart muscle | −3.27 | −3.79 | |
| *Synbranchus marmoratus* | White muscle | −3.69 | −3.05 | Heisler (1984) |
| | Heart muscle | −4.39 | −4.57 | |

[a]For details see text and Heisler (1984).
[b]Values given in mmol kg cell water$^{-1}$.

## Table VI

Intracellular Bicarbonate Production by Nonbicarbonate Buffering ($\Delta[HCO_3^-]_{NB_i}$) and Net Bicarbonate Transfer Processes ($\Delta[HCO_3^-]_{i \to e}$ and $\Delta[HCO_3^-]_{e \to w}$) with a 10°C Increase in Temperature[a,b]

| Species | Tissue ICS[c] | ICS water volume (ml) | $\Delta[HCO_3^-]_{NB_i}$ (1) | $\Delta[HCO_3^-]_i$ (2) | $\Delta[HCO_3^-]_{i \to e}$ (3)=(1)-(2) | Reference |
|---|---|---|---|---|---|---|
| *Scyliorhinus stellaris* (adult specimens) | White muscle | 357 | +0.19 | +0.07 | +0.12 | Heisler and Neumann (1980); Heisler *et al.* (1980) |
| | Red muscle | 35 | +0.26 | −0.01 | +0.27 | |
| | Other tissues[d] | 340 | (+0.18)[d] | (+0.06)[d] | (+0.12)[d] | |
| Total | | 732 | +0.63 | +0.12 | +0.51 | |
| *Synbranchus marmoratus* | White muscle | 600 | −2.21 | −0.38 | −1.83 | Heisler (1984) |
| | Other tissues[d] | 201 | (−0.74)[d] | (−0.13)[d] | (−0.61)[d] | |
| Total | | 801 | −2.95 | −0.51 | −2.44 | |

[a]For details see text and Heisler (1984).
[b]Values given in mmol kg body water⁻¹.
[c]ICS, intracellular space.
[d]Other tissues assumed to be white muscle.

known, the transmembrane bicarbonate-equivalent ion transfer of individual
tissues can be modeled (Eq. (6); for details, see Heisler, 1984).

The detailed information required for this type of analysis is available for
only two fish species (*Scyliorhinus*, Heisler and Neumann, 1980; Heisler *et
al.*, 1980; *Synbranchus*, Heisler, 1984). Per unit of volume, the amount of
bicarbonate transferred following temperature change is especially large in
red muscle of *Scyliorhinus* (Table V). However, also in heart muscle of
*Scyliorhinus* and in white and heart muscle of *Synbranchus*, amounts of
bicarbonate are transferred that even exceed the quantities transferred during hypercapnic disturbances of the acid–base status (see also Heisler, 1980,
1984; Toews *et al.*, 1983). The transfer in white muscle of *Scyliorhinus* is
much smaller (Table V), but because of the large relative volume of this
tissue, it still contributes significantly to the overall transfer (Table VI).

3. CONTRIBUTION OF TRANSMEMBRANE AND
   TRANSEPITHELIAL ION TRANSFER TO THE
   INTRACELLULAR AND EXTRACELLULAR ACID–BASE
   REGULATION

The contribution of ionic transfer processes to the compartmental pH
regulation can be delineated by modeling the respective compartment as a
closed buffer system. The change in pH with the change in temperature is
then dependent only on the characteristics of the nonbicarbonate buffers in
the system and the changes in $P_{CO_2}$. This can be performed by setting
$\Delta HCO_3^-{}_{i \to e}$ and $\Delta HCO_3^-{}_{e \to w}$ to zero in Eqs. (6) and (9), and expressing
$\Delta[HCO_3^-]$ of the respective compartment by pH and $P_{CO_2}$ values according
to the Henderson–Hasselbalch equation. The resulting equations (see also
Heisler and Neumann, 1980) can then be solved by computer iteration for
the change in pH. The $\Delta pH/\Delta t$ values calculated by this approach for the
extracellular space of six species and for intracellular compartments of two
species show (with the one exception of white muscle in *Scyliorhinus*) considerable differences as compared to the respective *in vivo* value (Table VII).
These data indicate that the closed buffer system concept propounded by
Reeves (1972) does not apply to fish.

This holds even in the air-breathing fish *Synbranchus*, which can escape
the limitations of water as gas exchange medium and would very well be
capable of adjusting arterial $P_{CO_2}$ such that no ionic transfer would be required, a measure that has been suggested as the only mechanism for intracellular and extracellular pH regulation with changes of temperature in
lower vertebrates (Reeves, 1972; Reeves and Malan, 1976).

**Table VII**

Changes of pH in Extracellular and Intracellular Body Compartments with Temperature ($\Delta pH/\Delta t$) in Vivo and in the Respective Compartments Modeled as Closed Buffer Systems (U °C$^{-1}$)

| Species | Body compartment[a] | $\Delta pH/\Delta t$ (in vivo) | $\Delta pH/\Delta t$ (Closed buffer system) | Reference |
|---|---|---|---|---|
| *Scyliorhinus stellaris* | White muscle ICS | −0.018 | −0.017 | Heisler *et al.* (1980) |
| | Red muscle ICS | −0.031 | −0.013 | |
| | Heart muscle ICS | −0.007 | −0.016 | |
| | ECS | −0.011 | −0.021 | |
| *Synbranchus marmoratus* | White muscle ICS | −0.009 | −0.017 | Heisler (1984) |
| | Heart muscle ICS | −0.003 | −0.016 | |
| | ECS | −0.017 | −0.003 | |
| *Salmo gairdneri* | ECS | −0.013 | −0.001 | Randall and Cameron (1973) |
| *Cynoscion arenarius* | ECS | −0.013 | −0.002 | Cameron (1978) |
| *Cyprinus carpio* | ECS | −0.012 | +0.002 | N. Heisler and P. Neumann (unpublished) |
| *Ictalurus punctatus* | ECS | −0.013 | −0.007 | Cameron and Kormanik (1982a) |

[a]ICS, intracellular space; ECS, extracellular space.

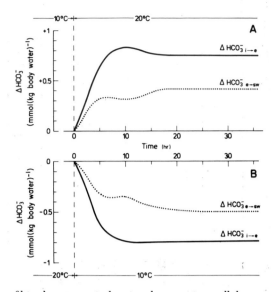

**Fig. 5.** Transfer of bicarbonate-equivalent ions between intracellular space and extracellular space $(\Delta HCO_3^-{}_{i \to e})$ and between extracellular space and environmental seawater $(\Delta HCO_3^-{}_{e \to sw})$ after changes in temperature in *Scyliorhinus stellaris*. (A) $10 \to 20°C$; (B) $20 \to 10°C$. (Data of Heisler, 1978.)

## 4. Model Calculations on the Kinetics of Intracellular pH Adjustment

It has not been possible to date to monitor directly the kinetics of intracellular pH adjustment after changes of temperature in fish. It is possible, however, to model roughly the time course of intracellular pH adjustment in white muscle on the basis of $P_{CO_2}$, pH, and $[HCO_3^-]$ as a function of time after changes of temperature, and the kinetics of transepithelial transfer, when the bicarbonate transfer kinetics are assumed to be the same in all tissue species (Heisler, 1984).

The acid–base relevant transfer processes between environmental water and extracellular space are not complete before 16 to 24 hr after changes of temperature in *Scyliorhinus* (Fig. 5), which is close to the time required for the extracellular pH to attain a new equilibrium (Fig. 2) (Heisler, 1978). The transmembrane transfer requires only about 12 hr, and accordingly, steady-state values for the intracellular pH of white muscle were attained at about the same time. Most of the change (~90%), however, was achieved within 30 min after the change in temperature (Fig. 6).

In *Synbranchus* the time course of the overall transmembrane bicarbonate transfer was similar even though about three times the amount was

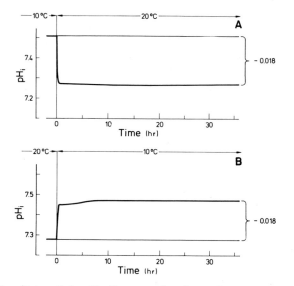

**Fig. 6.** Kinetics of intracellular pH adjustment after changes in temperature in white muscle of *Scyliorhinus stellaris*. (A) 10 → 20°C; (B) 20 → 10°C. (Data from Heisler, 1984.)

transferred (Fig. 7). Also adjustment of intracellular pH was complete within 10 hr, but $pH_i$ overshot considerably during the first hour and only slowly approached steady-state values (Fig. 8).

It is evident that adjustment of intracellular pH is a function of the transmembrane bicarbonate transfer and that a new steady state cannot be achieved before the transfer is complete. In white muscle of *Scyliorhinus*, however, the amount transferred is small enough per unit volume (Table V) to allow 90% of the change in pH to occur almost immediately. In red muscle

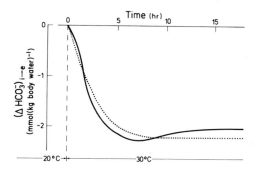

**Fig. 7.** Transfer of bicarbonate-equivalent ions between intracellular and extracellular space $(\Delta[HCO_3^-]_{i \to e})$ after a change in temperature in two specimens (solid and dotted line) of *Synbranchus marmoratus*. (Data of Heisler, 1983.)

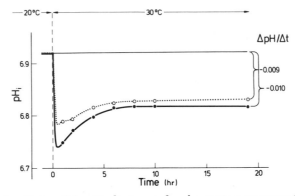

**Fig. 8.** Kinetics of intracellular pH adjustment after changes in temperature in white muscle in two specimens of *Synbranchus marmoratus*. (Data of Heisler, 1984.)

and heart muscle of *Scyliorhinus* the transmembrane bicarbonate transfer is much larger. According to the small relative volume of these tissues, the kinetics of $pH_i$ adjustment cannot be modeled with sufficient accuracy, but if the transfer rates are not higher than in white muscle by one order of magnitude, $pH_i$ adjustment in these tissues is expected to be a rather time-consuming process.

In *Synbranchus* white muscle, the extent of transmembrane transfer is much larger than in *Scyliorhinus* (Table VI). Therefore, the contribution of transfer processes is also much larger and accounts for the longer time to reach values close to the final steady state (Fig. 8). Since transepithelial ionic transfer in *Synbranchus* does not take place to a significant extent, the kinetics of both intracellular and extracellular pH adjustment are governed by the time course of transmembrane bicarbonate transfer.

A striking feature of the kinetics of acid–base regulation described earlier is that adjustment of pH after changes in temperature requires 10–24 hr (for intracellular and extracellular compartments, respectively); this is true even though the involved buffer mechanisms are instantaneous and the rate of bicarbonate transfer after changes of temperature between environmental water, extracellular space, and intracellular compartments is smaller by as much as one order of magnitude than those observed in other stress conditions (e.g., hypercapnia or lactacidosis, see later). Accordingly, the transfer mechanism cannot be the rate-limiting step of the process. The slow time course of the regulatory process is therefore probably the result of a slow adjustment of the acid–base setpoint values after changes of temperature and has to be considered as the result of an unknown acclimation process.

## B. Hypercapnia

Hypercapnia in fish may be induced by various mechanisms. Apparently they are all the result of changed environmental conditions. The most obvious reason for hypercapnia in fish is an elevated inspired $P_{CO_2}$ of the environmental water.

### 1. ENVIRONMENTAL HYPERCAPNIA

Fish have frequently been subjected to environmental hypercapnia in order to mimic elevations in $P_{CO_2}$ in their natural environment (see Section II,A,3), but also exposed to high ambient $P_{CO_2}$ in order to study the mechanisms of acid–base regulation in these animals.

When the environmental water $P_{CO_2}$ is elevated under experimental conditions in a step change, plasma $P_{CO_2}$ is also increased to a value between 1 and 4 mm Hg higher than the environmental $P_{CO_2}$ in a relatively short time period. The effect of such increased plasma $P_{CO_2}$ on the acid–base equilibrium has been studied in a number of water-breathing fish species (e.g., Lloyd and White, 1967; Cross et al., 1969; Cameron and Randall, 1972; Janssen and Randall, 1975; Eddy, 1976; Heisler et al., 1976a, 1980; Børjeson, 1976; Randall et al., 1976; Claiborne and Heisler, 1984).

The acid–base response to such an elevation of plasma $P_{CO_2}$ is rather uniform and characterized by an initial fall in pH after onset of hypercapnia, after which plasma pH starts to recover toward control values (Fig. 9). Finally, pH is almost completely compensated to less than $-0.1$ pH Units, in some cases less than $-0.05$ Units from the original in spite of continuing hypercapnia. This is the result of a large, up to fourfold increase in plasma bicarbonate concentration (Fig. 9), restoring the ratio $[HCO_3^-]/[CO_2]$ close to the control value. Also pH in those intracellular compartments that have been studied to date (white muscle, red muscle, and heart muscle of *Scyliorhinus stellaris*, Heisler et al., 1978; Heisler, 1980) and the mean whole-body intracellular pH of *Ictalurus punctatus* (Cameron, 1980) are compensated by elevated bicarbonate concentration to less than $-0.05$ pH Units from the control values.

Nonbicarbonate buffering of $CO_2$, which has been suggested by Cross et al. (1969) to be the source for the large amount of bicarbonate accumulated additionally in extracellular space and intracellular compartments, cannot be responsible. Since the nonbicarbonate buffer capacity (buffer value × volume) of the extracellular compartment (e.g., Albers and Pleschka, 1967; Holeton and Heisler, 1983; Toews et al., 1983) and the finally attained deviation of the extracellular pH from the control value are both small, the

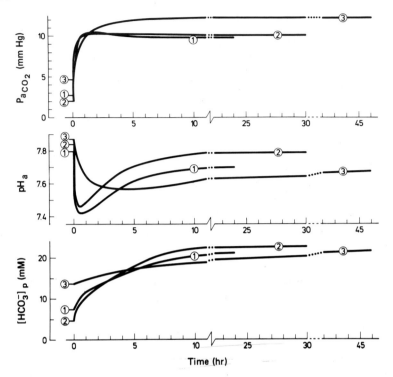

**Fig. 9.** Changes in arterial plasma $P_{CO_2}$, pH, and bicarbonate concentration in three fish species upon exposure to environmental hypercapnia. Species and data sources: (1) *Scyliorhinus stellaris* (Heisler *et al.*, 1976a); (2) *Conger conger* (Toews *et al.*, 1983); (3) *Cyprinus carpio* (Claiborne and Heisler, 1984).

contribution of extracellular nonbicarbonate buffering is negligible. Also, in the intracellular space the bicarbonate concentration is increased more than attributable to nonbicarbonate buffering (Heisler *et al.*, 1978; Heisler, 1980; Cameron, 1980; Cameron and Kormanik, 1982a), so that the increase in extracellular bicarbonate can also not be attributed to transfer from the intracellular space.

   If the accumulated bicarbonate is not produced in the metabolism or released from carbonate-containing structures intracorporally (which both are unlikely alternatives), it must have been gained from the environmental water. This hypothesis has already been tested by Cross *et al.* (1969). These authors, however, could not demonstrate any significant bicarbonate-equivalent transfer between environmental water and fish, probably because of their insensitive and inappropriate methodology (see Section II,B,1,a).

   More recently significant transfer of acid–base relevant ions has been demonstrated in two marine fish species, the elasmobranch *Scyliorhinus stellaris* (Heisler *et al.*, 1976a) and the teleost *Conger conger* (Toews *et al.*, 1984), and in the freshwater teleost fish *Cyprinus carpio* (Claiborne and Heisler, 1983). The fish were exposed to environmental hypercapnia of about 8 mm Hg combined with normoxia in a closed water recirculation system (see Section II,B,1,c), and the net transfer of bicarbonate-equivalent ions was determined by application of the Δ-bicarbonate measurement system (see Section II,C,1,c). Ammonia was determined with ammonia-sensitive electrodes (see Section II,C,2).

During the normocapnic control period the ureotelic *Scyliorhinus* excreted only small amounts of ammonia, which were almost balanced by release of bicarbonate-equivalent ions (Table III). The ammonia excretion remained constant on exposure to hypercapnia. The bicarbonate excretion, which is slightly positive under control conditions, increased for a short period (15 min) and then reversed into a considerable uptake of bicarbonate-equivalent ions. The net $H^+$ extrusion (net $\Delta H^+_{e \to w}$) as the difference between control and experimental sum of bicarbonate uptake and ammonia excretion leveled off at about 4.5 mmol kg body water$^{-1}$ (Fig. 10).

The ammoniotelic marine teleost fish *Conger conger* exhibited a sixfold higher control ammonia excretion than *Scyliorhinus*, which was also almost completely balanced by bicarbonate release (Table III). During hypercapnia the ammonia excretion was slightly reduced, which has to be considered as a maladjustment with respect to the net $H^+$ extrusion required for the compensatory bicarbonate accumulation in the animal (Fig. 10). The considerable control release of bicarbonate-equivalent ions was reversed, again with a short delay time, to a gain of bicarbonate from the environment (Toews *et al.*, 1983). About 5.3 mmol kg body water$^{-1}$ of bicarbonate was net taken up (net $\Delta H^+_{e \to w}$) before the fish returned to the control rate of bicarbonate release. The quantity of bicarbonate net gained from the environment was larger than in *Scyliorhinus* and corresponds to the higher degree of compensation of the extracellular pH achieved in *Conger* ($\Delta pH < -0.05$ U) as compared to *Scyliorhinus* ($\Delta pH < -0.1$ U).

In the freshwater teleost *Cyprinus carpio* the control ammonia release was similar to that in *Conger*, whereas the bicarbonate release was smaller, resulting in a higher $H^+$ ion-equivalent control excretion ($\Delta H^+_{excr}$, Table III).

Upon exposure to elevated $P_{CO_2}$ the bicarbonate release rate initially remained constant, whereas the ammonia release was increased. Later the ammonia excretion rate returned to control values, and bicarbonate was released to the environment at a lower rate. During the hypercapnic observation period there was a net gain of about 4.2 mmol kg body water$^{-1}$ of bicarbonate-equivalent ions from the environment (Claiborne and Heisler, 1984). Although the bicarbonate concentration in the plasma and the amount

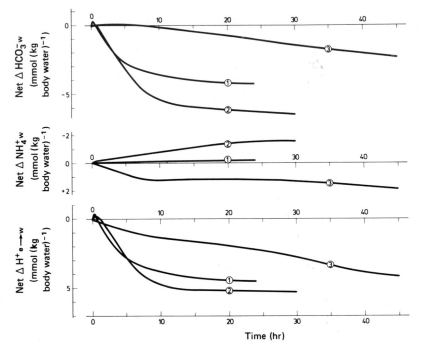

**Fig. 10.** Net changes in the amount of bicarbonate (net $\Delta HCO_3^-$ $_w$) and ammonium (net $(\Delta NH_4^+)_w$ in the environmental water, and the net amount of $H^+$ ions released (net $\Delta H^+_{e \to w}$) of three fish species on exposure to environmental hypercapnia. Species and data sources same as Fig. 9.

of bicarbonate gained were similar in all three species (Table VIII), the degree of extracellular pH compensation was much smaller in *Cyprinus* ($\Delta pH \sim -0.2$ U). This is because the original bicarbonate concentration in the plasma was about twice as high as in the two marine species. Accordingly much more bicarbonate is required to achieve complete pH compensation.

The time course of compensation and of net bicarbonate gained from the environment is considerably different between marine and freshwater fish. *Scyliorhinus* and *Conger* achieved almost complete compensation in about 8 to 10 hr (Heisler *et al.*, 1976a; Toews *et al.*, 1983), whereas in *Cyprinus* only partial compensation was achieved in 46 hr (Claiborne and Heisler, 1984), and almost complete compensation required 24 hr in *Ictalurus* (Cameron, 1980), and 22 or 72 hr in *Salmo gairdneri* (Eddy *et al.*, 1977; Janssen and Randall, 1975).

These extreme differences even between experiments on the same species with the underlying differences in ionic uptake rates are probably to only a minor extent related to species differences. One explanation of the

**Table VIII**

Extracellular Bicarbonate Concentrations (mM) during Control Conditions ($[HCO_3^-]_c$) and in Hypercapnia ($[HCO_3^-]_{hyp}$), and Transfer of Bicarbonate-Equivalent Ions between Intracellular and Extracellular Space ($\Delta HCO_3^-{}_{i \to e}$), and between Extracellular Space and Environmental Water ($\Delta HCO_3^-{}_{e \to w}$) during Hypercapnia[a]

| Species | Origin of hypercapnia | $[HCO_3^-]_c$ | $[HCO_3^-]_{hyp}$ | Exposure time (hr) | $\Delta HCO_3^-{}_{e \to w}$ | $\Delta HCO_3^-{}_{i \to e}$ | Reference |
|---|---|---|---|---|---|---|---|
| *Scyliorhinus stellaris* | Environmental | 7.6 | 20 | 8 | −4.5 | −1.1 | Heisler *et al.* (1976a) |
| *Conger conger* | Environmental | 5.0 | 22 | 10 | −4.9 | −1.4 | Toews *et al.* (1983) |
| | | | 22 | 30 | −5.3 | −1.8 | Toews *et al.* (1983) |
| *Cyprinus carpio* | Environmental | 13.0 | 22 | 48 | −4.2 | −1.7 | Claiborne and Heisler (1984) |
| *Scyliorhinus stellaris* | Hyperoxia induced | 5.3 | 20 | 25 | −5.2 | −1.35 | Heisler *et al.* (1981, 1984) |
| | | 5.3 | 24 | 144 | — | — | |
| *Synbranchus marmoratus* | Transition water to air breathing | 24 | 24 | 18 | −0.7 | −1.9 | Heisler (1982a) |
| | | 24 | 24 | 100 | — | — | Heisler (1982a) |

[a]Values given in mmol kg body water$^{-1}$.

differences between the time courses in the experiments of Eddy *et al.* and of Janssen and Randall in trout is based on the different water quality: the Vancouver, British Columbia, tap water used by Janssen and Randall is extremely soft water (D. J. Randall, personal communication) and has a correspondingly low pH (~5.5 in hypercapnia), which possibly slowed down the ionic uptake rates as a result of the much smaller availability of counterexchange ions (Heisler, 1982b). The same holds, to even much larger extent, for the ionic differences between seawater and low-concentrated fresh water (see Section V). Indeed, preliminary investigations have indicated that elevation of the water [NaHCO$_3$] to 3 m$M$ results in a twofold increase in the rate of net bicarbonate gain from the environment in carp (J. B. Claiborne and N. Heisler, unpublished).

The influence of the water pH (or, at constant $P_{CO_2}$, the bicarbonate concentration) on the time course of pH compensation during hypercapnia has also been demonstrated in *Scyliorhinus* (Heisler and Neumann, 1977; see also Heisler, 1980). With constant degree of hypercapnia (~8 mm Hg plasma $P_{CO_2}$), the maximal bicarbonate-equivalent uptake rate, 14–15 μmol (min kg body water)$^{-1}$, was achieved only when the water pH was higher than 7.2. At water pH values lower than 6.8, the uptake rate was gradually reduced. The best correlation was obtained between bicarbonate uptake rate and the arterial–seawater pH difference ($\Delta pH = pH_{pl} - pH_{sw}$), or, since $P_{CO_2}$ was constant, the bicarbonate concentration ratio. The uptake rate was almost constant and maximal at $\Delta pH$ from $-0.5$ to 0.6 and gradually decreased with rising pH, reaching zero at about $\Delta pH = 1.2$. Higher values for $\Delta pH$ resulted in net bicarbonate-equivalent ion loss.

The apparent lack of complete pH compensation in carp in contrast to the other investigated fish species can hardly be explained on the basis of the mechanism just described. The limiting factor in carp may instead be the maximal bicarbonate concentration that can be attained in the extracellular space by the bicarbonate-retaining and bicarbonate-resorbing structures. A review of literature data shows that during hypercapnia such a maximal threshold can be expected at a plasma bicarbonate concentration of about 25 to 32 m$M$ for fishes and amphibians (Heisler, 1985). The higher the initial bicarbonate concentration, the smaller is the possible compensatory increase during hypercapnia and thus restoration of pH toward the control value. In the aquatic salamander *Siren lacertina*, plasma [HCO$_3^-$] did not rise during hypercapnia above 25 m$M$, even though plasma pH remained reduced by more than 0.3 pH Units. Increasing the water [HCO$_3^-$] up to 8.6 m$M$ or even infusion of NaHCO$_3$ did not increase the hypercapnic steady-state plasma [HCO$_3^-$]. The infused bicarbonate was instead quantitatively released to the environment (Heisler *et al.*, 1982). Recent evidence on carp suggests that this fish species can also not maintain a bicarbonate concentration in plasma above about 25 m$M$, even during environmental hypercapnia

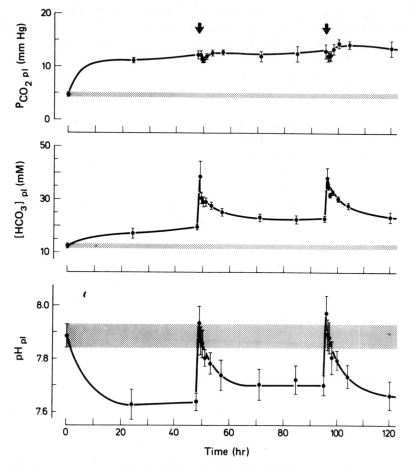

**Fig. 11.** Responses of arterial plasma $P_{CO_2}$, [$HCO_3{}^-$], and pH in carp (*Cyprinus carpio*) exposed to environmental hypercapnia (~8 mm Hg) and additionally infused with bicarbonate (arrows: 5 mmol kg$^{-1}$; see text).

of larger extent (5% $CO_2$) or following infusion of $NaHCO_3$ (5 mmol kg$^{-1}$) (Fig. 11) (Claiborne and Heisler, 1985). Accordingly, the pH compensation in carp during hypercapnia has to be considered as limited by the relatively high control plasma bicarbonate concentration.

## 2. HYPERCAPNIA INDUCED BY TRANSITION FROM WATER BREATHING TO AIR BREATHING

Fish are considered to regulate their ventilation primarily according to the oxygen demand; that is, ventilation is regulated to maintain a constant arterial $P_{O_2}$ (cf. Dejours, 1975). When facultative air-breathing fish terminate aquatic oxygen uptake through gills and skin because of too low environmental water $P_{O_2}$, the transition to exclusive air breathing regularly causes considerable increases in plasma $P_{CO_2}$ (e.g., DeLaney et al., 1974, 1977; Lenfant et al., 1966–1967; Heisler, 1982a). This is the result of reduced ventilation due to the much higher oxygen content of air and the largely reduced capacitance ratio of $CO_2/O_2$ in air as compared to water. An additional factor may be the low activity of carbonic anhydrase in the aerial gas exchange structure (cf. Randall et al., 1981).

The rise in arterial $P_{CO_2}$ as a result of the transition from water to air breathing is most pronounced in those species capable of exclusively breathing one or the other medium. In the South American freshwater teleost Synbranchus marmoratus, which is one of the rare representatives of this group, arterial $P_{CO_2}$ rises during transition from exclusive water breathing to exclusive air breathing from 5.6 to 26 mm Hg within 2 to 3 days (Heisler, 1982a). This rise in $P_{CO_2}$ is associated with a fall in plasma pH by more than 0.6 pH Units, which is not compensated by elevation of the plasma bicarbonate concentration (Table VIII). This lack of compensation of the considerable hypercapnic acidosis may be associated with the same factor preventing complete compensation of hypercapnic acid–base disturbances in carp, that is, that the control bicarbonate concentration of about 24 mM in Synbranchus cannot further be elevated because of the postulated threshold of the bicarbonate-retaining and -resorbing structures at about 25 mM. Thus, Synbranchus would behave very similarly in its acid–base regulation to the aquatic salamander species Siren lacertina and Amphiuma means (Heisler et al., 1982), which live in habitats with similarly adverse environmental conditions (low pH, $P_{O_2}$, and electrolyte concentration).

During the initial time period after the switch from water breathing to air breathing, the release of ammonia and bicarbonate, which is typical for ammoniotelic fish, is initially still maintained in Synbranchus, but then progressively reduced. This must be associated with a drastic reduction in ammonia production in order to avoid toxic effects in the organism. Whether or not this is a result of an overall reduction in nitrogen metabolism or a shift to other nitrogenous waste products (as shown for Periopthalmus, Gregory, 1977) is still unclear. The initiating factor is very likely the fall in air-breathing frequency and the associated gill water flush frequency (Heisler, 1982a), which results in a reduction of the gill–water contact time by more than a factor of 100 during steady-state air breathing as compared to water breathing.

The release of bicarbonate is reduced more than that of ammonia during the first 10 hr after the switch. This results in a net gain of 0.7 mmol kg body water$^{-1}$ of bicarbonate, which is, together with 0.8 mmol kg body water,$^{-1}$

produced by blood nonbicarbonate buffering, and 0.4 mmol kg body water$^{-1}$ from the extracellular bicarbonate pool, transferred to the intracellular space (Heisler, 1982a).

Although the extracellular compartment remains completely uncompensated, the intracellular compartments of white muscle and heart muscle are little affected by the severe hypercapnia. This is the result of an about 4.5-fold elevation of the intracellular bicarbonate concentration by nonbicarbonate buffering in the cells, and the relatively small amount of bicarbonate-equivalent ions transferred into the cell from the extracellular space (Table VIII).

The amount of bicarbonate transferred has a considerable effect on the intracellular pH compensation, and it suffices to allow tight regulation of the intracellular pH. This is only possible because of the initially low intracellular bicarbonate concentration (water breathing), whereas the same amount of bicarbonate distributed in the extracellular space on the background of 24 mM [$HCO_3^-$] would have resulted in only 20% compensation of the pH shift actually observed (Heisler, 1982a). This strategy of acid–base regulation with preference of the intracellular compartments over the extracellular compartments is similar to that in the aquatic salamander *Siren* (Heisler *et al.*, 1982) and can also be observed, but is less pronounced in other fish species, amphibians, reptiles, and even mammals.

Air-breathing fish generally do not appear to compensate the extracellular space. This has been demonstrated in *Channa argus* (Ishimatsu and Itazawa, 1983), which showed even decreases in plasma [$HCO_3^-$] associated with air breathing and the hypercapnia-induced fall in pH. Also, in *Amia calva* arterial total $CO_2$ content remained constant when $P_{CO_2}$ rose as a consequence of air exposure (Daxboeck *et al.*, 1981). In *Protopterus aethiopicus*, plasma bicarbonate concentration rose slightly during 5 days of air breathing in a mud burrow (DeLaney *et al.*, 1974), as well as during 7 months of air breathing also in a mud burrow (DeLaney *et al.*, 1977). This rise, however, has to be attributed to loss of water volume, which is indicated by the simultaneous overproportionate concentration increases of the other ions (DeLaney *et al.*, 1977).

All these data indicate that bicarbonate is at least not gained by ionic exchange with the environment, which would also be unlikely, since water contact is extremely limited. In contrast, all available data tend to indicate that with the developing hypercapnic acidosis during air breathing, bicarbonate (produced by extracellular nonbicarbonate buffering and from the extracellular pool) is shifted into the intracellular space in order to protect the intracellular pH. The available information, however, is still rather incomplete, and the mechanism of preferential intracellular pH protection in air-breathing fish species awaits further elucidation.

3. HYPEROXIA-INDUCED HYPERCAPNIA

The partial pressure of oxygen in natural waters is the result of various factors. Oxygen is consumed in the water by aerobic metabolism of microorganisms, various aquatic animal species, and the water flora during dark periods. In turn, oxygen is produced in the water by photosynthesis during light periods, enters the water by surface diffusion, and is transported into deeper water layers by convection. As a result of the light cycle-induced periodic changes in oxygen production in the water and limitations of water–air gas exchange by surface plant layers and convection-limiting thermostratification, $P_{O_2}$ may vary between zero and more than 400 mm Hg, under certain conditions even up to atmospheric pressure (e.g., Kramer *et al.*, 1978; Sioli, 1954, 1955, 1957).

Exposure of fish to hyperoxic water regularly results in an increase in plasma $P_{CO_2}$ (e.g., Dejours, 1972, 1973; Truchot *et al.*, 1980; Wood and Jackson, 1980; Wilkes *et al.*, 1981; Heisler *et al.*, 1981). This is correlated with the generally observed inverse relationship between gill ventilation and the oxygen content of the water (e.g., Saunders, 1962; Holeton and Randall, 1967; Davis and Cameron, 1970; Shelton, 1970; Davis and Randall, 1973; Randall and Jones, 1973; Eddy, 1974; Dejours *et al.*, 1977; Itazawa and Takeda, 1978), as a result of the primarily oxygen-oriented regulation of respiration in fishes (Dejours, 1975).

The fall in plasma pH resulting from the hyperoxia-induced rise in arterial $P_{CO_2}$ was, unlike the regulatory pattern during environmental hypercapnia, in most of the studied species not or only partially compensated by elevation of plasma bicarbonate even after long-term exposure (e.g., Bornarcin *et al.*, 1977; Dejours, 1973, 1975; Truchot *et al.*, 1980). Only in *Salmo gairdneri* (Wood and Jackson, 1980), *Catostomus commersoni* (Wilkes *et al.*, 1981), and *Scyliorhinus stellaris* (Heisler *et al.*, 1981) could complete or almost complete compensation be observed.

In *Scyliorhinus*, plasma pH deviated on the average from the control values in fact not more than −0.08 U throughout the whole experiment. When inspired $P_{O_2}$ was elevated in a step change fashion in a closed water recirculation system to about 500 mm Hg (see Fig. 12), gill ventilation was reduced to less than 40% of the control value within 2 min and fell to about 20% after 45 min (Fig. 12). As a result of this largely reduced ventilation rate, arterial $P_{CO_2}$ started to rise steeply and arterial pH fell by 0.08 pH Units. Then, however, gill ventilation was increased again by a factor of two over the next 2 hr, which slowed down the increase rate of arterial $P_{CO_2}$ considerably. With a short delay (15 min) after the initial rise in $P_{CO_2}$ and drop in plasma pH, the fish started to gain bicarbonate from the environment at about the same rate ($\sim$14 μmol (min kg body water)$^{-1}$), as has maximally been observed during environmental hypercapnia for the same

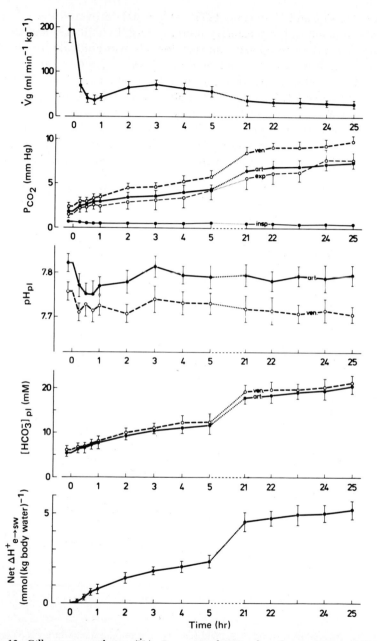

**Fig. 12.** Gill water ventilation ($\dot{V}_g$), $P_{CO_2}$, pH, [$HCO_3^-$], and net amount of $H^+$ ions transferred to the environmental water (Net $\Delta H^+_{e\to sw}$) of *Scyliorhinus stellaris* exposed to environmental hyperoxia ($P_{O_2}$ = 500 mm Hg). (Data of N. Heisler, G. F. Holeton, and D. P. Toews, unpublished.)

species (Heisler and Neumann, 1977). When pH recovered toward control values as a result of the secondary increase in gill ventilation and the compensatory bicarbonate uptake, the transfer rate was gradually reduced to 5 µmol (min kg body water)$^{-1}$ between hours 3 and 5, and to 2 µmol (min kg body water)$^{-1}$ after 21 hr. After the third hour of hyperoxia, gill ventilation slowly fell to about 15% of the control after 25 hr. This resulted in a further increase in $P_{CO_2}$ at a rate that could just be compensated by the transepithelial bicarbonate uptake, so that no major deviations of plasma pH occurred. When freely swimming fish specimens were investigated at the same level of environmental hyperoxia, it became evident that the process of

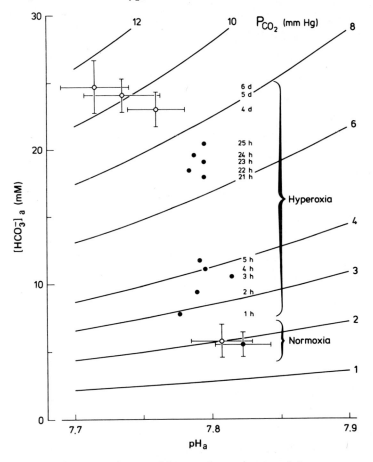

**Fig. 13.** Arterial plasma pH, $P_{CO_2}$, and [HCO$_3$$^-$] in normoxia and during environmental hyperoxia of up to 6 days duration (d, days; h, hours). Filled circles are values from animals kept in a closed water recirculation system (see Fig. 1); open circles are from freely swimming unrestrained fish. (Data of N. Heisler, G. F. Holeton, and D. P. Toews, unpublished.)

adaptation to the hyperoxic environment was not complete after 25 hr (Fig. 13). Arterial $P_{CO_2}$ further rose from about 7.5 to about 11 mm Hg after 6 days (Fig. 13). Plasma bicarbonate was also further elevated, very likely as a result of transepithelial ion transfer, although not to the extent required for complete pH compensation: $[HCO_3{}^-]$ only rose to 24 m$M$, and pH deviated after 6 days of hyperoxia by about 0.17 pH Units from the control values (Fig. 13). This demonstrates again that a maximal plasma bicarbonate level of more than 25 m$M$ can hardly be attained in water-breathing fish (Table VIII; see also Section IV,B,1 and 2).

The acid–base regulation during hyperoxia-induced hypercapnia appears to be not principally different from that in environmental hypercapnia. During both conditions changes in $P_{CO_2}$ are compensated by transepithelial ion transfer processes of similar extent (Table VIII; cf. Section IV,B,1).

The slower rate of $P_{CO_2}$ increase during hyperoxia enables the organism to avoid deviations in plasma pH, and, since the transepithelial transfer of bicarbonate is much larger than required for the compensation of the extracellular space alone, of also the intracellular pH (Table VIII). This is made possible by a tight adjustment of ventilation to the requirements of acid–base regulation. In contrast to normoxia, the oxygen demand of the animal can be supplied in hyperoxia without any limitations, and arterial $P_{O_2}$ is not tightly regulated by gill ventilation. During the first hours of hyperoxia, $P_{O_2}$ is much higher than required for full blood saturation (250–300 mm Hg), and falls with gill ventilation until after about 6 days of hyperoxia arterial $P_{O_2}$ has attained values well below 100 mm Hg in the range of values observed during normoxia.

These data suggest that during hyperoxia the oxygen-sensitive respiratory drive is subordinate to a pH-sensitive mechanism similar to that known for higher vertebrates. This mechanism, however, requires additional investigation.

## C. Strenuous Muscular Activity

Fish are usually equipped with a large amount of poorly perfused white musculature, which represents the major proportion of the body mass (see Section II,A,2,b) and a much smaller amount of well-perfused red muscle. The red muscles, which are mainly localized near to the outer body wall along the side lines, are considered to perform, exclusively with aerobic energy production, the normal cruising and positioning activity in fish. This is reflected by the low plasma lactate concentration of less than 1.5 m$M$ found in unstressed fish. In contrast, during emergency situations and predacious activities, recruitment of the barely perfused white muscles results in a large production of lactic acid as the main metabolic end product of anaerobic activity in vertebrates. Under such conditions up to 84 mmol kg tissue weight$^{-1}$ can be accumulated in white muscles of fish (Wardle, 1978).

At physiological pH in intracellular and extracellular compartments (see Section III,B) lactic acid, because of its low pK value (~3.9), is almost exclusively dissociated to equimolar quantities of $H^+$ and lactate ions and thus represents a considerable stress for the acid–base regulation of the animal. The dissociation products of lactic acid are gradually eliminated from the muscle cells. The lactate concentration in plasma rises to peak values of up to 30 m$M$ within 2 to 8 hr after strenuous exercise (see Fig. 14 and Table IX). Only in marine flatfish, the peak lactate concentrations are smaller by about a factor of 10 and usually do not exceed 2 m$M$ (Wood et al., 1977; Wardle, 1978; Turner et al., 1983). This difference was explained on the basis of nonrelease of lactate from muscle cells due to stress-induced high levels of circulating catecholamines (Wardle, 1978) or by the existence of a nonidentified metabolic end product of anaerobic glycolysis rather than lactic acid (Wood et al., 1977; Turner et al., 1983).

The acid–base disturbances in plasma induced by the efflux of $H^+$ from the muscle cells usually peak much earlier (~1 hr) than plasma lactate (Table IX). Piiper et al. (1972), however, pointed out that in spite of the apparently much faster efflux kinetics of $H^+$ ions, the excess of lactate in the extracellular space exceeded the amount of surplus $H^+$ ions several fold. This holds for all studied fish species, except for the marine flatfish (see Table IX). The apparent "$H^+$ ion deficit" could be explained by different distribution of $H^+$ ions and lactate between intracellular and extracellular body compartments of the fish (Piiper et al., 1972), or by net transfer of $H^+$ ions to the environmental water similar to the mechanism applied during hypercapnia and after temperature changes (see Section IV,A and B).

A quantitative evaluation of the mechanisms just cited has been performed in the elasmobranch Scyliorhinus stellaris (Holeton and Heisler, 1983), the marine teleost fish Conger conger (Holeton et al., 1985; see also Heisler, 1982b), and the freshwater trout Salmo gairdneri (Holeton et al., 1983). After electrical stimulation in a closed water recirculation system (see Section II,B,1,c; Fig. 1), the transfer of acid–base relevant ions and of ammonia between intracellular compartment, extracellular space, and environmental water was monitored for more than 30 hr after the imposed exercise stress.

All three fish species started immediately after exercise to transfer $H^+$ ions into the environmental water in considerable amounts (Fig. 15). The control excretion of ammonia in Scyliorhinus and Conger (Table III) remained unchanged, the control bicarbonate release (Table III) was reversed in both species to an uptake (Fig. 15). In contrast, in Salmo gairdneri net uptake of bicarbonate from the water contributed to only a small extent;

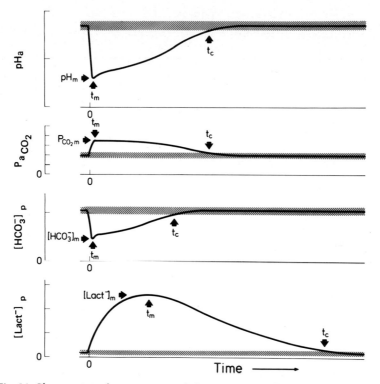

**Fig. 14.** Characteristic changes in arterial plasma pH, $P_{CO_2}$, [$HCO_3^-$], and [lactate] in fish after strenuous exercise. (Index "m" designates value and time of a parameter at maximal deflection from control values. $t_c$ represents the time when control values have been re-attained for the respective parameter. For values see Table IX.

most of the overall net $H^+$ excretion was the result of a considerable increase in ammonia release. In *Scyliorhinus* and *Salmo* the net $H^+$ extrusion continued at rather constant rate until the plasma bicarbonate concentration approached steady-state values (Fig. 14, Table IX), whereas in *Conger* the $H^+$ extrusion leveled off before complete normalization of the arterial acid–base status (Fig. 14, Table IX). The reason for this behavior in *Conger* is unknown. The $H^+$ ions net transferred to the environment in order to normalize the acid–base status during the first part of the recovery period are later returned into the fish for metabolization of lactic acid or resynthesis to glycogen.

The results of these studies indicate that a sizable proportion of the "hydrogen ion deficit" (Piiper *et al.*, 1972) observed in all studied fish species—except the flatfish—is due to temporary net transfer of $H^+$ ions to the ambient water. Taking this transfer into account, the amount of $H^+$ ions

Table IX

Characteristics of Changes in Acid–Base Status and Plasma Lactate Concentration in Fish after Strenuous Exercise[a]

| Species | $pH_a$ | | | | $P_{aCO_2}$ | | | | $[HCO_3^-]_p$ | |
|---|---|---|---|---|---|---|---|---|---|---|
| | $pH_c$ | $pH_m$ | $t_m$ (hr) | $t_c$ (hr) | $P_{CO_2c}$ (mm Hg) | $P_{CO_2m}$ (mm Hg) | $t_m$ (hr) | $t_c$ (hr) | $[HCO_3^-]_c$ (mM) | $[HCO_3^-]_m$ (mM) |
| *Scyliorhinus* | 7.78 | 7.20 | 1 | 18 | 1.3 | 3.0 | 0.25 | 3.5 | 3.8 | 1.24 |
| *stellaris* | 7.89 | 7.42 | 0.25 | (12) | 2.1 | 4.57 | 0.25 | 8 | 7.39 | 4.41 |
| | 7.83 | 7.19 | 0.50 | (18) | 2.02 | 5.1 | 0.5 | 12 | 6.84 | 3.18 |
| *Conger conger* | 7.86 | 7.42 | 0.25 | (12) | 1.95 | 4.75 | 0.25 | 11 | 5.15 | 3.30 |
| *Oncorhynchus* *nerka* | — | — | — | — | — | — | — | — | — | — |
| *Platichthys* *stellatus* | 7.90 | 7.52 | 0 | (8) | 2.0 | 7.0 | 0 | 4 | 6.9 | 4.6 |
| *Pleuronectus* *platessa* | — | — | — | — | — | — | — | — | — | — |
| *Hippoglossoides* | 7.80 | 7.32 | 0 | 12 | 2.2 | 5.3 | 0 | 2 | 5.1 | 3.2 |
| *elassodon* | 7.83 | 7.58 | 0 | 4 | 2.0 | 3.7 | 0 | 2 | 5.6 | 5.2 |
| *Salmo gairdneri* | 7.81 | 7.31 | 0 | 3.5 | 1.82 | 3.45 | 0.25 | 4 | 4.1 | 2.4 |
| | 7.88 | 7.34 | 0 | 8 | 3.1 | 7.8 | 0 | 4 | 8.3 | 3.6 |
| *Salvelinus* *namaycush* | — | — | — | — | — | — | — | — | — | — |
| | — | — | — | — | — | — | — | — | — | — |
| | — | — | — | — | — | — | — | — | — | — |

[a]Index c designates control values. For other symbols see Fig. 14 (for acid–base parameters only, measurements obtained from fish with indwelling catheters taken into account). Values in parentheses are interpolated.

[b]FW, fresh water; SW, seawater.

released from the intracellular compartments is actually for most of the recovery time period larger than the amount of lactate transferred to the extracellular space (Holeton and Heisler, 1983).

The efflux kinetics of $H^+$ ions and lactate ions from the muscle cells are apparently different. The rate constants for the efflux have been estimated for postexercise conditions in *Scyliorhinus* (Holeton and Heisler, 1983). It was found that the rate constant for the $H^+$ elimination from the muscle cells was at least 12 times larger than that for lactate. Since this factor is considerably different from the time course ratio observed in vivo, other factors must also be involved.

It has been argued that the slow efflux of lactic acid from muscle tissues in fish as compared to higher vertebrates was the result of partial, or even complete shutdown of perfusion to the lactic acid-loaded white muscles as a

Table IX (Continued)

| [HCO$_3^-$]$_p$ | | [Lact$^-$]$_p$ | | | | | | |
| $t_m$ (hr) | $t_c$ (hr) | [Lact$^-$]$_c$ (mM) | [Lact$^-$]$_m$ (mM) | $t_m$ (hr) | $t_c$ (hr) | Temperature (°C) | Environmental water[b] | Reference |
|---|---|---|---|---|---|---|---|---|
| 1 | 8 | 1.0 | 21 | 6 | 22 | 17 | SW | Piiper et al. (1972) |
| 0.5 | 8 | 1.3 | 8.7 | 4 | (18) | 16 | SW | Holeton and Heisler (1983) |
| 0.5 | 10 | 1.4 | 20.5 | 8 | >30 | 16 | SW | Holeton and Heisler (1983) |
| 0.25 | 10 | 0.6 | 13.4 | 2 | 23 | 17 | SW | Holeton et al. (1985) |
| — | — | 2.3 | 30.8 | 2.5 | 8 | 20 | SW | Black (1957c) |
| 1 | 4 | 0.3 | 1.8 | 4 | — | 9 | SW | Wood et al. (1977) |
| — | — | 0.8 | <2 | 4 | 6 | 9 | SW | Wardle (1978) |
| 2 | 12 | 0.2 | 0.8 | 2 | 8 | 11.5 | SW | Turner et al. (1983) |
| 0.5 | 4 | 0.2 | 0.3 | 0.5 | 0 | 11.5 | SW | Turner et al. (1983) |
| 0 | 3 | 0.9 | 13.6 | 2 | (18) | 15 | FW | Holeton et al. (1983) |
| 1 | 8 | 0.23 | 13.0 | 2 | (16) | 15 | FW | Turner et al. (1983) |
| — | — | 0.8 | 13 | 2 | 12 | 8–12 | FW | Black et al. (1959) |
| — | — | 0.6 | 17 | 2 | 12 | 11–12 | FW | Black et al. (1966) |
| — | — | 1.8 | 18.2 | 2 | 12 | 11–12 | FW | Black (1957a) |
| — | — | 1.4 | 19.4 | 2 | 16 | 11–12 | FW | Black (1957b) |

protection mechanism for the organism (Black et al., 1962; Stevens and Black, 1966; Wood et al., 1977; Wardle, 1978). This factor does not appear to be of any significance, since blood flow to the muscle tissues in trout was, during post-exercise recovery, increased considerably as compared to control values (Neumann et al., 1983). It was also shown by model calculations based on measured parameters that peak lactate concentration values in the blood of trout had to be expected by 2.5 min after exercise, if the process was exclusively perfusion limited (Neumann et al., 1983). Comparison with the actual time required (2 hr, Holeton et al., 1983) indicates that lactate efflux from muscle cells must, in contrast, be largely diffusion limited.

The H$^+$ efflux rate constant is so high (Holeton and Heisler, 1983) that perfusion limitation for H$^+$ ions may well be present. More important, however, for the slow removal from the muscle cells is the "equilibrium limitation" (Holeton and Heisler, 1983). According to the small volume and low buffer value of the extracellular space, transfer of only a small quantity of

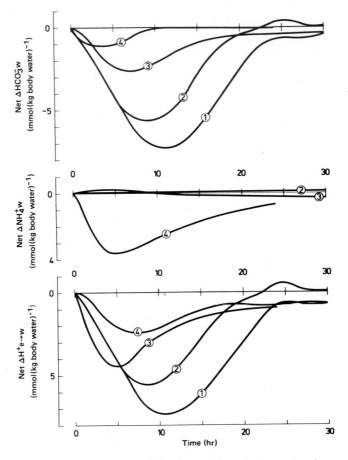

**Fig. 15.** Net changes in the amount of bicarbonate (Net $\Delta HCO_3^-{}_w$) and ammonium (net $\Delta NH_4^+{}_w$) in the environmental water, and the net amount of $H^+$ ions released (Net $\Delta H^+{}_{e \to w}$ of three fish species after strenuous exercise. Species and data sources: (1) and (2) *Scyliorhinus stellaris* (Holeton and Heisler, 1983); (3) *Conger conger* (G. F. Holeton, D. P. Toews, and N. Heisler, unpublished); (4) *Salmo gairdneri* (Holeton *et al.*, 1983).

$H^+$ ions suffices to lower the extracellular pH so much that a new equilibrium between intracellular and extracellular pH is achieved. This eliminates the driving force for further $H^+$ transfer.

The lactate efflux from the intracellular space of white muscle is governed by other factors. At equilibrium, lactate is distributed across the cell membrane either according to the membrane potential, or according to the intracellular/extracellular pH difference, depending on whether lactate is transferred across the cell membrane predominantly in ionized or non-ionized form. Thus, in contrast to $H^+$ ions, a large fraction of the lactate can be transferred to the extracellular space before equilibrium occurs.

These, and possibly other still unknown factors determine the distribution after strenuous exercise of $H^+$ and lactate among the intracellular compartments, extracellular space, and environmental water. The data on

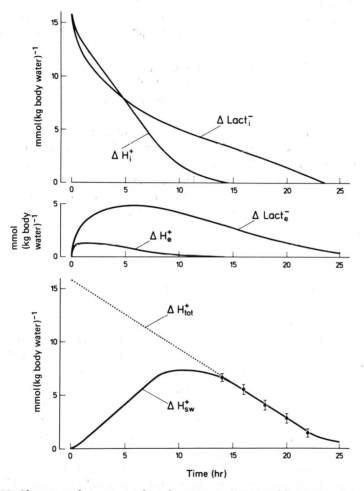

**Fig. 16.** Changes in the amounts of surplus $H^+$ ions ($\Delta H^+$) and lactate ions ($\Delta Lact^-$) in intracellular space (index i), extracellular space (index e), and environmental seawater (index sw). $\Delta H^+_{tot}$ is the total amount of surplus $H^+$ ions in fish and seawater, extrapolated from the regression of $\Delta H^+_{sw}$ between 14 and 24 hr. For details see Holeton and Heisler (1983).

*Scyliorhinus* exemplify all three studied fish species (Fig. 16) (Holeton and Heisler, 1983).

At the end of muscular activity $H^+$ ions are released from the intracellular space (fig. 16, $\Delta H_i^+$) at a high rate until saturation of the extracellular space occurs due to equilibrium limitation (fig. 16, plateau of the curve of $\Delta H_e^+$). Hydrogen ions are further removed at a rate of only ~25 μmol min$^{-1}$ kg body water$^{-1}$ according to the transfer of $H^+$ ions to the environment ($\Delta H^+_{e \to w}$, ~15 μmol min$^{-1}$ kg body water$^{-1}$) and the further aerobic metabolic processing ($\Delta H^+_{tot}$, ~10 μmol min$^{-1}$ kg body water$^{-1}$). Lactate, instead, is continuously diffusing out of the intracellular space, and more than 50% of the total amount may be present in the extracellular space. In contrast, at no time are more than 10% of surplus $H^+$ ions present in the extracellular compartment. The main proportion is continuously transferred from intracellular space via extracellular space to environmental water, and, when the extracellular load is reduced as a result of aerobic lactic acid processing, back into the extracellular space.

During the first hours after exercise most of the surplus $H^+$ ions are actually stored in the intracellular compartments as suggested by Piiper *et al.* (1972). Later all of the surplus $H^+$ ions are temporarily transferred into the environmental water. This important mechanism allows normalization of the acid–base status of the fish at a time when usually not more than 40–50% of the lactic acid as the original acid–base stress factor has been removed by aerobic metabolism.

## D. Acidic Environmental Water

Low environmental water pH is naturally observed in some electrolyte-poor water basins, such as in the Amazon, where the pH of most of the studied waters is between 4 and 5, sometimes up to 6 (Sioli, 1954, 1955, 1957). Industrial influences have artificially created during the last few decades wide areas with similarly acidic waters in the northeastern United States and Canada, and in central Europe and southern Scandinavia. Such low water pH values were rarely found in these areas until coal mine drainages and acid precipitation lowered the pH of originally neutral or slightly alkaline (pH 7–9) waters to values incompatible with sustained fish life (e.g., Parsons, 1968, 1976; Kinney, 1964; Leivestad *et al.*, 1976; Leivestad and Muniz, 1976; Beamish, 1976; Schofield, 1976).

The acid–base status in fish exposed to unnaturally low environmental water pH has been studied by a number of investigators. The results have been quite diverse and often contradictory. When brook trout (*Salvelinus fontinalis*) was exposed to environmental pH of 4.2 for 5 days (Diveley *et al.*, 1977) and brown trout (*Salmo trutta*) exposed to pH 4.0 for 8 days (Leivestad *et al.*, 1976), arterial pH was little affected, whereas rainbow trout (*Salmo gairdneri*) exposed to pH 3.15–4.5 or to pH 5.0 exhibited a significant fall in arterial plasma pH (Lloyd and Jordan, 1964; Janssen and Randall, 1975). In *Salvelinus fontinalis*, exposed to pH 3.0–3.3, arterial and venous blood pH and the bicarbonate concentration were also reduced (Packer and Dunson, 1970; Packer, 1979). Neville (1979a) reported that exposure of rainbow trout to pH 4.0 resulted in progressive acidosis of plasma pH, but when the fish were exposed to the same pH at elevated water $P_{CO_2}$ (~4.5 mm Hg), that is, at increased water bicarbonate concentration, no acidemia could be observed (Neville, 1979a,b,c).

The pattern of the associated behavior of the electrolyte status is similarly diverse, ranging from no effect at all to extreme disturbances in plasma and intracellular space (e.g., Lockhart and Lutz, 1977; Leivestad *et al.*, 1976; Leivestad and Muniz, 1976; McWilliams, 1980; Mudge and Neff, 1971; Beamish *et al.*, 1975; Neville, 1979c; Packer and Dunson, 1970, 1972; Swarts *et al.*, 1978; Dunson *et al.*, 1977).

It appears to be impossible to delineate a general pattern running through the huge body of literature data, so we instead will give a general description of the acid−base and associated responses of the ionoregulation by referring to the few studies that are more complete with respect to measured parameters and range of pH covered.

Exposure of carp (*Cyprinus carpio*) to environmental pH of 5.1 had little effect on acid−base and electrolyte status. Arterial $P_{CO_2}$, [Na$^+$], [Cl$^-$], and [K$^+$] remained unaffected, and plasma pH and [HCO$_3^-$] were shifted to only very slightly lower, new steady-state values (Ultsch *et al.*, 1981). Water pH of 4.0 had a larger effect. Plasma bicarbonate and pH, as well as plasma [Na$^+$] and [Cl$^-$], were considerably and steadily reduced and did not appear to attain new steady-state values even after 75 hr. Lowering the water pH to 3.5 led to little further changes in the electrolyte status, but to further deterioration of the acid−base status followed by death of the animals within a few hours. Death of the animals was closely correlated to arterial pH (pH prior to death: 6.91 ± 0.05).

Data obtained in rainbow trout at average water pH of 4.3 in soft water (McDonald *et al.*, 1980) are in rather close agreement with the observations on acid−base status and electrolyte status in carp at pH 4.0. In hard water, however, the acid−base disturbances at the same environmental pH of 4.3 were much larger, whereas the electrolyte disturbances were ameliorated as compared to the soft water experiments and the carp. The differences between trout in hard and soft water have been attributed to the differences in water calcium concentration (McDonald *et al.*, 1980). The calcium concentration in the environmental water of carp, however, was actually high (Ultsch *et al.*, 1981), whereas the concentration of $Na^+$ was comparable to that in the soft water used for the trout.

In carp at pH 5.1 the integrated net base loss (bicarbonate loss or $H^+$ uptake minus ammonia excretion) roughly accounted for the bicarbonate and pH changes in the fish, whereas at pH 4.0 the integrated net base loss during the 75 hr of exposure was at least three times larger than the amount attributable to bicarbonate changes and physicochemical buffering in the animal (Ultsch *et al.*, 1981). These data suggest that the loss of bicarbonate stimulates mobilization of carbonate from the bone structures of the fish. In experiments on trout in hard water at pH 4.3, the discrepancy between net base loss and the bicarbonate pool depletion and titration of nonbicarbonate buffers was not as pronounced during the experimental period as in carp (McDonald and Wood, 1981). The simultaneously observed increase in urinary phosphate excretion, however, suggests that similar mobilization processes are going on.

In trout exposed to pH 4.0 the urinary excretion of surplus $H^+$ ions was largely enhanced by about 10-fold in order to compensate for the net base loss occurring mainly at the branchial epithelium. About half of the branchial base loss was renally compensated. Most of the increase was due to a rise in ammonia excretion, which still did not contribute more than about 8% to the total ammonia release (McDonald and Wood, 1981).

A possible explanation of the concomitant disturbances of acid–base and ionoregulation is based on linked ion transfer mechanisms. The sodium uptake mechanism in the gills provided for compensation of the diffusional $Na^+$ loss is postulated to be an electroneutral $H^+/Na^+$ or $NH_4^+/Na^+$ ion exchange mechanism (e.g., Kerstetter *et al.*, 1970; Payan and Maetz, 1973; Maetz, 1973). The $H^+/Na^+$ exchange mechanisms would operate at $pH_w$ 4.0 against a 2500 times increased gradient for $H^+$ as compared to $pH_w$ 7.4, which may well slow down the mechanism and result in disturbances of both $Na^+$ and acid–base balance. Extrusion of $NH_4^+$ instead of $H^+$ by the same or a similar mechanism would then well support normalization of the active $Na^+$ uptake rate.

The total ammonia release in both carp and trout was actually increased by a factor of about 2 at pH 4.0, which was mainly due to an increase in branchial excretion (Ultsch *et al.*, 1981; McDonald and Wood, 1981). However, since water at pH 4 is an infinite sink for $NH_3$, it appears, according to experiments on the diffusion coefficient of $NH_3$ in trout gills (Cameron and Heisler, 1983), most likely that $NH_3$ passes through the epithelium by non-ionic diffusion similar to the conditions in the mammalian kidney (Rector, 1973). The $NH_3$ would be ionized within milliseconds and also increase the backpressure for $NH_4^+$. Thus, the reduced $Na^+$ uptake appears to be due to the backpressure-induced hindrance of counterion transfer, a hypothesis well supported by various data on the interference of low environmental pH with active $Na^+$ uptake in fish (e.g., Packer and Dunson, 1970; Bentley et al., 1976; Maetz et al., 1976).

In ammoniotelic freshwater fish, chloride lost by diffusion to the water is postulated to be regained in exchange with negatively charged endogenous ions, preferably $HCO_3^-$ (Maetz and Garcia-Romeu, 1964; De Renzis and Maetz, 1973; De Renzis, 1975; Kerstetter and Kirschner, 1972), which are produced when ammonia is ionized in the fish. At low environmental pH excretion of $HCO_3^-$ in exchange with $Cl^-$ would exacerbate the acid–base disturbances, especially in that no surplus $HCO_3^-$ is produced when $NH_3$ is eliminated by nonionic diffusion. If, then, in order to prevent this, the active $Cl^-$ uptake is reduced or shut down at already lowered plasma $[HCO_3^-]$ (Ehrenfeld and Garcia-Romeu, 1978), the fall in plasma $[Cl^-]$ could be explained also by passive loss exceeding active uptake, even when the passive loss is reduced according to the concomitant change in the transepithelial potential to more positive values (McWilliams and Potts, 1978). This hypothesis is supported by Neville's findings (1979c) that a 10-fold increase in water bicarbonate did not significantly improve the plasma $[Cl^-]$ regulation, but improved the acid–base balance, possibly by reducing the plasma/water diffusion ratio for bicarbonate.

There is, however, no satisfactory explanation for the effect of external calcium concentration. The transepithelial potential is largely dependent on the pH of the water (lowering pH shifts the potential to more positive values; McWilliams and Potts, 1978), and, at pH values higher than 5, also on the water calcium concentration (McWilliams and Potts, 1978). At the environmental pH of 4, however, the potential is independent of water calcium, so that this factor cannot account for the observed phenomenon.

The observed effect has been explained on the basis of quite a number of quantitatively different effects of calcium on various components of ion exchange mechanisms (McDonald et al., 1980). Lowering of the calcium concentration has an increasing effect on passive branchial ion permeabilities (Isaia and Masoni, 1976). The increase in passive $Na^+$ efflux must, in order to fit the data, be larger than the concomitant stimulation of the active $Na^+$ uptake (Eddy, 1975) by lowered $Ca^{2+}$. Then $Na^+$ is still lost in excess to the conditions in high-$Ca^{2+}$ water, whereas $H^+$ ions are extracted from the organism to a larger extent, thus ameliorating the acid–base disturbances. In addition, the $HCO_3^-/Cl^-$ ion exchange must be stimulated to a lesser extent than the increase in passive $Cl^-$ loss along the electrochemical gradient, and than the stimulation in $H^+/NH_4^+$ in order to result in the observed improvement in acid–base regulation in soft water as compared to hard water.

All these considerations to explain the effects of acid water pH for acid–base and ionoregulation are based on little experimental evidence, which has been obtained in not quite comparable conditions. Accordingly all models established still await verification by simultaneous determination of all implied variables.

## V. MECHANISMS AND SITES OF ACID–BASE RELEVANT TRANSEPITHELIAL ION TRANSFER PROCESSES

Transfer of $HCO_3^-$ and $OH^-$, or $H^+$ ions in the opposite direction, has identical effect on acid–base regulation as outlined previously (see Section II,C,1). When these ions are transferred across epithelia, however, they have to be accompanied by an oppositely charged co-ion, or they have to be transferred in exchange with another ion of the same charge in order to maintain electroneutrality. Since during steady-state conditions, and also during most stress situations, the osmoregulatory systems of the animal will attempt to maintain constant osmolarity and water spaces, ion transfer for the aim of acid–base regulation is most likely performed as a net 1:1 exchange against appropriate counterions. $Na^+$ has been proposed as counterion for $H^+$ and $NH_4^+$, which probably compete for the same carrier sites (e.g., Kerstetter et al., 1970; Payan and Maetz, 1973; Maetz, 1973; for reviews, see Maetz, 1974; Evans, 1979, 1980a), and $HCO_3^-$ transfer (and not ruled out yet, $OH^-$ transfer) has been attributed to the exchange with readily available $Cl^-$ ions (e.g., Maetz and Garcia-Romeu, 1964; De Renzis and Maetz, 1973; De Renzis, 1975; Kerstetter and Kirschner, 1972; Kormanik and Evans, 1979; for reviews, see Maetz, 1974; Evans, 1979, 1980a). Therefore, simultaneous determination of counterion fluxes may provide valuable additional information, but still does not allow one to distinguish absolutely between the various possibilities of acid–base relevant ion transfer. The problem is reflected by the fact that the theoretically expected and indispensable 1:1 stoichiometry could rarely be demonstrated in fish (for references, see Heisler, 1980; Maetz, 1974; Evans, 1979, 1980a). This is very likely because of methodological problems: fluxes of $H^+$, $HCO_3^-$, and $OH^-$ can be determined by only the lumped effect on the measured compartment, and also the release of $NH_4^+$ may well partially be the result of nonionic diffusion of $NH_3$. Only measurement of the transfer of ions not in thermodynamic equilibrium with other substances is unbiased by similar side effects, but may nonetheless interfere with the passive or active fluxes of ions other than $Na^+$ and $Cl^-$, like $SO_4^{2-}$, $HPO_4^{2-}$, $K^+$, $Mg^{2+}$, $Ca^{2+}$, and various organic cations and anions that have never been taken into account.

The application of semiquantitative ion flux inhibitors like SITS, DITS, and amiloride usually does not provide further evidence regarding acid–base regulation, according to problems with dosage-dependent side effects, incomplete blockade of transfer mechanisms, and the concomitant shifts to other mechanisms.

On this background, conclusions about the ion exchange mechanism involved in fish acid–base regulation must be speculative to a certain extent and should always be considered with a certain amount of discretion.

The ammonia production and elimination in those marine fish species that have been studied with respect to the role of ion transfer processes as mechanisms for acid–base regulation, was never increased during stress conditions (Table X). Accordingly, ammonium cannot have contributed as carrier for the removal of surplus $H^+$ ions. Thus, net bicarbonate gained from the environment very likely has been transferred by either $HCO_3^-$/ $Cl^-$ or $H^+/Na^+$ exchange.

During environmental hypercapnia and hyperoxia-induced hypercapnia in *Scyliorhinus*, the plasma osmolarity remained unchanged, indicating that the considerable increase in plasma bicarbonate concentration was balanced by a reduction in other anions. Since pH was almost completely restored, this could not be attributed to proteins, but must have been an equivalent decrease in $[Cl^-]$ (Heisler *et al.*, 1976a, and unpublished). Also, *Conger* maintained osmolarity extremely constant and actually decreased plasma $[Cl^-]$ equivalently (Toews *et al.*, 1983).

After strenuous exercise, osmolarity in the fish is largely increased as a result of production of lactic acid from the osmotically less active macromolecule glycogen. During recovery the osmolarity of both species was normalized more rapidly (and even reduced to values lower than the controls) when further aerobic processing of lactate occurred. This may result from a larger reduction in plasma chloride than can be attributed to lactate/chloride counterion exchange between intracellular compartments of muscle tissues and the extracellular space, an effect that could actually been confirmed by $[Cl^-]$ measurements in the plasma of *Conger* (Holeton *et al.*, 1985).

This suggests that $HCO_3^-$ is gained from the environment in ionic exchange with $Cl^-$, then reacts with surplus $H^+$ ions in the extracellular space, and is then eliminated as molecular $CO_2$ from the osmotic pool by diffusion into the water through the gills. This mechanism may later be reversed to eliminate bicarbonate and restore the $Cl^-$ pool of the animals. This conclusion is supported by findings of Kormanik and Evans (1979), who demonstrated by unidirectional flux measurements that in *Opsanus beta*, seawater bicarbonate is actually gained in exchange for $Cl^-$ originating in the fluids of the fish.

Although some evidence has been provided by osmoregulatory studies

## Table X

### Changes in Ammonia Excretion in Response to Acid–Base Stress Conditions

| Species | Stress factor | Ammonia excretion[a] | Reference |
|---|---|---|---|
| **Marine** | | | |
| *Scyliorhinus stellaris* | Temperature changes | ∅ | Heisler (1978) |
| | Hypercapnia | ∅ | Heisler et al. (1976a) |
| | Hyperoxia/hypercapnia | ∅ | Heisler et al. (1981) |
| | Exercise | ∅ | Holeton and Heisler (1983) |
| | HCl infusion | ∅ | N. Heisler and P. Neumann (unpublished) |
| | NaHCO₃ infusion | ∅ | N. Heisler and P. Neumann (unpublished) |
| | Acidic water | ∅ | Heisler and Neumann (1977) |
| *Squalus acanthias* | Hypercapnia | ∅ | Cross et al. (1969) |
| | Hypercapnia | + | Evans (1982) |
| *Opsanus beta* | Hypercapnia | − | Evans (1982) |
| *Conger conger* | Hypercapnia | − | Toews et al. (1983) |
| | Exercise | ∅ | Holeton et al. (1985) |
| **Freshwater** | | | |
| *Salmo gairdneri* | Exercise | + + | Holeton et al. (1983) |
| | Acidic water | + | McDonald and Wood (1981) |
| *Cyprinus carpio* | Acidic water | + | Ultsch et al. (1981) |
| | Hypercapnia | + | Claiborne and Heisler (1984) |
| | NH₄Cl infusion | + + | Claiborne and Heisler (1985) |
| *Synbranchus marmoratus* | Air breathing/hypercapnia | ∅ | Heisler (1982a) |
| *Ictalurus punctatus* | Temperature changes | ∅ | J. N. Cameron (personal communication) |
| | NaHCO₃ infusion | ∅ | Cameron and Kormanik (1982) |
| | NH₄Cl infusion | + + | Cameron and Kormanik (1982b) |

[a]Meaning of symbols: ∅, no change; +, increased excretion; + +, greatly increased excretion; −, reduced excretion.

that Na$^+$ influx is stimulated during NH$_4$Cl-induced acidosis (Evans, 1980b), and ammonia flux is increased during hypercapnia, usually marine fish species appear to avoid the osmotically and ionically inappropriate H$^+$/Na$^+$ and NH$_4$$^+$/Na$^+$ ion exchange mechanisms for acid—base regulation (see Table X). Extrusion of surplus H$^+$ ions directly via H$^+$/Na$^+$ exchange or via NH$_4$$^+$/Na$^+$ exchange would add to the influx of Na$^+$ into the animal, and stress the osmoregulation of the animal in addition to the Na$^+$ entering the fish by passive diffusion. At least in relatively impermeable marine species like *Conger* and *Scyliorhinus*, this factor would be a considerable additional stress for the ionoregulation.

Freshwater fish subjected to acid—base stress situations regularly reduce the rate of bicarbonate-equivalent ion release, or even reverse the bicarbonate-equivalent flux into a net uptake (see earlier). This has to be attributed to either HCO$_3$$^-$/Cl$^-$ or H$^+$/Na$^+$ ion exchange mechanisms. Differentiation between these possibilities is difficult because of the overlying effects of the ammonia release. Unidirectional ion fluxes with Na$^+$ and Cl$^-$ in the Arctic grayling (*Thymallus arcticus*) during hypercapnia and after changes of temperature (Cameron, 1976), and net ion fluxes in carp during hypercapnia (Claiborne and Heisler, 1984) and in trout after exercise (Holeton *et al.*, 1983), suggest that the HCO$_3$$^-$/Cl$^-$ exchange mechanism, and during hypercapnia and after exercise, also either one or both of the H$^+$/Na$^+$ and NH$_4$$^+$/Na$^+$ exchange mechanisms, were utilized.

At least in some of the experiments on freshwater fish, the ammonia elimination appears, in contrast to the conditions in marine fish, to be utilized in order to cope with acid—base disturbances. The ammonia release is increased as compared to control conditions in trout after exercise (Holeton *et al.*, 1983) and after exposure to acidic environmental water (McDonald and Wood, 1981), as well as in carp during hypercapnia (Claiborne and Heisler, 1984) and exposure to acidic water (Ultsch *et al.*, 1981). Also infusion of ammonium chloride into the channel catfish *Ictalurus punctatus* (Cameron and Kormanik, 1982b) and the carp, *Cyprinus carpio* (Claiborne and Heisler, 1985), causes a rise in ammonia release.

Production and elimination of ammonia appear, at first glance, to constitute an obvious mechanism to be exploited for freshwater fish acid—base regulation. The exchange of NH$_4$$^+$ against Na$^+$ is advantageous to the fish for compensation of passive diffusive loss of Na$^+$ and simultaneously provides elimination of the toxic nitrogenous waste product NH$_3$.

In mammals, the ammonia excretion may be enhanced during acid—base disturbances by a factor of 10 (Pitts, 1945). There, however, NH$_3$ is eliminated from the body fluids, and especially from the kidney tubule cells by nonionic diffusion into the tubule lumen according to its extremely high diffusivity (Rector, 1973) and serves intratubularly as a buffer for H$^+$ ions

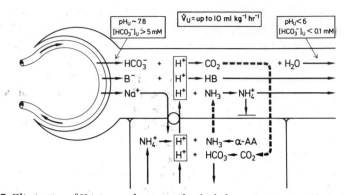

**Fig. 17.** Elimination of $H^+$ ions and ammonia by the kidney. $H^+$ ions are tubularly secreted in exchange with $Na^+$ and buffered with filtered nonbicarbonate buffers ($B^-$). $H^+$ ions also combine with filtered bicarbonate to $CO_2$, which diffuses back into the tubule cells for bicarbonate conservation. Ammonia predominantly enters the tubule lumen by nonionic diffusion according to its high diffusivity and buffers $H^+$ ions, which are to be excreted in surplus to filtered nonbicarbonate buffers. (Dashed pathways indicate nonionic diffusion; see also text.)

actively excreted in exchange with other positively charged ions (Pitts, 1964) (Fig. 17). This mechanism keeps the concentration ratio of $H^+$ ions between tubule cells and luminal fluid below the critical value of 800 to 1000, beyond which the tubule cells are incapable of further $H^+$ excretion (Pitts, 1973). This mechanism is only activated when the kidney release of other nonbicarbonate buffers (mainly phosphate) is, in relation to other demands of the organism and the type of nutrition, too small to buffer $H^+$ ions in sufficient quantity.

In fish, however, as indicated by the scarce number of studies on this subject, the kidneys usually contribute relatively little to the overall transfer of acid–base relevant ions. With a few exceptions, more than 94% of the transfer occurs at the gill epithelium with its extremely large surface area (Table XI). Other potential sites of ion exchange in fish (such as the rectal gland and abdominal pores of elasmobranchs, and the skin) play no significant role, at least in *Scyliorhinus*.

The conditions at the gills of fish are, however, quite different from those in the kidney. The gills are irrigated by a much larger flow in the range of 200 to 300 ml (min kg fish)$^{-1}$ (e.g., *Scyliorhinus* 240 ml kg$^{-1}$, Randall *et al.*, 1976), larger by at least a factor of $10^3$ compared to the maximal flow rate of urine in fish (Fig. 18; cf. Hunn, 1982), and more than $10^4$ compared to the urine flow rate in humans (~2 liters day$^{-1}$ 70 kg$^{-1}$). This large amount of water is capable of taking up $H^+$ ions at even maximal excretion rate (~15 μmol (min kg body weight)$^{-1}$ (see Section IV,B and C), without significant effect on the bicarbonate concentration of the water (Fig. 18), as long as the

**Table XI**

Contributions of Extrabranchial Sites to the Acid–Base Regulation in Response to Various Stress Situations[a]

| Species | Stress factor | Kidneys | Skin | Rectal gland + abdominal pores | Reference |
|---|---|---|---|---|---|
| *Squalus acanthias* | Hypercapnia | <1 | — | — | Cross *et al.* (1969) |
| | Hypercapnia | <1 | — | — | Evans (1982) |
| | $HCO_3^-$ infusion | <1 | — | — | Hodler *et al.* (1955) |
| | $HCO_3^-$ infusion | <1 | — | — | Murdaugh and Robin (1967) |
| *Scyliorhinus stellaris* | Hypercapnia | <1 | <1 | <1 | Heisler *et al.* (1976a) |
| | Exercise | <1 | — | <1 | Holeton and Heisler (1983) |
| | HCl infusion | <1 | <1 | <1 | N. Heisler and P. Neumann (unpublished) |
| | $HCO_3^-$ infusion | <1 | <1 | <1 | N. Heisler and P. Neumann (unpublished) |
| | Temperature changes | 3 | — | — | Heisler (1978) |
| *Opsanus beta* | Hypercapnia | <1 | — | — | Evans (1982) |
| *Parophrys vetulus* | HCl infusion | ~5 | — | — | McDonald *et al.* (1982) |
| *Ictalurus punctatus* | HCl infusion | <1 | — | — | Cameron (1980) |
| | Hypercapnia | 14 | — | — | Cameron (1980) |
| | Temperature changes | 45–55 | — | — | Cameron and Kormanik (1982a) |
| | $NH_4Cl$ infusion | ~30 | — | — | Cameron and Kormanik (1982b) |
| | $NaHCO_3$ infusion | ~27 | — | — | Cameron and Kormanik (1982b) |
| *Salmo gairdneri* | Exercise | 6 | — | — | Holeton *et al.* (1983) |
| | Lactic acid infusion | 5 | — | — | Kobayashi and Wood (1980) |
| | Hypoxia | 25 | — | — | Kobayashi and Wood (1980) |
| *Hoplias malabaricus* | Postoperative acidosis | No response | — | — | Cameron and Wood (1978) |
| | Hypercapnia | No response | — | — | Cameron and Wood (1978) |
| *Hoplerythrinus unitaeniatus* | Postoperative acidosis | No response | — | — | Cameron and Wood (1978) |
| | Hypercapnia | ~5 | — | — | Cameron and Wood (1978) |
| *Synbranchus marmoratus* | Air breathing/hypercapnia | 4 | — | — | Heisler (1982a) |

[a]Contributions given as percentage of total. For the quite variable experimental approaches, see the original references.

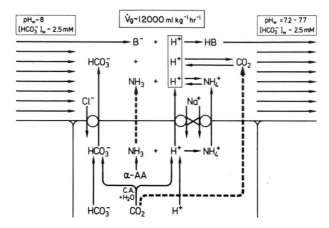

**Fig. 18.** Elimination of $H^+$ ions and ammonia by the gills. Since flow and buffer capacity of the water are extremely high as compared to the urine in the kidney tubules, ammonia is not required to buffer $H^+$ ions (cf. Fig. 17). Accordingly, production of ammonia for the sole purpose of acid–base regulation is inefficient. (Dashed pathways indicate nonionic diffusion; see also text.)

water bicarbonate concentration is not extremely low. The ionic exchange of $NH_4^+$ against sodium is accordingly not advantageous over $H^+/Na^+$, but would energetically be comparatively inefficient, if valuable amino acids had to be broken down to form ammonia in addition to the steady-state production for the sole purpose of acid–base regulation.

At low environmental water pH, elimination of $H^+$ ions in exchange with sodium may be largely impaired because of the too high $H^+$ ion concentration ratio between plasma and water (see Section IV,D). Under these conditions elimination of ammonia is actually increased by about a factor of two (Ultsch *et al.*, 1981; McDonald and Wood, 1981). However, since the diffusivity of $NH_3$ is extremely high (see later) and the acid water is an infinite sink for $NH_3$, the ammonia is very likely transferred by nonionic diffusion similar to the conditions in the mammalian kidney (Rector, 1973) and thus does not contribute to acid–base regulation (Ultsch *et al.*, 1981). This hypothesis is supported by the fact that unidirectional sodium influx at pH 4 is reduced close to zero (McWilliams and Potts, 1978). Since the increased ammonia production is utilized in the renal elimination pathway to only a minor extent (McDonald and Wood, 1981), its production in surplus to the steady-state rate has to be attributed to disregulation of metabolic pathways, or to acid hydrolysis.

Nonionic diffusion of ammonia through the gill epithelium may also be an important mechanism of elimination in other examples of increased ammonia release in freshwater fish. Ammonium chloride infusion results in a considerable acidosis in *Ictalurus* (Cameron and Kormanik, 1982b), in *Salmo gairdneri* (Cameron and Heisler, 1983), and in *Cyprinus* (Claiborne and Heisler, 1985), which cannot have been caused by the neutral ammonium salt. The acidosis can only be attributed to nonionic diffusion of $NH_3$ into the water, leaving HCl behind in the plasma of the fish. Rough estimates based on the extent of acidosis and the amount of ammonia released to the water (Eq. (4)) suggest that at least half of the ammonia was eliminated by nonionic diffusion.

Analysis of the transfer of $Na^+$ and $Cl^-$ accompanying the acid–base relevant transfer processes allows some additional conclusions. The net amount of $Na^+$ taken up from the environment by *Salmo gairdneri* after strenuous exercise is considerably smaller than the amount of $NH_4^+$ that occurred in the water. On the basis of a 1:1 ion exchange ratio (to be claimed for conservation of electroneutrality), at least this difference has to be attributed to nonionic elimination of $NH_3$. This is in accordance with the changes in $HCO_3^-$ and $Cl^-$ in the water: more chloride was excreted to the water than water bicarbonate was reduced. The difference may well be explained by ionization of $NH_3$ to $NH_4^+$ and the accompanying formation of bicarbonate (Holeton *et al.*, 1983).

Since $Na^+$ is the postulated counterion for both $H^+$ and $NH_4^+$, the difference between the changes in $Na^+$ and $NH_4^+$ represents the lower limit of nonionic $NH_3$ elimination (about one-third of the total), which would be valid, if no $H^+/Na^+$ ion exchange would take place. However, the data would also fit the assumption that ammonia was exclusively eliminated by nonionic diffusion, and all sodium was exchanged against $H^+$ ions.

In carp exposed to hypercapnia, similar discrepancies between $Na^+$ and $NH_4^+$ changes in the environmental water could be observed, which, during certain time periods, indicated that considerable fractions of the net $Na^+$ transfer has to be correlated with $H^+/Na^+$ transfer, and that more than 50% of the ammonia was eliminated by nonionic diffusion. It is even possible that a major fraction of the ammonia eliminated in excess of the control rate was not additionally produced but was released by nonionic diffusion according to the lowered environmental water pH (>1 Unit) (Claiborne and Heisler, 1984).

Recently the basis for nonionic ammonia elimination in trout was reevaluated by determination of the solubility of $NH_3$ in plasma and water, and measurement of the plasma–water diffusion gradients (Cameron and Heisler, 1983). It was found that the apparent diffusion coefficient of $NH_3$ was comparable to that of other respiratory gases ($CO_2$, $O_2$), which would be expected from literature data on aqueous diffusion coefficients (Radford, 1964) when no active transfer was involved. According to the high water solubility of $NH_3$, which is about $10^3$ larger than for $CO_2$, only about 55 $\mu$torr partial pressure difference are sufficient to provide elimination of the control ammonia production. Cameron and Heisler (1983) concluded that at least in trout with normal plasma concentration and relatively low water concentration, ammonia was exclusively eliminated by nonionic diffusion, whereas active $NH_4^+$ extrusion in exchange with $Na^+$ only takes place at elevated environmental ammonia concentration.

The transfer factor determined for nonionic ammonia elimination ($\dot{M}_{NH_3}/\Delta P_{NH_3} = 0.102$ $\mu$mol (min kg $\mu$torr)$^{-1}$) by Cameron and Heisler (1983) is similar to the slope of the ammonia elimination as a function of partial pressure difference of $NH_3$ between plasma and water presented by Maetz (1973; Fig. 10, $\dot{M}_{NH_3}/\Delta P_{NH_3} = 0.079$ $\mu$mol (min kg $\mu$torr)$^{-1}$). This relatively close agreement between two different approaches confirms the evaluation of Holeton et al. (1983), who concluded, based on the data of Maetz (1973), that nonionic elimination of ammonia in trout after exercise could well be higher than the lower limit determined by the discrepancy between $Na^+$ and $NH_4^+$ transfer (see earlier). If, as has been reported for Cyprinus carpio and Tilapia mossambica (Driedzic and Hochachka, 1976; Kutty, 1972), the ammonia production of muscle fibers during anaerobic activity is enhanced and plasma ammonia accordingly increased (Driedzic and Hochachka, 1976), the nonionic elimination rate may rise to considerable values (Fig. 19).

It seems that the role of ammonia excretion for the acid–base regulation in fish has been overestimated. Ammonia has to be considered also as a respiratory gas, which is transferred under normal conditions by mechanisms similar to those of $CO_2$ and $O_2$. The ionic active transfer of $NH_4^+$ is clearly present as a mechanism (as for the $CO_2$–bicarbonate buffer system the ionic exchange mechanism for $HCO_3^-$), but it appears to be reserved for situations with extraordinarily high environmental ammonia concentrations. In marine teleost fish, which are usually rather permeant to ions, $NH_4^+$ may also be eliminated by ionic diffusion, having then the same effect as potentially acid–base relevant ion as when actively transferred (e.g., Claiborne et al., 1982; Goldstein et al., 1982). However, since marine fish do not appear to increase their ammonia release during stress conditions, this transfer pathway seems to have only minor significance for the acid–base regulation.

Accordingly, the importance of the other two main ion exchange mechanisms $HCO_3^-/Cl^-$ and $H^+/Na^+$ is much greater. Unfortunately, as outlined earlier, the elimination of ammonia interferes largely with the effects of these mechanisms on the environmental water. Measurements of the changes in acid–base status of the environmental water and various body compartments lead to rather complete quantitative analyses of the overall acid–base relevant ion transfer processes, but do not clearly distinguish between the involved mechanisms. However, the significance of both net

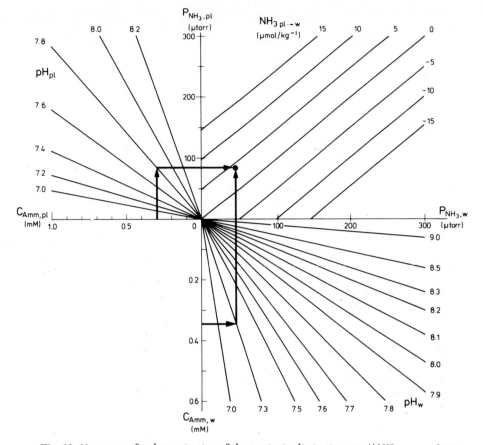

**Fig. 19.** Nomogram for determination of the nonionic elimination rate ($\Delta NH_{3\ pl\rightarrow w}$) of $NH_3$ from total ($NH_3 + NH_4^+$) concentration ($C$), and pH in plasma and environmental water of trout. (Based on data of Cameron and Heisler, 1983.)

and unidirectional counterion fluxes is impaired by possible competition of various acid—base relevant and other ions for the same carrier sites (e.g., at least the two $NH_4^+$ and $H^+$ for exchange with $Na^+$, and $HCO_3^-$ and $OH^-$ for the exchange with $Cl^-$), and the methodological limitations of these methods, which allow only semiquantitative analysis. Since removal of certain ions from the environmental water and the application of ion transfer inhibitors are charged with various limitations and have to be expected to shift the acid—base relevant transfer from the affected mechanism to other processes, these methods can only be utilized for the demonstration of principles. Further elucidation of physiologically applied mechanisms can thus only be expected on the basis of simultaneous determination of the complete pattern of acid–base variables, net ion fluxes, and unidirectional tracer fluxes.

## REFERENCES

Albers, C. (1970). Acid-base balance. *In* "Fish Physiology" (W. S. Hoar and D. J. Randall, eds.), Vol. 4, pp. 173–208. Academic Press, New York.

Albers, C., and Pleschka, K. (1967). Effects of temperature on $CO_2$ transport in elasmobranch blood. *Respir. Physiol.* **2**, 261–273.

Arillo, A., Margiocco, C., Melodia, F., Mensi, P., and Schenone, G. (1981). Ammonia toxicity mechanism in fish: Studies on rainbow trout (*Salmo gairdneri* Rich.). *Ecotoxicol. Environ. Saf.* **5**, 316–328.

Aster, P. L., Casey, C. A., Vorhaben, J. E., and Campell, J. W. (1982). Hepatic ammoniogenesis in the catfish, *Ictalurus punctatus*. *Am. Zool.* **21**, 1029.

Babak, E. (1907). Über die funktionelle Anpassung der äußeren Kiemen beim Sauerstoffmangel. *Zentralbl. Physiol.* **21**, 97–99.

Ball, I. R. (1967). The relative susceptibilities of some species of fresh-water fish to poisons. I. Ammonia. *Water Res.* **1**, 767–775.

Bartels, H., and Wrbitzky, R. (1960). Bestimmung des $CO_2$-Absorptionskoeffizienten zwischen 15 und 38°C in Wasser und Plasma. *Pfluegers Arch.* **271**, 162–168.

Beamish, R. J. (1976). Acidification of lakes in Canada by acid precipitation and the resulting effects on fishes. *In* Proceedings of the First International Symposium on Acid Precipitation and the Forest Eco-System. *USDA For. Serv. Tech. Rep.* **23**, 479–498.

Beamish, R. J., Lockhart, W. L., Van Loon, J. C., and Harvey, H. H. (1975). Long-term acidification of a lake and resulting effects on fishes. *Ambio* **4**, 98–102.

Benadé, A. J. S., and Heisler, N. (1978). Comparison of efflux rates of hydrogen and lactate ions from isolated muscles in vitro. *Respir. Physiol.* **32**, 369–380.

Bentley, P. J., Maetz, J., and Payan, P. (1976). A study of the unidirectional fluxes of Na and Cl across the gills of the dogfish *Scyliorhinus canicula* (Chondrichthyes). *J. Exp. Biol.* **64**, 629–637.

Bilinski, E. (1960). Biosynthesis of trimethylammonium compounds in aquatic animals. I. Formation of trimethylamine oxide and betaine from $C^{14}$-labelled compounds by lobster (*Homarus americanus*). *J. Fish. Res. Board Can.* **17**, 895–902.

Bilinski, E. (1961). Biosynthesis of trimethylammonium compounds in aquatic animals. II. Role of betaine in the formation of trimethylamine oxide by lobster (*Homarus americanus*). *J. Fish. Res. Board Can.* **18**, 285–286.

Bilinski, E. (1962). Biosynthesis of trimethylammonium compounds in aquatic animals. III. Choline metabolism in marine crustacea. *J. Fish. Res. Board Can.* **19**, 505–510.

Black, E. C. (1957a). Alterations in the blood level of lactic acid in certain salmonid fishes following muscular activity. I. Kamloops trout, *Salmo gairdneri*. *J. Fish. Res. Board Can.* **14**, 117–134.

Black, E. C. (1957b). Alterations in the blood level of lactic acid in certain salmonid fishes following muscular activity. II. Lake trout, *Salvelinus namaycush*. *J. Fish. Res. Board Can.* **14**, 645–649.

Black, E. C. (1957c). Alterations in the blood level of lactic acid in certain salmonid fishes following muscular activity. III. Sockeye salmon *Oncorhynchus nerka*. *J. Fish. Res. Board Can.* **14**, 807–814.

Black, E. C., Chiu, W. G., Forbes, F. D., and Hanslip, A. (1959). Changes in pH, carbonate and lactate of the blood of yearling Kamloops trout, *Salmo gairdneri*, during and following severe muscular activity. *J. Fish. Res. Board Can.* **16**, 391–402.

Black, E. C., Connor, A. R., Lam, K. C., and Chiu, W. G. (1962). Changes in glycogen, pyruvate and lactate in rainbow trout (*Salmo gairdneri*) during and following muscular activity. *J. Fish. Res. Board Can.* **19**, 409–436.

Black, E. C., Manning, G. T., and Hayashi, K. (1966). Changes in levels of hemoglobin, oxygen, carbon dioxide and lactate in venous blood of rainbow trout (*Salmo gairdneri*) during and following severe muscular activity. *J. Fish. Res. Board Can.* **23**, 783–795.

Børjeson, H. (1976). Some effects of high carbon dioxide tension on juvenile salmon (*Salmo salar L.*). *Acta Univ. Ups.* **383**, 1–35.

Bornancin, M., DeRenzis, G., and Maetz, J. (1977). Branchial Cl$^-$ transport, anion stimulated ATPase and acid-base balance in *Anguilla anguilla* adapted to freshwater: Effects of hyperoxia. *J. Comp. Physiol.* **117**, 313–322.

Boutilier, R. G., Randall, D. J., Shelton, G., and Toews, D. P. (1979a). Acid-base relationships in the blood of the toad, *Bufo marinus*. I. The effects of environmental $CO_2$. *J. Exp. Biol.* **82**, 331–344.

Boutilier, R. G., Randall, D. J., Shelton, G., and Toews, D. P. (1979b). Acid-base relationships in the blood of the toad, *Bufo marinus*. II. The effects of dehydration. *J. Exp. Biol.* **82**, 345–355.

Boutilier, R. G., Randall, D. J., Shelton, G., and Toews, D. P. (1979c). Acid-base relationships in the blood of the toad, *Bufo marinus*. III. The effect of burrowing. *J. Exp. Biol.* **82**, 357–365.

Bower, C. E., and Bidwell, J. P. (1978). Ionization of ammonia in seawater: Effects of temperature, pH and salinity. *J. Fish. Res. Board Can.* **35**, 1012–1016.

Cameron, J. N. (1971). Rapid method for determination of total carbon dioxide in small blood samples. *J. Appl. Physiol.* **31**, 632–634.

Cameron, J. N. (1974). Evidence for the lack of by-pass shunting in the teleost gills. *J. Fish. Res. Board Can.* **31**, 211–213.

Cameron, J. N. (1976). Branchial ion uptake in arctic grayling: Resting values and effects of acid–base disturbance. *J. Exp. Biol.* **64**, 711–725.

Cameron, J. N. (1978). Regulation of blood pH in teleost fish. *Respir. Physiol.* **33**, 129–144.

Cameron, J. N. (1980). Body fluid pools, kidney function, and acid–base regulation in the freshwater catfish *Ictalurus punctatus*. *J. Exp. Biol.* **86**, 171–185.

Cameron, J. N., and Heisler, N. (1983). Studies of ammonia in rainbow trout: Physico-chemical parameters, acid-base behaviour and respiratory clearance. *J. Exp. Biol.* **105**, 107–125.

Cameron, J. N., and Kormanik, G. A. (1982a). Intracellular and extracellular acid-base status as a function of temperature in the freshwater channel catfish, *Ictalurus punctatus*. *J. Exp. Biol.* **99**, 127–142.

Cameron, J. N., and Kormanik, G. A. (1982b). The acid-base responses of gills and kidneys to infused acid and base loads in the channel catfish, *Ictalurus punctatus*. *J. Exp. Biol.* **99**, 143–160.

Cameron, J. N., and Randall, D. J. (1972). The effect of increased ambient $CO_2$ on arterial $CO_2$-tension, $CO_2$ content and pH in rainbow trout. *J. Exp. Biol.* **57**, 673–680.

Cameron, J. N., and Wood, C. M. (1978). Renal function and acid-base regulation in two Amazonian erythrinid fishes: *Hoplias malabaricus*, a water breather, and *Hoplerythrinus unitaeniatus*, a facultative air breather. *Can. J. Zool.* **56**, 917–930.

Claiborne, J. B., and Heisler, N. (1984). Acid-base regulation in the carp (*Cyprinus carpio*) during and after exposure to environmental hypercapnia. *J. Exp. Biol.* **108**, 25–43.

Claiborne, J. B., and Heisler, N. (1985). In preparation.

Claiborne, J. B., Evans, D. N., and Goldstein, L. (1982). Fish branchial $Na^+/NH_4^+$ exchange is via basolateral $Na^+$-$K^+$-activated ATPase. *J. Exp. Biol.* **96**, 431–434.

Cross, C. E., Packer, B. S., Linta, J. M., Murdaugh, H. V., Jr., and Robin, E. D. (1969). $H^+$

buffering and excretion in response to acute hypercapnia in the dogfish *Squalus acanthias*. *Am. J. Physiol.* **216**, 440–452.

Davis, J. C., and Cameron, J. N. (1970). Water flow and gas exchange at the gills of rainbow trout, *Salmo gairdneri*. *J. Exp. Biol.* **54**, 1–18.

Davis, J. C., and Randall, D. J. (1973). Gill irrigation and pressure relationships in rainbow trout, *Salmo gairdneri*. *J. Fish. Res. Board Can.* **30**, 99–104.

Daxboeck, C., Barnard, D. K., and Randall, D. J. (1981). Functional morphology of the gills of the bowfin, *Amia calva* L. with special reference to their significance during air exposure. *Respir. Physiol.* **43**, 349–364.

Dejours, P. (1972). Action des changements de $P_{O_2}$ de l'eau sur la ventilation et l'équilibre acide-base du sang chez quelques poissons téléostéens. *J. Physiol. (Paris)* **65**, 386A.

Dejours, P. (1973). Problems of control of breathing in fishes. *In* "Comparative Physiology" (L. Bolis, K. Schmidt-Nielsen, and S. H. P. Maddrell, eds.), pp. 117–133. North-Holland Publ., Amsterdam.

Dejours, P. (1975). "Principles of Comparative Respiratory Physiology." North-Holland Publ., Amsterdam.

Dejours, P., Toulemond, A., and Truchot, J. P. (1977). The effects of hyperoxia on the breathing of marine fishes. *Comp. Biochem. Physiol. A* **58A**, 409–411.

DeLaney, R. G., Lahiri, S., and Fishman, A. P. (1974). Aestivation of the African Lungfish *Protopterus aethiopicus*: Cardiovascular and respiratory functions. *J. Exp. Biol.* **61**, 111–128.

DeLaney, R. G., Lahiri, S., Hamilton, R., and Fishman, A. P. (1977). Acid-base balance and plasma composition in the aestivating lungfish (*Protopterus*). *Am. J. Physiol.* **232**, R10–R17.

De Renzis, G. (1975). The branchial chloride pump in the goldfish *Carassius auratus*: Relationship between $Cl^-/HCO_3^-$ and $Cl^-/Cl^-$ exchanges and the effect of thiocyanate. *J. Exp. Biol.* **63**, 587–602.

De Renzis, G., and Maetz, J. (1973). Studies on the mechanism of the chloride absorption by the goldfish gill: Relation with acid–base regulation. *J. Exp. Biol.* **59**, 339–358.

Dively, J. L., Mudge, J. E., Neff, W. H., and Antony, A. (1977). Blood $P_{O_2}$, $P_{CO_2}$ and pH changes in brook trout (*Salvelinus fontinalis*) exposed to sublethal levels of acidity. *Comp. Biochem. Physiol. A* **57A**, 347–351.

Driedzic, W. R., and Hochachka, P. W. (1976). Control of energy metabolism in fish white muscle. *Am. J. Physiol.* **230**, 579–582.

Dunson, W. A., Swarts, F., and Silvestri, M. (1977). Exceptional tolerance of low pH of some tropical blackwater fish. *J. Exp. Zool.* **201**, 157–162.

Eddy, F. B. (1974). Blood gases in the tench (*Tinca tinca*) in well-aerated and oxygen-deficient waters. *J. Exp. Biol.* **60**, 71–83.

Eddy, F. B. (1975). The effect of calcium on gill potentials and on sodium and chloride fluxes in the goldfish, *Carassius auratus*. *J. Comp. Physiol.* **96**, 131–142.

Eddy, F. B. (1976). Acid-base balance in rainbow trout (*Salmo gairdneri*) subjected to acid stresses. *J. Exp. Biol.* **64**, 159–171.

Eddy, F. B., Lomholt, J. P., Weber, R. E., and Johansen, K. (1977). Blood respiratory properties of rainbow trout (*Salmo gairdneri*) kept in water of high $CO_2$ tension. *J. Exp. Biol.* **67**, 37–47.

Edsall, J. T., and Wyman, J. (1958). "Biophysical Chemistry." Academic Press, New York.

Ehrenfeld, J., and Garcia-Romeu, F. (1978). Coupling between chloride absorption and base excretion in isolated skin of *Rana esculenta*. *Am. J. Physiol.* **235**, F33–F39.

Emerson, K., Russo, R. C., Lund, R. W., and Thurston, R. V. (1975). Aqueous ammonia equilibrium calculation: Effect of pH and temperature. *J. Fish. Res. Board Can.* **32**, 2379–2383.

Evans, D. H.(1977). Further evidence for Na/NH₄ exchange in marine teleost fish. *J. Exp. Biol.* **70**, 213–220.

Evans, D. H. (1979). Fish. *In* "Comparative Physiology of Osmoregulation in Animals" (G. M. O. Maloiy, ed.), Vol. 1, pp. 305–390. Academic Press, New York.

Evans, D. H. (1980a). Kinetic studies of ion transport by fish gill epithelium. *Am. J. Physiol.* **238**, R224–R230.

Evans, D. H. (1980b). Na⁺/NH₄⁺ exchange in the marine teleost, *Opsanus beta:* Stoichiometry and role in Na⁺ balance. *In* "Epithelial Transport in the Lower Vertebrates" (B. Lahlou, ed.), pp. 197–205. Cambridge Univ. Press, London and New York.

Evans, D. H. (1982). Mechanisms of acid extrusion by two marine fishes: The teleost, *Opsanus beta,* and the elasmobranch, *Squalus acanthias. J. Exp. Biol.* **97**, 289–299.

Fromm, P. O. (1963). Studies on renal and extrarenal excretion in a freshwater teleost, *Salmo gairdneri. Comp. Biochem. Physiol.* **10**, 121–128.

Garey, W. F. (1972). Determination of normal blood pH of fishes. *Respir. Physiol.* **14**, 180–181.

Glass, M. L., Boutilier, R. G., and Heisler, N. (1984). Effects of body temperature on respiration, blood gases and acid-base status in the turtle *Chrysemys picta bellii. J. Exp. Biol.* (in press).

Goldstein, L., Forster, G., and Fanelli, G. M., Jr. (1964). Gill blood flow and ammonia excretion in the marine teleost, *Myoxocephalus scorpius. Comp. Biochem. Physiol.* **12**, 489–499.

Goldstein, L., Claiborne, J. B., and Evans, D. E. (1982). Ammonia excretion by the gills of two marine teleost fish: The importance of NH₄⁺ permeance. *J. Exp. Zool.* **219**, 395–397.

Gregory, R. B. (1977). Synthesis and total excretion of waste nitrogen by fish of the *Periopthalmus* (mud skipper) and Scartelaos families. *Comp. Biochem. Physiol. A* **57A**, 33–36.

Harvey, H. W. (1974). "The Chemistry and Fertility of Sea Waters." Cambridge Univ. Press, London and New York.

Heisler, N. (1978). Bicarbonate exchange between body compartments after changes of temperature in the larger spotted dogfish (*Scyliorhinus stellaris*). *Respir. Physiol.* **33**, 145–160.

Heisler, N. (1980). Regulation of the acid-base status in fishes. *In* "Environmental Physiology of Fishes" (M. A. Ali, ed.), pp. 123–162. Plenum, New York.

Heisler, N. (1982a). Intracellular and extracellular acid-base regulation in the tropical freshwater teleost fish *Synbranchus marmoratus* in response to the transition from water breathing to air breathing. *J. Exp. Biol.* **99**, 9–28.

Heisler, N. (1982b). Transepithelial ion transfer processes as mechanisms for fish acid-base regulation in hypercapnia and lactacidosis. *Can. J. Zool.* **60**, 1108–1122.

Heisler, N. (1984). Role of ion transfer processes in acid-base regulation with changes of temperature in fishes. *Am. J. Physiol.* (in press).

Heisler, N. (1985). Acid—base equilibrium comparative. *In* "Acid–Base Regulation in Animals" (N. Heisler, ed.). Elsevier, Amsterdam (in press).

Heisler, N., and Neumann, P. (1977). Influence of sea water pH upon bicarbonate uptake induced by hypercapnia in an elasmobranch (*Scyliorhinus stellaris*). *Pflugers Archiv.* **368**, Suppl. R.19.

Heisler, N., and Neumann, P. (1978). The role of buffering in the regulation of intracellular pH after changes of temperature in the larger spotted dogfish (*Scyliorhinus stellaris*). *Pflügers Arch.* **377**, Suppl., R. 17.

Heisler, N., and Neumann, P. (1980). The role of physico-chemical buffering and of bicarbonate transfer processes in intracellular pH regulation in response to changes of temperature in the larger spotted dogfish (*Scyliorhinus stellaris*). *J. Exp. Biol.* **85**, 99–110.

Heisler, N., and Piiper, J. (1971). The buffer value of rat diaphragm muscle tissue determined by $P_{CO_2}$ equilibration of homogenates. *Respir. Physiol.* **12**, 169–178.

Heisler, N., and Piiper, J. (1972). Determination of intracellular buffering properties in rat diaphragm muscle. *Am. J. Physiol.* **222**, 747–753.

Heisler, N., Weitz, H., and Weitz, A. M. (1976a). Hypercapnia and resultant bicarbonate transfer processes in an elasmobranch fish. *Bull. Eur. Physiopathol. Respir.* **12**, 77–85.

Heisler, N., Weitz, H., and Weitz, A. M. (1976b). Extracellular and intracellular pH with changes of temperature in the dogfish *Scyliorhinus stellaris. Respir. Physiol.* **26**, 249–263.

Heisler, N., Weitz, H., Weitz, A. M., and Neumann, P. (1978). Comparison of the acid-base regulation in hypercapnia between a marine elasmobranch fish (*Scyliorhinus stellaris*) and a facultative air breathing fish (*Synbranchus marmoratus*). *Physiologist* **21**, 52.

Heisler, N., Neumann, P., and Holeton, G. F. (1980). Mechanisms of acid-base adjustment in dogfish (*Scyliorhinus stellaris*) subjected to long-term temperature acclimation. *J. Exp. Biol.* **85**, 89–98.

Heisler, N., Holeton, G. F., and Toews, D. P. (1981). Regulation of gill ventilation and acid-base status in hyperoxia-induced hypercapnia in the larger spotted dogfish (*Scyliorhinus stellaris*). *Physiologist* **24**, 305.

Heisler, N., Forcht, G., Ultsch, G. F., and Anderson, J. F. (1982). Acid-base regulation in response to environmental hypercapnia in two aquatic salamanders, *Siren lacertina* and *Amphiuma means. Respir. Physiol.* **49**, 141–158.

Heisler, N., *et al.* (1984). In preparation.

Hicks, G. H., and Stiffler, D. F. (1980). The effects of temperature and hypercapnia on acid-base status of larval *Ambystoma tigrinum. Fed. Proc., Fed. Am. Soc. Exp. Biol.* **36**, 1060.

Hillaby, B. A., and Randall, D. J. (1979). Acute ammonia toxicity and ammonia excretion in rainbow trout (*Salmo gairdneri*). *J. Fish. Res. Board Can.* **36**, 621–629.

Hodler, J., Heinemann, H. O., Fishman, A. P., and Smith, H. W. (1955). Urine pH and carbonic anhydrase activity in the marine dogfish. *Am. J. Physiol.* **183**, 155–162.

Holeton, G. F., and Heisler, N. (1983). Contribution of net ion transfer mechanisms to the acid-base regulation after exhausting activity in the larger spotted dogfish (*Scyliorhinus stellaris*). *J. Exp. Biol.* **103**, 31–46.

Holeton, G. F., and Randall, D. J. (1967). The effect of hypoxia upon the partial pressure of gases in the blood and water afferent and efferent to the gills of rainbow trout. *J. Exp. Biol.* **46**, 317–327.

Holeton, G. F., Neumann, P., and Heisler, N. (1983). Branchial ion exchange and acid-base regulation after strenuous exercise in rainbow trout (*Salmo gairdneri*). *Respir. Physiol.* **51**, 303–318.

Holeton, G. F., Toews, D. P., and Heisler, N. (1985). In preparation.

Hunn, J. B. (1982). Urine flow rate in freshwater salmonids: A review. *Prog. Fish-Cult.* **44**, 119–125.

Isaia, J., and Masoni, A. (1976). The effects of calcium and magnesium on water and ionic permeability in the seawater adapted eel *Anguilla anguilla* L. *J. Comp. Physiol.* **109**, 221–233.

Ishimatsu, A., and Itazawa, Y. (1983). Blood oxygen levels and acid-base status following air exposure in a air-breathing fish, *Channa argus:* The role of air ventilation. *Comp. Biochem. Physiol. A* **74A**, 787–793.

Itazawa, Y., and Takeda, T. (1978). Gas exchange in the carp gills in normoxic and hypoxic conditions. *Respir. Physiol.* **35**, 263–269.

Jackson, D. C., and Heisler, N. (1982). Plasma ion balance of submerged anoxic turtles at 3°C: The role of calcium lactate formation. *Respir. Physiol.* **49**, 159–174.

Jackson, D. C., and Heisler, N. (1983). Intracellular and extracellular acid–base and electrolyte status of submerged anoxic turtles at 3°C. *Respir. Physiol.* **53**, 187–201.

Janssen, R. G., and Randall, D. J. (1975). The effect of changes in pH and $P_{CO_2}$ in blood and water on breathing in rainbow trout, *Salmo gairdneri. Respir. Physiol.* **25**, 235–245.

Kerstetter, F. H., and Kirschner, L. B. (1972). Active chloride transport by the gills of rainbow trout (*Salmo gairdneri*). *J. Exp. Biol.* **56**, 263–272.

Kerstetter, F. H., Kirschner, L. B., and Rafuse, D. D. (1970). On the mechanism of sodium ion transport by the irrigated gills of rainbow trout (*Salmo gairdneri*). *J. Gen. Physiol.* **56**, 342–359.

Kinney, E. (1964). Extent of acid mine pollution in United States affecting fish and wildlife. *U.S. Dep. Int. Bur. Sport, Fish. Wildl., Circ.* **191**, 1–27.

Kobayashi, K. A., and Wood, C. M. (1980). The response of the kidney of the freshwater rainbow trout to true metabolic acidosis. *J. Exp. Biol.* **84**, 227–244.

Kormanik, G. A., and Evans, D. II. (1979). $IICO_3{}^-$-stimulated $Cl^-$ efflux in the gulf toadfish acclimated to sea-water. *J. Exp. Zool.* **208**, 13–16.

Kramer, D. L., Lindsey, C. C., Moodie, G. E. E., and Stevens, E. D. (1978). The fishes and aquatic environment of the central Amazon basin, with particular reference to respiratory patterns. *Can. J. Zool.* **56**, 717–729.

Krogh, A. (1939). "Osmotic Regulation in Aquatic Animals." Cambridge Univ. Press, London and New York.

Kutty, M. N. (1972). Respiratory quotient and ammonia excretion in *Tilapia mossambica. Mar. Biol.* **16**, 126–133.

Leivestad, H., and Muniz, I. (1976). Fish kill at low pH in a Norwegian river. *Nature (London)* **259**, 391–392.

Leivestad, H., Hendry, G., Muniz, I., and Snevik, E. (1976). Effects of acid precipitation on fresh water organisms. *In* "Impact of Acid Precipitation on Forest and Fresh Water Ecosystems in Norway" (F. Braekke, ed.), Res. Rep. 6/76, pp. 87–111. SNSF Project Nisk, 1432 AaS-NLN, Norway.

Lenfant, C., Johansen, K., and Grigg, G. G. (1966/1967). Respiratory properties of blood and pattern of gas exchange in the lungfish *Neoceratodus forsteri* (Krefft). *Respir. Physiol.* **2**, 1–21.

Lloyd, R., and Jordan, D. H. M. (1964). Some factors affecting the resistance of rainbow trout (*Salmo gairdneri*) to acid waters. *Int. J. Air Water Pollut.* **8**, 393–403.

Lloyd, R., and White, W. R. (1967). Effect of high concentrations of carbon dioxide on the ionic composition of rainbow trout blood. *Nature (London)* **216**, 1341–1342.

Lockhart, W. L., and Lutz, A. (1977). Preliminary biochemical observations of fishes inhabiting an acidified lake in Ontario, Canada. *Water, Air, Soil Pollut.* **7**, 317–332.

McDonald, D. G., and Wood, C. M. (1981). Branchial and renal acid and ion fluxes in the rainbow trout, *Salmo gairdneri* at low environmental pH. *J. Exp. Biol.* **93**, 101–118.

McDonald, D. G., Hobe, H., and Wood, C. M. (1980). The influence of calcium on the physiological responses of the rainbow trout, *Salmo gairdneri*, to low environmental pH. *J. Exp. Biol.* **88**, 109–131.

McDonald, D. G., Walker, R. L., Wilkes, P. R. H., and Wood, C. M. (1982). $H^+$ excretion in the marine teleost *Parophrys vetulus. J. Exp. Biol.* **98**, 403–414.

McWilliams, P. G. (1980). Acclimation to an acid medium in the brown trout *Salmo trutta. J. Exp. Biol.* **88**, 267–280.

McWilliams, P. G., and Potts, W. (1978). The effects of pH and calcium concentrations on gill potentials in the brown trout (*Salmo trutta*). *J. Comp. Physiol.* **126**, 277–286.

Maetz, J. (1978). $Na^+/NH_4{}^+$, $Na^+/H^+$ exchanges and $NH_3$ movement across the gill of *Carassius auratus. J. Exp. Biol.* **58**, 255–275.

Maetz, J. (1974). Adaptation to hyper-osmotic environments. *Biochem. Biophys. Perspect. Mar. Biol.* **1**, 91–149.

Maetz, J., and Garcia-Romeu, F. (1964). The mechanism of sodium and chloride uptake by the gills of a fresh water fish, *Carassius auratus*. II. Evidence for $NH_4^+/Na^+$ and $HCO_3^-/Cl^-$ exchanges. *J. Gen. Physiol.* **47**, 1209–1227.

Maetz, J., Payan, P., and De Renzis, G. (1976). Controversial aspects of ionic uptake in freshwater animals. *In* "Perspectives in Experimental Biology" (P. Spencer Davies, ed.), Vol. 1, pp. 77–91. Pergamon, Oxford.

Maffly, R. H. (1968). A conductometric apparatus for measuring micromolar quantities of carbon dioxide. *Anal. Biochem.* **23**, 252–262.

Markham, A. E., and Kube, K. A. (1941). The solubility of carbon dioxide and nitrous oxide in aqueous solution. *J. Am. Chem. Soc.* **63**, 449.

Morris, S., and Taylor, A. C. (1983). Diurnal and seasonal variations in physico-chemical conditions within intertidal rockpools. *Estuarine Coastal Shelf Sci.* **17**, 339–355.

Mudge, J. E., and Neff, W. H. (1971). Sodium and potassium levels in serum of acid-exposed brook trout, *Salvelinus fontinalis*. *Proc. Pa. Acad. Sci.* **45**, 101–103.

Murdaugh, H. V., Jr., and Robin, E. D. (1967). Acid-base metabolism in the dogfish shark. *In* "Sharks, Skates and Rays" (W. Gilbert, R. F. Mathewson, and D. P. Rall, eds.), pp. 249–264. Johns Hopkins Press, Baltimore, Maryland.

Neumann, P., Holeton, G. F., and Heisler, N. (1983). Cardiac output and regional blood flow in gills and muscles after exhaustive exercise in rainbow trout (*Salmo gairdneri*). *J. Exp. Biol.* **105**, 1–14.

Neville, C. M. (1979a). Sublethal effects of environmental acidification on rainbow trout (*Salmo gairdneri*). *J. Fish. Res. Board Can.* **36**, 84–87.

Neville, C. M. (1979b). Ventilatory response of rainbow trout (*Salmo gairdneri*) to increased $H^+$ ion concentration in blood and water. *Comp. Biochem. Physiol. A* **63A**, 373–376.

Neville, C. M. (1979c). Influence of mild hypercapnia on the effects of environmental acidification on rainbow trout (*Salmo gairdneri*). *J. Exp. Biol.* **83**, 345–349.

Nicol, S. C., Glass, M. L., and Heisler, N. (1983). Comparison of directly determined and calculated plasma bicarbonate concentration in the turtle *Chrysemys picta bellii* at different temperatures. *J. Exp. Biol.* **107**, 521–525.

Ott, M. E., Ultsch, G. R., and Heisler, N. (1980). A re-evaluation of the relationship between temperature and the critical oxygen tension in freshwater fishes. *Comp. Biochem. Physiol.* **67A**, 337–340.

Packer, R. K. (1979). Acid–base balance and gas exchange in brook trout (*Salvelinus fontinalis*) exposed to acidic environments. *J. Exp. Biol.* **79**, 127–134.

Packer, R., and Dunson, W. (1970). Effects of low environmental pH on blood pH and sodium balance of brook trout. *J. Exp. Zool.* **174**, 65–72.

Packer, R., and Dunson, W. (1972). Anoxia and sodium loss associated with the death of brook trout at low pH. *Comp. Biochem. Physiol. A* **41A**, 17–26.

Parsons, J. D. (1968). The effects of acid-strip-mine effluents on the ecology of stream. *Arch. Hydrobiol.* **65**, 25–50.

Parsons, J. D. (1976). Effects of acid-mine wastes on aquatic ecosystems. *USDA US Dept. Agriculture For. Serv. Gen. Tech. Rep. NE* **NE-23**, 1074.

Payan, P., and Maetz, J. (1973). Branchial sodium transport mechanisms in *Scyliorhinus canicula*: Evidence for $Na^+/NH_4^+$ and $Na^+/H^+$ exchanges and for a role of carbonic anhydrase. *J. Exp. Biol.* **58**, 487–502.

Pequin, L. (1962). Les teneurs en azote ammoniacal du sang chez la carpe (*Cyprinus carpio* L.). *C. R. Hebd. Seances Acad. Sci.* **255**, 1795–1797.

Pequin, L., and Serfaty, A. (1963). L'excretion ammoniacale chez un téléostéen dulcicole: *Cyprinus carpio* L. *Comp. Biochem. Physiol.* **10**, 315–324.

Peyraud, C., and Serfaty, A. (1964). Le rythme respiratoire de la carpe (*Cyprinus carpio* L.) et ses relations avec le taux de l'oxygène dissous dans le biotope. *Hydrobiologia* **23**, 165–178.

Piiper, J., and Schumann, D. (1967). Efficiency of $O_2$ exchange in the gills of the dogfish, *Scyliorhinus stellaris*. *Respir. Physiol.* **2**, 135–148.

Piiper, J., Meyer, M., and Drees, F. (1972). Hydrogen ion balance in the elasmobranch *Scyliorhinus stellaris* after exhausting activity. *Respir. Physiol.* **16**, 290–303.

Pitts, R. F. (1945). The renal regulation of acid-base balance with special reference to the mechanism for acidifying the urine. *Science* **102**, 49–81.

Pitts, R. F. (1964). Renal production and excretion of ammonia. *Am. J. Med.* **36**, 720–742.

Pitts, R. F. (1973). Production and excretion of ammonia in relation to acid–base regulation. *In* "Handbook of Physiology" (J. Orloff and R. W. Berliner, eds.), Sect. 8, pp. 455–496. Amer. Physiol. Soc., Washington, D.C.

Radford, E. P. (1964). The physics of gases. *In* "Handbook of Physiology" (W. O. Fenn and H. Rahn, eds.), Sect. 3, Vol. I, pp. 125–152. Am. Physiol. Soc., Washington, D.C.

Rahn, H. (1966a). Development of gas exchange. Evolution of the gas transport system in vertebrates. *Proc. R. Soc. Med.* **59**, 493–494.

Rahn, H. (1966b). Aquatic gas exchange: Theory. *Respir. Physiol.* **1**, 1–12.

Rahn, H. (1967). Gas transport from the environment to the cell. *In* "Development of the Lung" (A. V. S. de Reuck and R. Porter, eds.), pp. 3–23. Churchill, London.

Rahn, H., and Baumgardner, F. W. (1972). Temperature and acid-base regulation in fish. *Respir. Physiol.* **14**, 171–182.

Randall, D. J., and Cameron, J. N. (1973). Respiratory control of arterial pH as temperature changes in rainbow trout. *Am. J. Physiol.* **225**, 997–1002.

Randall, D. J., and Jones, D. R. (1973). The effect of deafferentation of the pseudobranch on the respiratory response to hypoxia and hyperoxia in the trout (*Salmo gairdneri*). *Respir. Physiol.* **17**, 291–301.

Randall, D. J., Heisler, N., and Drees, F. (1976). Ventilatory response to hypercapnia in the larger spotted dogfish *Scyliorhinus stellaris*. *Am. J. Physiol.* **230**, 590–594.

Randall, D. J., Burggren, W. W., Farrell, A. P., and Haswell, M. S. (1981). "The Evolution of Air Breathing in Vertebrates." Cambridge Univ. Press, London and New York.

Rector, F. C. (1973). Acidification of the urine. *In* "Handbook of Physiology" (J. Orloff and R. W. Berliner, eds.), Sect. 8, pp. 431–454. Am. Physiol. Soc., Washington, D.C.

Reeves, R. B. (1972). An imidazole alphastat hypothesis for vertebrate acid-base regulation: Tissue carbon dioxide content and body temperature in bullfrogs. *Respir. Physiol.* **14**, 219–236.

Reeves, R. B., and Malan, A. (1976). Model studies of intracellular acid-base temperature responses in ectotherms. *Respir. Physiol.* **28**, 49–63.

Saunders, R. L. (1962). The irrigation of the gills in fishes. *Can. J. Zool.* **40**, 817–862.

Schenone, G., Arillo, A., Margiocco, C., Melodia, F., and Mensi, P. (1982). Biochemical bases for environmental adaptation in goldfish (*Carassius auratus*): Resistence to ammonia. *Ecotoxicol. Environ. Saf.* **6**, 479–488.

Schofield, G. (1976). Acid precipitation: Effects on fish. *Ambio* **5**, 228–230.

Shelton, G. (1970). The regulation of breathing. *In* "Fish Physiology" (W. S. Hoar and D. J. Randall, eds.), Vol. 4, pp. 293–359. Academic Press, New York.

Sioli, H. (1954). Beiträge zur regionalen Limnologie des Amazonasgebietes. II. Der Rio Arapiuns. *Arch. Hydrobiol.* **49**, 448–518.

Sioli, H. (1955). Beiträge zur regionalen Limnologie des Amazonasgebietes. III. Über einige Gewässer des oberen Rio Negro-Gebietes. *Arch. Hydrobiol.* **50**, 1–32.

Sioli, H. (1957). Beiträge zur regionalen Limnologie des Amazonasgebietes. IV. Limnologische Untersuchungen in der Region der Eisenbahnlinie Belem-Braganca (Zona bragantina) im Staate Para, Brasilien. *Arch. Hydrobiol.* **53**, 161–222.

Smith, H. W. (1929a). The composition of the body fluids of elasmobranchs. *J. Biol. Chem.* **81,** 407–419.

Smith, H. W. (1929b). The excretion of ammonia and urea by the gills of fish. *J. Biol. Chem.* **81,** 727–742.

Soivio, A., Westman, D. C., and Nyholm, K. (1972). Improved method of dorsal aorta catherization: Haematological effects followed for three weeks in rainbow trout (*Salmo gairdneri*). *Finn. Fish. Res.* **1,** 11–21.

Soivio, A., Nyholm, K., and Westman, D. C. (1975). A technique for repeated sampling of blood of individual resting fish. *J. Exp. Biol.* **63,** 207–217.

Solarzano, L. (1969). Determination of ammonia in natural waters by the phenolhypochlorite method. *Limnol. Oceanogr.* **14,** 799–801.

Stevens, D. E. (1968). The effect of exercise on the distribution of blood to various organs in rainbow trout. *Comp. Biochem. Physiol.* **25,** 615–625.

Stevens, D. E., and Black, E. C. (1966). The effect of intermittent exercise on carbohydrate metabolism in rainbow trout (*Salmo gairdneri*). *J. Fish. Res. Board Can.* **23,** 471–485.

Swarts, F. A., Dunson, W. A., and Wright, J. E. (1978). Genetic and environmental factors involved in increased resistence of brook trout to sulfuric acid solutions and mine acid polluted waters. *Trans. Am. Fish. Soc.* **107,** 651–677.

Thurston, R. V., and Russo, R. C. (1981). Ammonia toxicity to fishes. Effect of pH on the toxicity of the un-ionized ammonia species. *Environ. Sci. Technol.* **15,** 837–840.

Toews, D. P., and Heisler, N. (1982). The effects of hypercapnia on intracellular and extracellular acid-base status in the toad *Bufo marinus*. *J. Exp. Biol.* **97,** 77–86.

Toews, D. P., Holeton, G. F., and Heisler, N. (1983). Regulation of the acid-base status during environmental hypercapnia in the marine teleost fish *Conger conger*. *J. Exp. Biol.* **107,** 9–20.

Tomasso, J. R., Boudie, C. A., Simco, B. A., and Davis, K. B. (1980). Effects of environmental pH and calcium on ammonia toxicity in channel catfish. *Trans. Am. Fish. Soc.* **109,** 229–234.

Truchot, J. P., and Duhamel-Jouve, A. (1980). Oxygen and carbon dioxide in the marine intertidal environment: Diurnal and tidal changes in rock pools. *Respir. Physiol.* **39,** 241–254.

Truchot, J. P., Toulmond, A., and Dejours, P. (1980). Blood acid-base balance as a function of water oxygenation: A study at two different ambient $CO_2$ levels in the dogfish, *Scyliorhinus canicula*. *Respir. Physiol.* **41,** 13–28.

Turner, J. D., Wood, C. M., and Hobe, H. (1983). Physiological consequences of severe exercise in the inactive bethic flathead sole (*Hippoglossoides elassodon*); a comparison with the active pelagic rainbow trout (*Salmo gairdneri*). *J. Exp. Biol.* **104,** 269–288.

Ultsch, G. R., Ott, M. E., and Heisler, N. (1980). Standard metabolic rate, critical oxygen tension, and aerobic scope for spontaneous activity of trout (*Salmo gairdneri*) and carp (*Cyprinus carpio*) in acidified water. *Comp. Biochem. Physiol. A* **67A,** 329–335.

Ultsch, G. R., Ott, M. E., and Heisler, N. (1981). Acid-base and electrolyte status in carp (*Cyprinus carpio*) exposed to low environmental pH. *J. Exp. Biol.* **93,** 65–80.

Van Dam, L. (1938). "On the Utilization of Oxygen and Regulation of Breathing in Some Aquatic Animals." Volharding, Grøningen.

Van Slyke, D. D. (1917). Studies of acidosis. II. A method for the determination of carbon dioxide and carbonates in solution. *J. Biol. Chem.* **30,** 347–368.

Van Slyke, D. D., Sendroy, A. B., Hastings, A. B., and Neill, J. M. (1928). Studies of gas and electrolyte equilibria in blood. X. The solubility of carbon dioxide at 38°C in water, salt solution, serum and blood cells. *J. Biol. Chem.* **78,** 765.

Waddell, W. J., and Butler, T. C. (1959). Calculation of intracellular pH from the distribution of

5,5-dimethyl-2,4-oxazolidinedione (DMO). Application to skeletal muscle of the dog. *J. Clin. Invest.* **38**, 720–729.

Walsh, P. J., and Moon, T. W. (1982). The influence of temperature on extracellular and intracellular pH in the American eel, *Anguilla rostrata* (Le Suer). *Respir. Physiol.* **50**, 129–140.

Wardle, C. S. (1978). Non-release of lactic acid from anaerobic swimming muscle of plaice *Pleuronectes platessa* L: A stress reaction. *J. Exp. Biol.* **77**, 141–155.

Weast, R. G., ed. (1975–1976). "Handbook of Chemistry and Physics," 56th ed. CRC Press, Cleveland, Ohio.

Wilkes, P. R. H., Walker, R. L., McDonald, D. G., and Wood, C. M. (1981). Respiratory, ventilatory, acid-base and ionoregulatory physiology of the white sucker *Catostomus commersoni:* The influence of hyperoxia. *J. Exp. Biol.* **91**, 239–254.

Wood, C. M., and Jackson, E. B. (1980). Blood acid-base regulation during environmental hyperoxia in the rainbow trout (*Salmo gairdneri*). *Respir. Physiol.* **42**, 351–372.

Wood, C. M., McMahon, B. R., and McDonald, D. G. (1977). An analysis of changes in blood pH following exhausting activity in the starry flounder, *Platichthys stellaris. J. Exp. Biol.* **69**, 173–185.

Wood, J. D. (1958). Nitrogen excretion in some marine teleosts. *Can. J. Biochem. Physiol.* **36**, 1237–1242.

# APPENDIX: PHYSICOCHEMICAL PARAMETERS FOR USE IN FISH RESPIRATORY PHYSIOLOGY*

*ROBERT G. BOUTILIER*
*THOMAS A. HEMING*
*GEORGE K. IWAMA*

Department of Zoology
University of British Columbia
Vancouver, British Columbia, Canada

## I. INTRODUCTION

The intent of this appendix is to bring together certain physicochemical parameters that are applicable to the blood plasma of fish. Inasmuch as the aquatic environment represents an extension of the extracellular space via the gills, attention will also be given to physical principles governing the acid–base chemistry, ammonia equilibrium, and solubility of respiratory gases in waters of known ionic composition. Every effort has been made to provide clear distinctions in the text between data that have been measured and those that have been extrapolated. In addition, brief annotations of the

*The authors were supported by N.S.E.R.C. (Canada).

FISH PHYSIOLOGY, VOL. XA

prevailing experimental conditions and measurement devices are reported whenever possible, along with source references to the original data. In many instances, we have constructed best-fit empirically determined equations to describe published data, if not already reported in th;' form.

It is evident that the use of tabulated "constants" of blood plasma and water are of only marginal value compared to experimentally measured values. It is our hope that the following compilation of data will only be used in studies where conditions are such that they restrict direct measures of the parameters in question.

## II. PHYSICOCHEMICAL PROPERTIES OF PLASMA

### A. Carbon Dioxide Equilibria

In clinical situations, it is common practice to assess acid–base status by measuring the pH and $CO_2$ partial pressure of an anaerobically obtained blood sample and then to calculate plasma bicarbonate concentration using the Henderson–Hasselbalch Equation. Such estimates require values for the solubility of carbon dioxide in plasma ($\alpha_{CO_2}$) and the apparent $pK$ ($pK_{app}$) of the dissociation of carbonic acid, both of which are affected by temperature and ionic strength; $pK_{app}$ is also influenced by pH in a nonlinear fashion (Albers, 1970; see also Chapter 6, this volume). Determinations of these "constants" have been performed several times for human blood plasma, and they have become well defined (Severinghaus, 1965), albeit under the narrow conditions of temperature, pH, and ionic strength of mammalian plasma. The absence of such experimentally determined values in fish species has led to widespread use of these human plasma constants in determining either $P_{CO_2}$ or $[HCO_3{}^-]$.

In one regard, it is fortuitous that the ionic strength of human plasma approximates that of many teleosts, since this has lent considerable support to the applicability of the $\alpha_{CO_2}$ and $pK_{app}$ values of human plasma to fish. However, at temperatures frequently encountered in studies on fish, these values must be extrapolated from the much higher temperatures at which the actual experiments were performed. Human plasma $CO_2$ solubility measurements, for example, have only been made for the temperature range of 15 to 38°C (Bartels and Wrbitzky, 1960; Austin *et al.*, 1963), whereas the $pK_{app}$ values widely used in fish studies are based on linear extrapolation of determinations made at only two temperatures (24 and 37.5°C by Severinghaus *et al.*, 1956). In point of fact, many of these "constants" lack any experimental verification at the physiological temperatures of a large number of fishes.

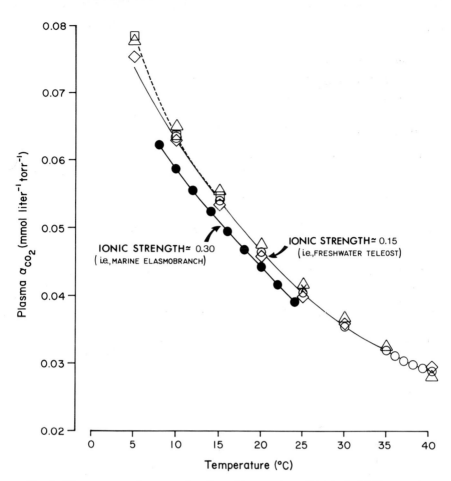

**Fig. 1.** Plasma $\alpha_{CO_2}$ values as a function of temperature. (A) (glyph ◯) Human plasma (Severinghaus, 1965). These data appear as tabulated values in Severinghaus (1965), who recalculated the original data of Bartels and Wrbitzky (1960) and Austin *et al.* (1963). The latter authors measured the $\alpha_{CO_2}$ of human plasma at several intervals spanning the 15–38°C temperature range. As plotted, the original measurements have been extrapolated to 10 and 40°C. The original measurements were carried out on serum acidified with concentrated lactic acid (2 vol% solution; Austin *et al.*, 1963) and equilibrated with $CO_2$ at various temperatures for at least 30 min. Plasma $CO_2$ contents were determined by the manometric method of Peters and Van Slyke (1932). Consideration was given to the water, lipid, protein, and electrolyte composition of the serum, all of which influence solubility (see Chapter 6, this volume, for discussion). The tabulated values given by Severinghaus (1965) have been modeled with the formula:

$$\alpha_{CO_2} = 0.0869 - 0.0028(T) + 4.6143 \times 10^{-5}(T^2)$$
$$- 2.8889 \times 10^{-7}(T^3), \text{ mmol liter}^{-1} \text{ torr}^{-1}$$

where $T$ is the temperature in °C ($r^2 = 0.9999$). Predictions of $\alpha_{CO_2}$ by this equation agree with the tabled data of Severinghaus (1965) to the fourth decimal place. We have extrapolated the regression line in Fig. 1 to 5°C only for comparison purposes. (B) (glyph ●) Dogfish plasma (Pleschka and Wittenbrock, 1971). Plasma $CO_2$ solubilities were experimentally determined on plasma obtained from dogfish species, *Scyliorhinus canicula* and *S. stellaris* (Pleschka and

## 1. Carbon Dioxide Solubility of Plasma

Carbon dioxide solubilities of plasma in fishes and mammals are given in Fig. 1 as a function of temperature.

## 2. The Apparent pK of Carbonic Acid in Plasma

The following nomograms and/or equations have been constructed from data in studies in which simultaneous determinations of plasma pH, $CO_2$

---

**Fig. 1.** (*Continued*)

Wittenbrock, 1971). Because measurements made over the 10–22°C range did not differ between species, the data were pooled; extrapolation to 8 and 25°C was made in the original article. Blood samples were obtained from the caudal artery of animals anesthetized by immersion in a 2% ethylurethane solution. After the osmolality of the separated plasma was measured, the sample was then acidified to pH <4 with diluted lactic acid (1% solution), and its osmolality redetermined at the lower value. The initial level of plasma osmolality was then restored by adding known amounts of NaCl. This procedure was required because of the precipitation of plasma proteins if concentrated lactic acid was added to the plasma. Samples treated in this fashion were equilibrated in tonometers at 10, 13, 16, 19, and 22°C with gas mixtures containing 19.8% $CO_2$ + 20% $O_2$—balance $N_2$ for 1 to 3 hr. Total $CO_2$ contents in 0.1-ml plasma samples were determined by the micromanometric technique of Van Slyke and Plazin (1961). Partial pressures of $CO_2$ in the equilibration gases were calculated according to prevailing atmospheric pressure conditions, corrected for temperature and water vapor pressure; these values were taken as the plasma $P_{CO_2}$ levels. The data were reported in the original article as Bunsen's solubility coefficients, which express the milliliters of gas, measured at 0°C, 760 torr, dissolved per milliliter of solution when the latter is under 1 atm of pressure. Bunsen coefficients were transformed to mmol liter$^{-1}$ torr$^{-1}$ using the equation

$$\alpha_{CO_2} \text{ (mmol liter}^{-1} \text{ torr}^{-1}) = \frac{\text{ml } CO_2/\text{ml plasma}}{(22.262 \times 0.76)}$$

See Van Slyke *et al.* (1928) and Radford (1964). An empirically determined formula to describe these data is

$$\alpha_{CO_2} = 0.1131 - 1.3847 \times 10^{-2}(T) + 1.4995 \times 10^{-3}(T^2) - 8.8008 \times 10^{-5}(T^3)$$
$$+ 2.4998 \times 10^{-6}(T^4) - 2.7369 \times 10^{-8}(T^5), \text{ mmol liter}^{-1} \text{ torr}^{-1}$$

where $T$ is the temperature in °C. (C) (glyph □) Rainbow trout plasma (Boutilier *et al.*, 1984). Separated plasma was obtained by centrifugation of blood samples obtained from dorsal aorta cannulae of *Salmo gairdneri* that had been chronically catheterized and left to recover in darkened boxes (10°C running water) for at least 48 hr. Plasma samples were acidified slowly to pH ≈2.5 (without apparent protein denaturation) by additions of pure lactic acid and then equilibrated with 40% $CO_2$ + 20.09% $O_2$—balance $N_2$ (Wösthoff pumps, Böchum, Federal Republic of Germany) at 5, 10, and 15°C. Carbon dioxide partial pressures were calculated from prevailing barometric pressure, water vapor pressure, and temperature readings taken over the same time course as equilibration of each sample. Total $CO_2$ contents were measured by the method of Cameron (1971) with a resolution of 0.05 mmol liter$^{-1}$. The water content and ionic tensions $(P_{CO_2})$, and total $CO_2$ content $(C_{CO_2})$ have allowed for calculation of $pK_{app}$ by rearrangement of the Henderson–Hasselbalch equation:

$$pK_{app} = pH - \log \frac{C_{CO_2}}{\alpha_{CO_2} P_{CO_2}} - 1$$

where $\alpha_{CO_2}$ has either been measured directly or taken from published values (Figs. 2 and 3). This apparent $pK$ has no thermodynamic meaning and

---

**Fig. 1.** (*Continued*)

strength of each plasma sample was also determined. A formula describing these data has been constructed:

$$\alpha_{CO_2} = 1.0064 \times 10^{-1} - 5.4431 \times 10^{-3}(T) + 2.1776 \times 10^{-4}(T^2)$$
$$- 4.9731 \times 10^{-6}(T^3) + 4.5288 \times 10^{-8}(T^4), \text{ mmol liter}^{-1} \text{ torr}^{-1}$$

where $T$ is temperature in °C. The equation predicts the mean measured values at 5, 10, and 15°C to within 0.0001 mmol liter$^{-1}$ torr$^{-1}$. (D) (glyph $\diamond$) Modeled human plasma (Reeves, 1976a). Solubility values for $CO_2$ were calculated from the data on human plasma reported by Severinghaus [see (A) and Fig. 1], according to the polynomial equation:

$$\alpha_{CO_2} = 0.0907 - 0.3373 \times 10^{-2}(T) + 0.6749 \times 10^{-4}(T^2)$$
$$- 5.4076 \times 10^{-7}(T^3), \text{ mmol liter}^{-1} \text{ torr}^{-1}$$

where $T$ is temperature in °C. Values for $\alpha_{CO_2}$ plotted in Fig. 1 using this formula do not predict the tabulated data in Severinghaus with the same degree of accuracy as the equation given in (A). (E) (glyph $\triangle$) Generalized model (see Chapter 6). Carbon dioxide solubility values were calculated on the basis of a formula derived by Heisler, based on literature data of Van Slyke *et al.* (1928), Markham and Kube (1941), and Bartels and Wrbitzky (1960), among others, and experimentally determined measurements on various body fluids in a number of species by Heisler and co-workers (see Chapter 6 for details). The formula

$$\alpha_{CO_2} = 0.1008 - 29.80 \times 10^{-3}[M] + (1.218 \times 10^{-3}[M]$$
$$- 3.639 \times 10^{-3})T$$
$$- (19.57 \times 10^{-6}[M] - 69.59 \times 10^{-6})T^2 - (71.71 \times 10^{-9}[M]$$
$$- 559.6 \times 10^{-9})T^3, \text{ mmol liter}^{-1} \text{ torr}^{-1}$$

is valid (between $T$ of 0 and 40°C) for pure water, salt solutions, and various body fluids. $[M] =$ molarity of dissolved species (mol liter$^{-1}$). $\alpha$ refers to the volume of water in a solution; solubility per liter of solution requires appropriate correction for the volume of proteins and salts. $[M]$ of the values plotted in Fig. 1 was taken as 0.3 mol liter$^{-1}$ (i.e., freshwater teleost molarity). (F) (glyph $\times$) Lungfish plasma (DeLaney *et al.*, 1977). Separated plasma samples (of 94.12% water content) were acidified by the addition of lactic acid (1% solution) and equilibrated at 25°C with 100% $CO_2$. Total $CO_2$ contents were determined by manometric analysis of 3-ml samples according to Van Slyke *et al.* (1928). The $\alpha_{CO_2}$ value reported (0.0410 mmol liter$^{-1}$ torr$^{-1}$) falls in close proximity to the values obtained from Severinghaus and Heisler for 25°C (Fig. 1).

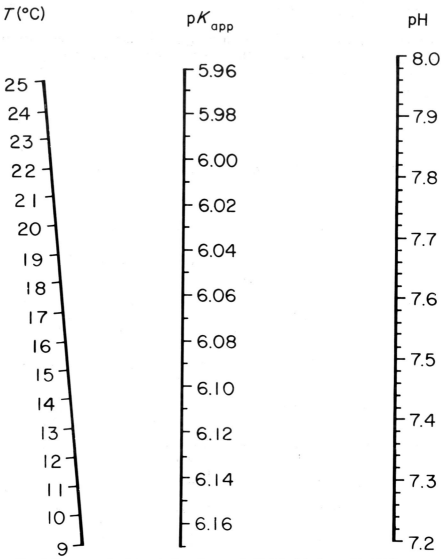

**Fig. 2.** Nomogram for determination of apparent p$K$ of dogfish plasma. Straight lines connecting values for temperature and pH lead to the corresponding p$K_{app}$. Nomogram constructed from the pH versus p$K_{app}$ data illustrated in Albers and Pleschka (1967) and Albers (1970) for dogfish plasma. Individual data points were digitized on a HP Graphics Tablet and HP9816 microprocessor to allow for reconstruction of $\Delta pK_{app}/\Delta pH$ regression lines and subsequent development of an accurate mathematical description of the data by a polynomial expression. In constructing the nomogram, regression lines for 9, 17, and 25°C (Albers and Pleschka, 1967) were independently assessed, giving $\Delta pK_{app}/\Delta pH$ slopes of $-0.0841$, $-0.1020$, and $-0.1066$, respectively. The averaged $\Delta pK_{app}/\Delta pH$ slope of the regression lines given by Albers and Pleschka was $-0.097$; that by our digitizing procedure was $-0.0976$. The formula describing the nomogram is

is strictly for use in the Henderson–Hasselbalch Equation when the apparent bicarbonate concentration is measured gasometrically or titrimetrically (i.e., $[HCO_3^-] + [CO_3^{2-}] + [NaCO_3^-] + [protein–NH–COO^-]$). Detailed discussions of the physicochemical boundary conditons applicable to

---

**Fig. 2.** (*Continued*)

$$pK_{app} = 6.4996 + \log T(0.3648 - 0.0521 \text{ pH}) - 0.0353 \text{ pH} - 0.0074T$$

where $T$ is temperature in °C.

The original data of Albers and Pleschka (1967) were experimentally determined by tonometry of blood samples taken from *Scyliorhinus canicula* and *S. stellaris*. Samples were equilibrated against gas mixtures containing 0.2–1.5% $CO_2$ in $O_2$ or $N_2$ at temperatures of 9, 17, and 25°C. Plasma $CO_2$ tensions were taken to be those of the equilibration gases (compressed gas cylinders analyzed to ±0.001% SD) after appropriate corrections for temperature, barometric pressure, and water vapor. Total $CO_2$ contents of 1-ml true plasma samples were measured by the Van Slyke apparatus (Van Slyke and Neill, 1924). The pH was measured using a Radiometer capillary glass electrode (Type G297/G II) calibrated with three National Bureau of Standards buffers. The $\alpha_{CO_2}$ of plasma was taken to be 73% of pure water values for each respective temperature. A subsequent study (Pleschka and Wittenbrock, 1971) recalculated some of the earlier $pK_{app}$) values of Albers and Pleschka (1967) using new experimentally determined $\alpha_{CO_2}$ measurements on dogfish plasma (see Fig. 1B) and found slightly different $pK_{app}$ estimates. At a pH of 7.40, the recalculated $pK_{app}$ values were 6.192 (9°C), 6.150 (17°C), and 6.077 (25°C). Unfortunately, in our treatment of the data, we were unable to correct the original data of Albers and Pleschka (1967) using the more recent $\alpha_{CO_2}$ measurements because of uncertainty about the $\alpha_{CO_2}$ values used in the original article (i.e., 73% of an unspecified distilled water value). Using the polynomial function given by the equation above, we calculate $pK_{app}$ values at pH 7.40 of 6.152 (9°C), 6.113 (17°C), and 6.024 (25°C). The deviations from the recalculated $pK_{app}$ values reported by Pleschka and Wittenbrock (1971) are respectively +0.040, +0.037, and +0.053.

Pleschka and Wittenbrock (1971) stated that their new $pK_{app}$ estimates (see earlier) were based on an $\alpha_{CO_2}$ value "scarcely different" from those observed (but not reported) by Albers and Pleschka (1967) for 81% $CO_2$ solubility of distilled water. However, the data of Albers and Pleschka (1967) used in construction of the nomogram and equation (given earlier) were based on the lower $\alpha_{CO_2}$ value they used and reported as corresponding to 73% solubility of distilled water. Estimates of $pK_{app}$ from the nomogram and/or equation just given can be adjusted for these new $\alpha_{CO_2}$ values (Pleschka and Wittenbrock, 1971) as follows, from the Henderson–Hasselbalch Equation:

$$\alpha_{CO_2} = \frac{C_{CO_2}}{P_{CO_2}(10^{pH - pK_{app}} + 1)}$$

Given that the new solubility value ($\alpha_{CO_2}'$) is 81/73 or 1.1096 times greater than the solubility values used by Albers and Pleschka (1967), it follows that

$$\alpha_{CO_2}' = \frac{1.1096\ C_{CO_2}}{P_{CO_2}(10^{pH - pK_{app}} + 1)}$$

where 1.1096 is a constant correction over the temperature range. $\alpha_{CO_2}'$ is then substituted these "operational" $pK_{app}$ values are developed elsewhere (Siggaard-Andersen, 1962; Albers, 1970; see also Chapter 6, this volume).

1. In addition to the data reported in Figs. 2 and 3, Delaney *et al.* (1977) experimentally determined the plasma $pK_{app}$ of lungfish (*Protopterus aethiopicus*) as a function of pH at 25°C. The data are described by the equation $pK_{app} = 6.606 - 0.061$ pH for 25°C only. Values of $pK_{app}$ were estimated by simultaneously measuring the pH, $P_{CO_2}$ (Radiometer System), and total $CO_2$ content (apparatus of Van Slyke *et al.*, 1928) of separated plasma samples after tonometry with gas mixtures containing 2.0, 3.5, 7.0, and 8.0% $CO_2$ in air. Values for $\alpha_{CO_2}$ were experimentally determined at the same temperature (DeLaney *et al.*, 1977; Fig. 1F).

2. Many studies have reported $pK_{app}$ values for human plasma based on either experimental data or theoretical interpretations of published works (e.g., Severinghaus *et al.*, 1956; Siggaard-Andersen, 1962; Austin *et al.*, 1963; Mitchell *et al.*, 1965; Severinghaus, 1965; Rispens *et al.*, 1968). The values of Severinghaus (1965) have been recommended for use with freshwater fish (Albers, 1970) and have been widely used. These tabulated data of Severinghaus (1965) have been described by polynomial expressions in Table I.

3. Reeves (1976a,b), conducting studies on temperature-induced changes in acid–base status of mammalian blood, reassessed the $pK_{app}$ data of Severinghaus (1965), according to the equation $pK_1' = f(I,T)$, where $pK_1'$ is the first dissociation constant of carbonic acid (not $pK_{app}$) as a function (f) of ionic strength (*I*) and temperature (*T*). This equation was originally used by Harned and Bonner (1945), who determined $pK_1'$ in aqueous solutions of sodium chloride (0–1 *M*; over 0 to 50°C) using an electromotive cell without liquid junction. Reeves (1976a) compared the shapes of the Harned and Bonner curves with the original data of Severinghaus *et al.* (1956), who had

---

**Fig. 2.** (*Continued*)
back into another rearrangement of the Henderson–Hasselbalch equation to obtain the adjusted apparent dissociation constant ($pK_{app}^{adj}$) as follows,

$$pK_{app}^{adj} = pH - \log \left( \frac{C_{CO_2}}{\alpha_{CO_2}' P_{CO_2}} - 1 \right)$$

$$= pH - \log \left( \frac{10^{pH - pK_{app}} + 1}{1.1096} - 1 \right)$$

Estimates of $pK_{app}$ at pH 7.40 from the nomogram in Fig. 2 adjusted in this fashion, are 6.200 (9°C), 6.161 (17°C), and 6.071 (25°C), values quite similar to those corrected to new experimentally obtained $\alpha_{CO_2}$ measurements by Pleschka and Wittenbrock (1971).

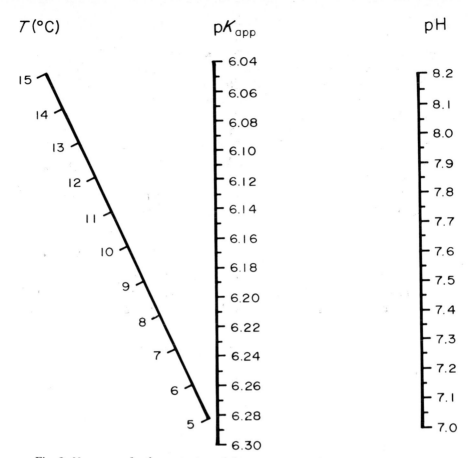

Fig. 3. Nomogram for determination of the apparent p$K$ of rainbow trout true plasma. Straight lines connecting values for temperature and pH lead to respective p$K_{app}$. Nomogram constructed from in vitro experimental determinations on rainbow trout blood by Boutilier *et al.* (1984). Arterial blood, obtained from *Salmo gairdneri* that had been chronically catheterized in the dorsal aorta and left to recover in darkened boxes (10°C running water) for at least 48 hr, was tonometered for 45 min at 5, 10, and 15°C against gas mixtures containing 0.25, 0.50, 1.00, and 2.00% $CO_2$—balance air. Blood $CO_2$ tensions were measured using Radiometer $CO_2$ electrodes selected for extreme stability and preconditioned to the expected $P_{CO_2}$ values as recommended by Boutilier *et al.* (1978) and Bateman *et al.* (1980). Carbon dioxide electrode time constants were used to back-interpolate blood measurements based on the bracketing of each sample with calibration gases delivered by Wösthoff pumps. The pH of true plasma was determined with Radiometer glass capillary electrodes (Type G297/G 2) calibrated with Radiometer precision buffers before and after each measurement. Total $CO_2$ contents of anaerobically obtained true plasma were measured by the method of Cameron (1971). Plasma $CO_2$ solubility coefficients were experimentally determined (Fig. 1), as was the [Na$^+$], [K$^+$], [Cl$^-$[, [Mg$^{2+}$], [Ca$^{2+}$], [lactate$^-$], and water content of each sample. The nomogram was constructed from the empirically determined formula:

$$p K_{app} = 6.4755 T^{-0.0187} + \log T(1.1704 - 0.1672 \text{ pH}) + 0.1073 \text{ pH} - 0.7511$$

assumed a linear relationship to exist between $pK_{app}$ and temperature, based on measurements made at only two temperatures. Making the assumption that $pK_{app}$ would change with temperature in the same fashion as $pK_1'$, Reeves (1976a) used the Harned and Bonner equation to fit a curvilinear line to the original data of Severinghaus et al. (1956) and presented the following polynomial equation: $pK_1' = 6.3852 - 1.3288 \times 10^{-2}(T) + 1.7364 \times 10^{-4}(T^2) - 6.0084 \times 10^{-7}(T^3)$ where $T$ is temperature (°C). This equation makes no correction for the known pH dependence of $pK_{app}$ (Severinghaus et al., 1956; Table I).

4. Siggaard-Andersen (1976) has derived a formula for calculation of apparent $pK$ values in plasma in which the effects of both temperature and pH on $pK_{app}$ are taken into account. The polynomial expression:

$$pK_{app} = 6.125 - \log(1 + 10^{pH - 8.7}) - 0.0026(t - 37) + 0.00012(t - 37)^2$$

where $t$ = temperature in °C, was derived on the basis of data found in Harned and Davis (1943), Maas et al. (1971) and Siggaard-Andersen (1962).

5. Heisler (Chapter 6, this volume) has developed a generally applicable

---

**Fig. 3.** (*Continued*)

where $T$ is temperature in °C, valid only for the 5–15°C range.

It is important to note that values obtained by this formula or read from the nomogram are based on pH measurements of whole blood and total $CO_2$ measurements of true plasma separated anaerobically at room temperature. Strictly speaking, plasma $pK_{app}$ values, such as those obtained from the experiments on separated plasma by Severinghaus et al. (1956) and others, will apply only to experimental protocols that allow for the measurement of pH and $C_{CO_2}$ of true plasma separated at the same temperature as the blood pool under investigation. When these conditions were met for trout blood equilibrated in vitro, it was found that the pH of whole blood was always lower than that of its true plasma (Boutilier et al., 1984). Moreover, this difference became magnified as temperature decreased, a finding similar to that of Siggaard-Andersen (1976). These factors can be explained at least in part by the rupture of red cells at the electrolyte bridge between blood and saturated KCl and the subsequent effects this has on the liquid junction potential of the pH measuring apparatus (Siggaard-Andersen, 1976; Boutilier et al., 1984). Also, if cool blood is allowed to warm toward room temperature before and during centrifugation for true plasma, $CO_2$ may leave the red blood cells and result in a higher total $CO_2$ relative to the amount contained in the original blood plasma. Ongoing red blood cell metabolism may also contribute to small errors in $CO_2$ measurements. Since these strict experimental precautions are rarely adhered to in normal practice, the use of $pK_{app}$ values obtained from gasometric determinations on separated plasma samples is not entirely appropriate. For these reasons and perhaps other factors not yet accounted for, the $\Delta pK_{app}/\Delta T$ slope of this nomogram is higher than that expected on the basis of the thermodynamic constants of the system. Clearly, the operational $pK$ of the first dissociation step must be larger than the thermodynamic one because of overlap with the second step and with the formation of $NaCO_3^-$ complexes. The presence of red blood cells in our determinations and their effects on pH and $C_{CO_2}$ measurements limit the operational usage of this nomogram to data collected under similar experimental conditions.

## Table I

Coefficients of Third-Order Polynomials to Describe Severinghaus's Tabulated Values (1965) of Human Plasma $pK_{app}$ as a Function of pH for Temperatures 10–40°C[a]

| Temperature (°C) | $pK_{app}$ f(pH) |
|---|---|
| 10 | $16.817 - 4.2990(\text{pH}) + 5.9286 \times 10^{-1}(\text{pH}^2) - 2.7778 \times 10^{-2}(\text{pH}^3)$ |
| 15 | $5.8195 + 1.8251 \times 10^{-1}(\text{pH}) - 1.7858 \times 10^{-2}(\text{pH}^2) + 3.7037 \times 10^{-8}(\text{pH}^3)$ |
| 20 | $5.4202 + 2.7765 \times 10^{-1}(\text{pH}) - 2.3813 \times 10^{-2}(\text{pH}^2) + 1.6667 \times 10^{-7}(\text{pH}^3)$ |
| 25 | $5.6555 + 1.9894 \times 10^{-1}(\text{pH}) - 1.7859 \times 10^{-2}(\text{pH}^2) + 8.3333 \times 10^{-8}(\text{pH}^3)$ |
| 30 | $16.043 - 4.1441(\text{pH}) + 5.8452 \times 10^{-1}(\text{pH}^2) - 2.7778 \times 10^{-2}(\text{pH}^3)$ |
| 35 | $5.0457 + 3.3739 \times 10^{-1}(\text{pH}) - 2.6192 \times 10^{-2}(\text{pH}^2) + 8.3333 \times 10^{-8}(\text{pH}^3)$ |
| 40 | $-5.6993 + 4.7494(\text{pH}) - 6.3214 \times 10^{-1}(\text{pH}^2) + 2.7778 \times 10^{-2}(\text{pH}^3)$ |

[a]These equations can be used to reproduce the tabulated values of Severinghaus (1965) to the third decimal place, and cover intermediate pH values. Note, however, that the table in Severinghaus (1965) was constructed from original data collected by Severinghaus et al. (1956) over the 6.6–8.0 pH range but only at two temperatures. Severinghaus et al. (1956) conducted in vitro experiments on human serum or plasma of freshly drawn blood. Samples were tonometered for 45 min at 37.5°C and at precisely known temperatures between 23 and 25°C against various gas mixtures containing 0.857–23.45% $CO_2$ in air (gas cylinders analyzed by Scholander or Haldane apparatus, Scholander, 1947). Plasma $P_{CO_2}$ values were taken to be equal to those of the equilibration gas tensions. Total $CO_2$ contents were determined by the manometric apparatus of Van Slyke over a plasma pH range of 6.6 to 8.0.

Based on revised National Bureau of Standards buffer values, correction for the KCl to buffer liquid junction potentials and revised $CO_2$ solubility values (see Fig. 1), the original $pK_{app}$ measurements by Severinghaus et al. (1956) have been recalculated by Severinghaus (1965) and extrapolated to cover the 10–40°C temperature range. The original assumption of linearity of $pK_{app}$ versus temperature between 24 and 37.5°C (Severinghaus et al., 1956) and subsequent extrapolation of these data outside the measured range (Severinghaus, 1965) have both been questioned by Reeves (1976a); see Section II,A,2, paragraph 3, for discussion.

formula based on theoretical considerations as well as experimentally determined $pK_{app}$ values in water, salt solutions, and biological fluids from numerous animals. The formula is considered valid for measurements of pH with low-temperature glass electrodes and double-electrolyte bridge reference electrodes (calibrated with buffers having similar ionic strength to the measured sample), and for measurements of apparent bicarbonate concentration ($[HCO_3^-]_{app}$) by gasometric techniques (i.e., $[HCO_3^-]_{app} = $ total $CO_2 - \alpha_{CO_2}P_{CO_2}$) where $\alpha_{CO_2}$ values are obtained from a corresponding general formula by Heisler (Chapter 6), shown in Fig. 1E. The generalized formula of Heisler for calculation of $pK_{app}$ is

$$pK_{app} = 6.583 - 13.41 \times 10^{-3}(T) + 228.2 \times 10^{-6}(T^2) - 1.516 \times 10^{-6}(T^3) - 0.341I^{0.323}$$
$$- \log\{1 + 0.00039[Pr] + 10^{pH} - 10.64 + 0.011T + 0.737I^{0.323}$$
$$\times (1 + 10^{1.92 - 0.01T} - 0.737I^{0.323} + \log[Na^+] + (-0.494I + 0.651)(1 + 0.0065[Pr]))\}$$

where $T$ is the temperature (°C, 0–40°), $I$ is the ionic strength of nonprotein ions, $[Na^+]$ is sodium concentration in mol liter$^{-1}$, and $[Pr]$ is protein concentration in g liter$^{-1}$.

## B. Oxygen Solubility of Plasma

Scarcely any measurements are available for the solubility of $O_2$ ($\alpha_{O_2}$) in the plasma of fishes. Two notable exceptions are the studies on flounder by Wood *et al.* (1979a,b) and that on the Antarctic icefish (Ruud, 1954). In both instances the total $O_2$ capacity of plasma (or blood lacking hemoglobin in the case of the icefish) was determined at a single temperature. Detailed studies on the $\alpha_{O_2}$ of human blood and plasma as a function of temperature have been performed (Christoforides and Hedley-Whyte, 1969, 1976; Christoforides *et al.*, 1969; Roughton and Severinghaus, 1973), and some of these data are detailed in Table II along with comparative values for fish.

## III. PHYSICOCHEMICAL PROPERTIES OF WATER

Steady-state conditions seldom exist in natural waters because of dynamic interactions between the water body and surrounding systems. In addition, no two bodies of water are similar because of geologically based influences and those that are a consequence of biotic factors including industrial and recreational usage by humans. Given these manifold interactions of the water system, it is not possible to ascribe a single set of constants that will apply with absolute accuracy to any particular aquatic environment. More recently, the convention has been to describe water systems such as distilled water, seawater of defined chlorinity and salinity, or artifically prepared seawater, in attempts to arrive at some standard conditions (see Skirrow, 1975). The following data are therefore defined within a much more narrow range of boundary conditions than can be expected in nature.

## A. Carbon Dioxide Solubility of Water

Most determinations of $\alpha_{CO_2}$ in water are made using solutions that have been acidified so that the measured total $CO_2$ approximates to $C_{CO2(aqueous)}$

## Table II

Oxygen Solubility of Human Plasma ($\alpha_{O_2}$, in $\mu$mol liter$^{-1}$ torr$^{-1}$) as a Function of Temperature[a,b]

| Temperature (°C) | $\alpha_{O_2}$ | Temperature (°C) | $\alpha_{O_2}$ | Temperature (°C) | $\alpha_{O_2}$ |
|---|---|---|---|---|---|
| 0 | 2.5854 | 14 | 1.8098 | 28 | 1.4455 |
| 1 | 2.5149 | 15 | 1.7745 | 29 | 1.4220 |
| 2 | 2.4502 | 16 | 1.7393 | 30 | 1.3985 |
| 3 | 2.3856 | 17 | 1.7040 | 31 | 1.3750 |
| 4 | 2.3210 | 18 | 1.6746 | 32 | 1.3515 |
| 5 | 2.2622 | 19 | 1.6511 | 33 | 1.3280 |
| 6 | 2.2035 | 20 | 1.6276 | 34 | 1.3103 |
| 7 | 2.1447 | 21 | 1.6041 | 35 | 1.2927 |
| 8 | 2.0859 | 22 | 1.5806 | 36 | 1.2751 |
| 9 | 2.0331 | 23 | 1.5571 | 37 | 1.2574 |
| 10 | 1.9861 | 24 | 1.5336 | 38 | 1.2457 |
| 11 | 1.9390 | 25 | 1.5101 | 39 | 1.2339 |
| 12 | 1.8920 | 26 | 1.4866 | 40 | 1.2222 |
| 13 | 1.8509 | 27 | 1.4631 | | |

[a]Experimental measurements were performed using two methods. In the direct method, hemoglobin-free plasma samples (of known protein concentration) were equilibrated with analyzed 99.99% $O_2$ in tonometer flasks thermostatted at 18, 28, and 37°C. Samples of 5 ml were withdrawn and transferred anaerobically into the extraction chamber of the Van Slyke apparatus (Van Slyke and Neill, 1924) for measurement. By the indirect method, plasma samples were equilibrated with room air at temperatures ranging from 18 to 40°C and then warmed or cooled to 38°C in a sealed syringe followed by measurement for partial pressure of oxygen (Radiometer electrode). In both methods, water samples were treated in the same fashion. These measurements revealed human plasma to have a constant relative solubility with respect to water (i.e., plasma $\alpha_{O_2}$:water $\alpha_{O_2}$ is constant). On this basis, Christoforides et al. (1969) interpolated plasma $\alpha_{O_2}$ values for the 0–40°C temperature range. Their values were expressed as Bunsen solubility coefficients (i.e., milliliters of gas reduced to standard temperature and pressure that is dissolved at complete equilibrium in 1.0 ml of liquid exposed to 1 conventional atm). We transformed the Bunsen coefficients to $\mu$mol liter$^{-1}$ torr$^{-1}$ using the equation, $\alpha_{O_2}$ ($\mu$mol liter$^{-1}$ torr$^{-1}$) = Bunsen coefficient/22.393 × 0.00076 (Van Slyke et al., 1928), where 22.393 equals the volume (in liters STPD) occupied by 1 mol of oxygen (Radford, 1964). Subsequent measurements of $O_2$ solubility coefficients of human blood revealed that $\alpha_{O_2}$ blood:$\alpha_{O_2}$ water is also constant for blood of a given hemoglobin concentration (Christoforides and Hedley-Whyte, 1969, 1976); values for 0–40°C were interpolated based on actual blood and water measurements at 0, 6, 12, 18, 28, and 37°C and are tabulated therein.

(continued)

**Table II** *(Continued)*

Plasma $\alpha_{O_2}$ values (of 2.048 $\mu$mol liter$^{-1}$ torr$^{-1}$ at 9°C) for the starry flounder, *Platichthys stellatus*, were interpreted from the studies by Wood *et al.* (1979a,b), who equilibrated plasma samples with room air and measured the resulting total $O_2$ content in volumes percent (Lex-$O_2$-Con Apparatus, especially calibrated for low $O_2$ contents). Ruud (1954) similarly determined the total $O_2$ content of blood lacking hemoglobin from the Antarctic icefish, *Chaenocephalus aceratus*, using the microgasometric method of Roughton and Scholander (1943). Ruud (1954) found the $\alpha_{O_2}$ of that blood to be 1.930 $\mu$mol liter$^{-1}$ torr$^{-1}$. The basic assumptions used in converting the volumes percent measurements to $\alpha_{O_2}$ (in $\mu$mol liter$^{-1}$ torr$^{-1}$) has been a room temperature of 20°C, a value of 22.393 (liters STPD) for 1 mol of oxygen (Radford, 1964), an equilibration gas tension of 155 torr $P_{O_2}$, and a cubic volume expansion factor of 0.0003 for plasma and for blood lacking hemoglobin.

[b]After Christoforides *et al.* (1969).

**Table III**

Values for the $CO_2$ Solubility of Water ($\alpha_{CO_2}$, mmol liter$^{-1}$ torr$^{-1}$) as a Function of Temperature and Salinity[a,b]

| Temperature (°C) | Salinity (‰) | | | | | |
|---|---|---|---|---|---|---|
| | 0 | 10 | 20 | 30 | 35 | 40 |
| −1 | — | — | 0.09570 | 0.09083 | 0.08847 | 0.08620 |
| 0 | 0.10208 | 0.09689 | 0.09197 | 0.08730 | 0.08507 | 0.08287 |
| 1 | 0.09813 | 0.09317 | 0.08846 | 0.08397 | 0.08183 | 0.07974 |
| 2 | 0.09439 | 0.08964 | 0.08512 | 0.08083 | 0.07876 | 0.07675 |
| 3 | 0.09086 | 0.08629 | 0.08196 | 0.07784 | 0.07587 | 0.07393 |
| 4 | 0.08750 | 0.08312 | 0.07896 | 0.07501 | 0.07312 | 0.07126 |
| 5 | 0.08432 | 0.08011 | 0.07612 | 0.07233 | 0.07050 | 0.06872 |
| 6 | 0.08129 | 0.07725 | 0.07342 | 0.06978 | 0.06803 | 0.06632 |
| 8 | 0.07567 | 0.07196 | 0.06842 | 0.06507 | 0.06345 | 0.06187 |
| 10 | 0.07061 | 0.06717 | 0.06391 | 0.06080 | 0.05930 | 0.05784 |
| 12 | 0.06601 | 0.06284 | 0.05982 | 0.05693 | 0.05555 | 0.05420 |
| 14 | 0.06184 | 0.05891 | 0.05611 | 0.05345 | 0.05216 | 0.05091 |
| 16 | 0.05805 | 0.05533 | 0.05274 | 0.05026 | 0.04907 | 0.04791 |
| 18 | 0.05459 | 0.05208 | 0.04967 | 0.04737 | 0.04626 | 0.04518 |
| 20 | 0.05145 | 0.04911 | 0.04687 | 0.04474 | 0.04371 | 0.04270 |
| 22 | 0.04857 | 0.04639 | 0.04432 | 0.04233 | 0.04137 | 0.04043 |
| 24 | 0.04593 | 0.04391 | 0.04197 | 0.04013 | 0.03924 | 0.03836 |
| 26 | 0.04351 | 0.04163 | 0.03983 | 0.03812 | 0.03728 | 0.03646 |
| 28 | 0.04129 | 0.03954 | 0.03787 | 0.03626 | 0.03549 | 0.03472 |
| 30 | 0.03925 | 0.03762 | 0.03607 | 0.03457 | 0.03384 | 0.03313 |
| 32 | 0.03737 | 0.03586 | 0.03441 | 0.03301 | 0.03233 | 0.03167 |

Table III (Continued)

| Temperature (°C) | Salinity (‰) | | | | | |
|---|---|---|---|---|---|---|
| | 0 | 10 | 20 | 30 | 35 | 40 |
| 34 | 0.03563 | 0.03422 | 0.03287 | 0.03158 | 0.03095 | 0.03033 |
| 36 | 0.03404 | 0.03272 | 0.03146 | 0.03025 | 0.02966 | 0.02909 |
| 38 | 0.03255 | 0.03134 | 0.03016 | 0.02904 | 0.02849 | 0.02795 |
| 40 | 0.03118 | 0.03005 | 0.02896 | 0.02791 | 0.02739 | 0.02689 |

[a]The original data set was experimentally determined by Murray and Riley (1971), who measured $\alpha_{CO_2}$ at eight temperatures over the 1–35°C range in both distilled water and acidified seawater of five different salinities (up to 38‰). The seawater was collected from the Atlantic, south of the Canary Isles (surface water of salinity 36‰), and immediately filtered through a 0.45-μm filter. Measurements were made by stripping carbon dioxide from a precisely known volume of $CO_2$-saturated water and determining the liberated carbon dioxide gravimetrically (see Murray and Riley, 1971, for additional details of saturation and degassing/absorption apparatus). Mathematical treatment of the measured values was used to construct comprehensive tables by interpolation of intermediate values (Murray and Riley, 1971). In a subsequent article, Weiss (1974) pointed out that Murray and Riley (1971) did not acidify their distilled water samples (for $\alpha_{CO_2}$ determination) and also did not allow for nonideal behavior in the gas phase. Accordingly, Weiss (1974) corrected the original data set of Murray and Riley (1971) and expressed $\alpha_{CO_2}$ values as $10^{-2}$ mol liter$^{-1}$ atm$^{-1}$ (in Skirrow, 1975) via the equation

$$\ell n\ \alpha_{CO_2} = -58.0931 + 90.5069(100/T_k) + 22.2940\ \ell n(T_k/100)$$
$$+ (0.027766 - 0.025888(T_k/100) + 0.0050578(T_k/100)^2)S$$

where $T_k$ is temperature in degrees Kelvin (°C + 273.15) and $S$ is the salinity in parts per thousand. We transformed the tabulated data of $\alpha_{CO_2}$ (given by this equation as $10^{-2}$ mol liter$^{-1}$ atm$^{-1}$) in Table A9.26 of Skirrow (1975) to the units, mmol liter$^{-1}$ torr$^{-1}$, for presentation herein.

[b]Data were obtained from tabulated values in Riley and Skirrow (1975).

$+ C_{H_2CO_3} = C_{CO_2(total)}$ (see Skirrow, 1975, for discussion). Of the several sets of distilled water and saltwater data reported (Lit. in Skirrow, 1975), the comparatively recent gravimetric determinations by Murray and Riley (1971), with refinements by Weiss (1974), are shown in Table III. For comparative purposes, the distilled water and full-strength seawater values in Table III have been plotted in Fig. 4 along with measurements made by manometric and gasometric techniques (Austin et al., 1963; Randall et al., 1976; Boutilier et al., 1984; see Table IV).

Note also that the generalized formula by Heisler (Fig. 1E) is valid for pure water and various salt concentrations up to the concentration of salt in seawater (Fig. 4).

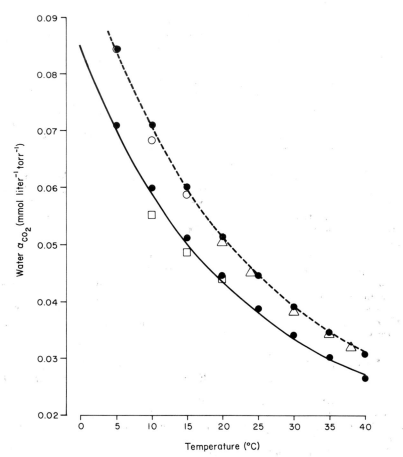

**Fig. 4.** Carbon dioxide solubility of pure water and seawater as a function of temperature. (A) $\alpha_{CO_2}$ in distilled water (broken line) and 35‰ salinity seawater (solid line), as determined gravimetrically by Murray and Riley (1971), revised by Weiss (1974), and given in this chapter in Table III (in mmol liter$^{-1}$ torr$^{-1}$). See footnote to Table III for additional details. (B) (glyph ◯) $\alpha_{CO_2}$ in distilled water samples that were acidified (to pH $\simeq 2.5$) and then equilibrated with 40% $CO_2$ + 20.09% $O_2$—balance $N_2$ (Wösthoff pumps) at 5, 10, and 15°C (Boutilier et al., 1984). The same methodology was used as in Fig. 1C. (C) (glyph △) $\alpha_{CO_2}$ of distilled water samples acidified with hydrochloric acid to a concentration of 0.1 N, and equilibrated with carbon dioxide (mean concentration of 99.7% as determined by manometric method of Peters and Van Slyke, 1932) at five temperatures in the 20–38°C range. Total $CO_2$ contents after 30 to 60 min were determined using the manometric Van Slyke apparatus with a 50-ml chamber (Austin et al., 1963). (D) (glyph ☐) $\alpha_{CO_2}$ of seawater ($\simeq 35$‰ salinity) collected in the Bay of Naples and equilibrated with $CO_2$ at 10, 15, and 20°C. Total $CO_2$ contents determined by Van Slyke apparatus (Randall et al., 1976). See Table IV for additional details. (E) $\alpha_{CO_2}$ values for distilled water and seawater generated from the formula of Heisler (see Fig. 1E) using 0 and 0.5580 respectively for the molarity of dissolved species ([M]) in moles per liter.

**Table IV**

Values for $\alpha_{CO_2}$ f $(T)$ and p$K_{app}$ $f$ (pH, $T$) of Seawater ($\approx$35‰
Salinity) Samples Collected in the Bay of Naples, Italy, in the
Spring of 1973[a]

| Temperature ($^\circ$C) | $CO_2$ solubility (mmol liter$^{-1}$ torr$^{-1}$) | Apparent p$K$ as a function of pH |
|---|---|---|
| 10 | 0.0553 ± 0.001 | −0.0983 × pH + 6.72 |
| 15 | 0.0487 ± 0.002 | −0.0979 × pH + 6.68 |
| 20 | 0.0414 ± 0.001 | −0.1003 × pH + 6.67 |

[a]The samples were transported frozen to the Max-Planck-Institut
(Göttingen, Federal Republic of Germany), for measurements by Randall *et al.* (1976). Samples of seawater were equilibrated at several $P_{CO_2}$
levels and then measured for pH, $P_{CO_2}$, and total $CO_2$ by glass electrode, Radiometer gas electrodes, and Van Slyke apparatus, respectively. The apparent p$K$ was then estimated by appropriate rearrangement of the Henderson–Hasselbalch Equation using the
corresponding determinations of seawater $CO_2$ solubility ($\alpha_{CO_2}$). The
nonbicarbonate buffer value of this seawater was less than 0.2 mEq
liter$^{-1}$ pH$^{-1}$; [Na$^+$] = 524 ± 4.5 mEq liter$^{-1}$; [K$^+$] = 11 ± 0.4 mEq
liter$^{-1}$; [Phosphate]$_{inorg}$ = 0.32 ± 0.04 mmol liter$^{-1}$; [Phosphate]$_{org}$
= <0.2 mmol liter$^{-1}$ (means ± 1 SD throughout; Randall *et al.*, 1976;
D. J. Randall and N. Heisler, personal communication).

## B. pH Measurements and p$K$ of Carbonic Acid in Water

On the basis of fundamental thermodynamic properties, there are many
limitations associated with the measurement of pH, one practical example
being the large differences often occurring between the ionic strength of the
unknown and standard buffer solutions (Skirrow, 1975; see also Chapter 6,
this volume). Although rigorous theoretical interpretation is considered by
some to be possible only for dilute aqueous solution of buffers and simple salt
solutions with ionic strength ($I$, mol liter$^{-1}$) of 0.1 or less (Bates, 1973), it is
possible to make reproducible measurements of seawater ($I$ = 0.70) for use
with correspondingly determined apparent dissociation constants of carbonic
acid. This was carried out by Randall *et al.* (1976) for seawater samples
collected in the Bay of Naples, Italy (Table IV). Since that time, Heisler
(personal communication) has made similar measurements of Naples seawater on three different occasions, using improved methodology for the measurement of pH in seawater (see Chapter 6). These more recent measure-

Table V

Solubility of Oxygen in Water ($\alpha_{O_2}$ in $\mu$mol liter$^{-1}$ torr$^{-1}$) as a Function of of Temperature and Salinity[a]

| Temperature (°C) | Salinity (‰) | | | | | |
|---|---|---|---|---|---|---|
| | 0 | 10 | 20 | 30 | 35 | 40 |
| −1 | — | — | 2.5837 | 2.4117 | 2.3299 | 2.2509 |
| 0 | 2.8839 | 2.6920 | 2.5142 | 2.3477 | 2.2716 | 2.1954 |
| 1 | 2.8060 | 2.6197 | 2.4503 | 2.2894 | 2.2132 | 2.1398 |
| 2 | 2.7312 | 2.5533 | 2.3866 | 2.2313 | 2.1579 | 2.0872 |
| 3 | 2.6592 | 2.4868 | 2.3257 | 2.1760 | 2.1053 | 2.0347 |
| 4 | 2.5899 | 2.4231 | 2.2676 | 2.1234 | 2.0527 | 1.9848 |
| 5 | 2.5262 | 2.3650 | 2.2150 | 2.0736 | 2.0057 | 1.9406 |
| 6 | 2.4628 | 2.3071 | 2.1627 | 2.0240 | 1.9589 | 1.8966 |
| 8 | 2.3470 | 2.1996 | 2.0636 | 1.9332 | 1.8736 | 1.8141 |
| 10 | 2.2400 | 2.1038 | 1.9732 | 1.8511 | 1.7943 | 1.7375 |
| 12 | 2.1415 | 2.0136 | 1.8913 | 1.7747 | 1.7206 | 1.6666 |
| 14 | 2.0518 | 1.9293 | 1.8153 | 1.7070 | 1.6528 | 1.6044 |
| 16 | 1.9705 | 1.8534 | 1.7449 | 1.6421 | 1.5936 | 1.5450 |
| 18 | 1.8925 | 1.7837 | 1.6806 | 1.5833 | 1.5346 | 1.4888 |
| 20 | 1.8230 | 1.7196 | 1.6191 | 1.5273 | 1.4842 | 1.4383 |
| 22 | 1.7595 | 1.6587 | 1.5666 | 1.4773 | 1.4341 | 1.3938 |
| 24 | 1.6993 | 1.6039 | 1.5143 | 1.4305 | 1.3900 | 1.3496 |
| 26 | 1.6419 | 1.5520 | 1.4679 | 1.3866 | 1.3489 | 1.3083 |
| 28 | 1.5906 | 1.5061 | 1.4246 | 1.3459 | 1.3109 | 1.2702 |
| 30 | 1.5456 | 1.4607 | 1.3846 | 1.3085 | 1.2733 | 1.2353 |
| 32 | 1.5009 | 1.4215 | 1.3479 | 1.2743 | 1.2419 | 1.2066 |
| 34 | 1.4593 | 1.3853 | 1.3113 | 1.2433 | 1.2107 | 1.1781 |
| 36 | 1.4242 | 1.3497 | 1.2812 | 1.2157 | 1.1829 | 1.1531 |
| 38 | 1.3894 | 1.3174 | 1.2514 | 1.1884 | 1.1584 | 1.1284 |
| 40 | 1.3581 | 1.2886 | 1.2251 | 1.1646 | 1.1343 | 1.1071 |

[a]These data are based on original measurements by Carpenter (1966) and Murray and Riley (1969). Carpenter (1966) made his measurements on distilled water and seawater (0–35.6‰ Cl), using a modified Winkler method, and observed that the $O_2$ solubility was not a linear function of chlorinity as had been earlier assumed. Carpenter's values were corroborated by Murray and Riley (1969), who made measurements using both chemical (modified Winkler technique) and physical (gasometric apparatus) methodology. Both data sets were fitted by Weiss (1970) to a thermodynamically consistent equation. Tabulated values, generated by this equation (Riley and Skirrow, 1975), were transformed from cm$^3$ dm$^{-3}$ atm$^{-1}$ (at 20.95% $O_2$, 100% relative humidity, and 760 torr) to $\mu$mol liter$^{-1}$ torr$^{-1}$ by the formula

$$\alpha_{O_2} \ (\mu\text{mol liter}^{-1} \text{ torr}^{-1}) = \frac{(AP_{corr}\, f_{O_2}^{-1})(\text{ml } O_2 \text{ liter}^{-1})}{R \times 0.76} \quad 1000$$

where $AP_{corr}$ is the atmospheric pressure (in torr), corrected for water vapor pressure at the appropriate temperature, $f_{O_2}$ the fractional $O_2$ concentration (in torr $P_{O_2}$), and $R$ the volume of oxygen (22.393 liters) per mol STPD (Radford, 1964).

ments have been incorporated into the generalized $pK_{app}$ formula of Heisler shown on page 414 of this chapter. This formula is valid for distilled water and various saltwater concentrations up to and including full strength seawater.

A procedure that avoids conventional pH measurement (where $I$ of standard solution is dilute with respect to test solution) has been proposed by Hansson (1972), where organic buffers [e.g., tris(hydroxymethyl)aminomethane plus its hydrochloride] are made up in synthetic seawater (see Skirrow, 1975). The hydrogen ion activity scale employed by Hansson (1972) is not based on the infinitely dilute convention of physicochemical investigations; rather, its expressed purpose is that of determining the acidity constants of carbonic acid. For this purpose, the activity scale is such that the total hydrogen ion concentration is effectively the same as its activity over the range of $H^+$ concentrations used. This is not the case for the National Bureau of Standards (NBS) scale. Skirrow (1975) points out that "for this reason, pH values based on the Hansson scale have a more distinct meaning than do those determined on, say, the NBS scale." Consequently, several sets of tabulated data for the $pK_1'$ and $pK_2'$ of carbonic acid in water as a function of chlorinity (salinity) and temperature have been reported (see Riley and Skirrow, 1975); based on the NBS pH scale (e.g., Lyman, 1956) and on the Hansson pH scale (Hansson, 1972).

## C. Oxygen Solubility of Water

Oxygen solubilities of distilled water and natural water are found in Table V, along with the conditions of measurement and treatment of the data.

## IV. AMMONIA

### A. Ammonia Solubility Values: Water and Plasma

Cameron and Heisler (1983) have determined the ammonia solubilities of water and rainbow trout plasma as a function of temperature, using the Bunsen coefficients of Washburn (1928) and experimentally determined relative solubilities in various solutions at 15°C. The relative solubility of ammonia in a 10-mmol $NH_4Cl$ liter$^{-1}$ distilled water solution was taken as the 100% value. Relative solubilities were then determined for this ammonia-containing distilled water solution made up to 150 or 500 m$M$ NaCl and for the plasma of *Salmo gairdneri*. Measurements of the partial pressure of ammonia gas in the solutions ($P_{NH_3}$) were made using a HNU Systems ammonia electrode.

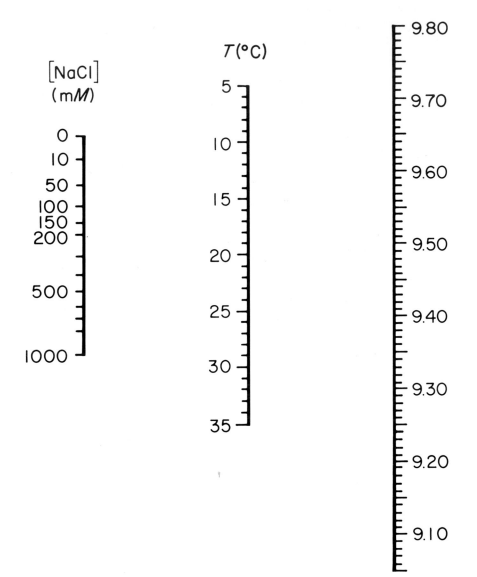

**Fig. 5.** Nomogram for determination of pK for ammonia ($pK_{amm}$) in water. Straight lines connecting known sodium chloride concentration and temperature lead to the respective $pK_{amm}$ (from Cameron and Heisler, 1983). Plasma from *Salmo gairdneri* at 15°C behaves as a solution whose ionic strength is equivalent to 370 mM NaCl. Experimental determination of $pK_{amm}$ was carried out on the following solutions: 10 mM $NH_4Cl$, 10 mM $NH_4Cl$–150 mM NaCl, 10 mM $NH_4Cl$–500 mM NaCl, and in plasma of rainbow trout. Note that zero NaCl point on the nomogram refers to a solution of 10 mmol liter$^{-1}$ of $NH_4Cl$ in distilled water. Procedure was to fill a thermostatted 17.5-ml cuvet with the solution and then titrate the fluid with 1 N

Using the tabulated values for ammonia solubility in water and rainbow trout plasma as determined by Cameron and Heisler (1983), we have constructed polynomial equations that accurately describe the solubility of ammonia as a function of temperature ($T$, 0–30°C range).

$$\text{Water } \alpha_{NH_3} = 66.0930 - 1.4738(T) + 0.0234(T^2)$$
$$- 0.0006(T^3) + 8.3621 \times 10^{-6}(T^4)$$

$$\text{Plasma } \alpha_{NH_3} = 68.3400 - 1.5244(T) + 0.0243(T^2)$$
$$- 0.0007(T^3) + 8.6419 \times 10^{-6}(T^4)$$

The coefficient of correlation ($r$) of each equation equals 1.0000.

## B. Ammonia Equilibrium p$K$ and p$K'$

A number of workers have concerned themselves with the movement of ammonia:nitrogen ($NH_3$ gas and the conjugate acid, $NH_4^+$) across the gills of fishes. To quantify the relationship between $NH_3$ and $NH_4^+$ movement across biological membranes, an exact value for the p$K'$ of the ammonia equilibrium (p$K'_{amm}$) is required at the prevailing temperature and ionic strength of the fluid under investigation (Fig. 5, Table VI; see review by Kormanik and Cameron, 1981).

### 1. Ion Effects on p$K$

The equilibrium condition of an acid–base reaction, and hence its equilibrium constant, is influenced by the ionic strength of the solution. The magnitude of this ion effect can be estimated using formulations derived from the Debye–Hückel theory (Stumm and Morgan, 1970) or more empirically derived expressions. The operational equilibrium constant ($K'$) is related to $K$, the constant valid when ionic strength is zero (infinite dilution), and the appropriate activity coefficients ($f$) as follows:

$$K' = K\frac{f\text{HB}}{f\text{B}}$$

where B is any base of any charge, and HB is the corresponding acid. In dilute solutions (ionic strength $<0.1$ $M$) of a simple electrolyte, ion activities

---

**Fig. 5.** (*Continued*)

NaOH and/or 1 $N$ HCl within the 7.5–11.5 pH range. The cuvet was fitted with a pH electrode (Ingold Type 401; or Ingold Type U402-S7 for higher ionic strengths) and a HNU Systems ammonia electrode. The p$K_{amm}$ was estimated from the pH/$P_{NH_3}$ data pairs and titrant volumes. See Cameron and Heisler (1983) for further details on procedure and plasma sample preparation.

**Table VI**

pK Values for Ammonia in Aqueous Solutions of Zero
Salinity at Temperatures 0–30°C[a]

| Temperature (°C) | $pK_{amm}$ | Temperature (°C) | $pK_{amm}$ |
|:---:|:---:|:---:|:---:|
| 0 | 10.0826 | 16 | 9.5297 |
| 1 | 10.0461 | 17 | 9.4972 |
| 2 | 10.0099 | 18 | 9.4649 |
| 3 | 9.9740 | 19 | 9.4328 |
| 4 | 9.9384 | 20 | 9.4010 |
| 5 | 9.9030 | 21 | 9.3693 |
| 6 | 9.8678 | 22 | 9.3379 |
| 7 | 9.8329 | 23 | 9.3067 |
| 8 | 9.7983 | 24 | 9.2757 |
| 9 | 9.7639 | 25 | 9.2448 |
| 10 | 9.7297 | 26 | 9.2143 |
| 11 | 9.6958 | 27 | 9.1839 |
| 12 | 9.6621 | 28 | 9.1537 |
| 13 | 9.6287 | 29 | 9.1237 |
| 14 | 9.5955 | 30 | 9.0939 |
| 15 | 9.5625 | | |

[a]Values significant only to three figures, but are given by
Emerson *et al.* (1975) to four figures, to minimize round-off
errors in subsequent calculations. These data were compiled
after critical evaluation of the literature on the ammonia
equilibrium in water, and are the result of new calculations
made by Thurston *et al.* (1974) and Emerson *et al.* (1975) on
the previously reported data set of Bates and Pinching
(1949, 1950). Bates and Pinching determined $pK_{amm}$ at 5°C
intervals from 0 to 50°C, and interpolated $pK_{amm}$ for intermediate temperatures using calculated coefficients for two
equation forms (suggested earlier by Harned and Robinson,
1940). Emerson *et al.* (1975) found that the coefficients reported by Bates and Pinching were slightly in error and
attributed this to the determination of values very close to
zero being greatly magnified by comparatively small round-
ing-off errors in calculations. The values given here were
calculated by Emerson *et al.* (1975) using double-precision
maths ($\simeq$15 significant figures) with the equation

$$pK_{amm} = 0.09018 + 2729.92/T_k$$

where $T_k$ is the temperature in degrees Kelvin ($T_k = °C +
273.15$). Because this equation was empirically determined,
Emerson *et al.* state that "extrapolations to temperatures
above 50°C or below 0°C should not be made." Calculations

**Table VI** (*Continued*)

of the fraction of $NH_3$ ($f_{NH_3}$) using these values of $pK_{amm}$ can be mathematically obtained by

$$f_{NH_3} = 1/(10^{pK_{amm}-pH} + 1)$$

Tabulated values of this fraction, expressed as percentages, can be found in Emerson *et al.* (1975) for 1°C increments (0–30°C range) and 0.5 pH Unit intervals (6–10 pH range). For a comprehensive table of $f_{NH_3}$ (0–40°C, 0.2°C increments; pH 5.00–12.00, 0.01-Unit increments), see Thurston *et al.* (1979).

can be calculated adequately from theoretical expressions of the Debye–Hückel limiting law (see Stumm and Morgan, 1970, for further details). However, limiting laws for ion activities cannot be applied satisfactorily to electrolyte mixtures of unlike charge types, such as blood plasma. In complex electrolyte mixtures with ionic strengths of 0.1 to 0.5 $M$, the Davies approximation for ion activity is most applicable (Stumm and Morgan, 1970). Using this empirically derived approximation, the operational equilibrium constant ($pK'$) can be calculated from the constant valid when ionic strength is zero ($pK$):

$$pK' = pK + A \left( Z\frac{2}{HB} - Z\frac{2}{B} \right)\left( \frac{I}{1 + \sqrt{I}} - 0.2I \right)$$

where $A = 1.82 \times 10^6(\epsilon T)^{-1.5}$, $\epsilon$ is the dielectric constant of water at 1 atm (see *Handbook of Chemistry and Physics, 1979–1980*), $T$ is temperature in degrees Kelvin (273.15 + °C), $I$ (ionic strength) = 0.5 $\Sigma C_i Z_i^2$, $C$ = molar

**Table VII**

Comparison of the Ammonia $pK'$ Values ($pK'_{amm}$) Calculated Using the Davies Approximation Formula with Experimentally Determined $pK'_{amm}$ Values in Various Solutions[a]

| Solution | Measured $pK'_{amm}$ | Calculated $pK'_{amm}$ |
|---|---|---|
| 150 m$M$ NaCl, 15°C | 9.576[b] | 9.604 |
| 500 m$M$ NaCl, 15°C | 9.677[b] | 9.661 |
| Trout plasma, 15°C | 9.648[b] | 9.605 |
| Mammalian plasma, 37°C | 9.150[c] | 8.933 |

[a]See Section IV,B for details. Deviations between measured and calculated $pK'_{amm}$ values are greatest for the plasma.
[b]Data of Cameron and Heisler (1983).
[c]Data of Bromberg *et al.* (1960).

concentration of each ion in solution, Z = charge. Application of this equation, using the pK values for the ammonia equilibrium in infintely dilute solutions (Table VI; see Emerson *et al.*, 1975), compares the pK' calculated $(pK'_{amm})$ with that measured $(pK'_{amm})$ in NaCl solutions and plasma in Table VII.

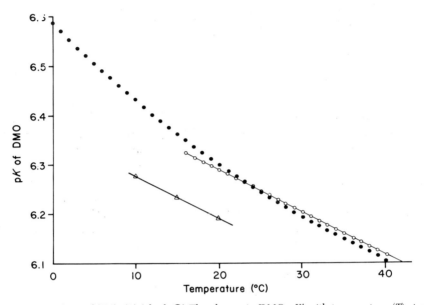

**Fig. 6.** $pK'_{DMO}$ f $(T,I)$. (A) (glyph ●) The change in DMO pK' with temperature $(T)$ at an ionic strength $(I)$ of 0.125. Data of Dr. E. D. Robin, Stanford University Medical Center (Stanford, California), together with additional titration determinations at 25 and 37°C, were compiled by Malan *et al.* (1976) and described mathematically for the 0–40°C range as

$$pK'_{DMO} = 19.941 - 0.0811T_k + 0.0001179T_k^2$$

where $T_k$ is temperature in degrees Kelvin ($T_k$ = °C + 273.15). (B) (glyph ○) $pK'_{DMO}$ at temperatures between 18 and 44°C determined electrometrically by NaOH titration of a solution containing 20 μmol DMO in 20 ml 0.9% NaCl. Regression equation of 15 determinations given as

$$pK'_{DMO} = 6.464 - 0.00874T$$

where $T$ is temperature in °C (from Albers *et al.*, 1971). (C) (glyph △) pK' values of DMO determined photometrically in solutions having the same ionic strength as dogfish plasma ($I$ = 0.3), at 10, 15, and 20°C. Linear regression analysis of the mean measurements reported in Heisler *et al.* (1976) gives the following equation:

$$pK'_{DMO} = 6.3605 - 0.0085T$$

where $T$ is temperature in °C.

## V. pK' of DMO

As in many groups of animals, estimates of intracellular pH in tissues of fish are made possible by measurements of the distribution, between plasma and cells, of an injected quantity of the weak acid, 5,5-dimethyl-2,4-ox-azolidinedione: the DMO method (Waddell and Butler, 1959). Computation of an intracellular pH requires several independently determined experimental quantities, in addition to the dissociation constant of DMO for the prevailing experimental conditions. pK' values for DMO as a function of temperature have been determined in solutions having similar ionic strengths to teleost and elasmobranch plasma; these are illustrated in Fig. 6 along with mathematical descriptions of the data.

### ACKNOWLEDGMENTS

We express our appreciation to Dr. D. J. Randall for his invaluable advice and assistance throughout the preparation of this appendix, and to Drs. J. N. Cameron, N. Heisler, and C. M. Wood for allowing us access to unpublished data.

### REFERENCES

Albers, C. (1970). Acid base balance. In "Fish Physiology" (W. S. Hoar and D. J. Randall, eds.), Vol. 4, p. 173. Academic Press, New York.

Albers, C., and Pleschka, K. (1967). Effect of temperature on $CO_2$ transport in elasmobranch blood. *Respir. Physiol.* **2,** 261–273.

Albers, C., Usinger, W., and Spaich, P. (1971). Effect of temperature on the intracellular $CO_2$ dissociation curve and pH. *Respir. Physiol.* **11,** 211–222.

Austin, W. H., Lacombe, E., Rand, P. W., and Chatterjee, M. (1963). Solubility of carbon dioxide in serum from 15 to 38 C. *J. Appl. Physiol.* **18,** 301–304.

Bartels, H., and Wrbitzky, R. (1960). Bestimmung des $CO_2$-Absorptionskoeffizienten zwischen 15 und 38°C in wasser und plasma. *Pfluegers Arch. Gesamte Physiol. Menschen Tiere* **271,** 162–168.

Bateman, N. T., Musch, T. I., Smith, C. A., and Dempsey, J. A. (1980). Problems with the gas-calibrated $P_{CO_2}$ electrode. *Respir. Physiol.* **41,** 217–226.

Bates, R. G. (1973). "Determination of pH; Theory and Practise." Wiley, New York.

Bates, R. G., and Pinching, G. D. (1949). Acidic dissociation constant of ammonium ion at 0° to 50°C, and the base strength of ammonia. *J. Res. Natl. Bur. Stand. (U.S.)* **42,** 419–430.

Bates, R. G., and Pinching, G. D. (1950). Dissociation constant of aqueous ammonia at 0° to 50° from e.m.f. studies of the ammonium salt of a weak acid. *J. Am. Chem. Soc.* **72,** 1393–1396.

Boutilier, R. G., Randall, D. J., Shelton, G., and Toews, D. P. (1978). Some response characteristics of $CO_2$ electrodes. *Respir. Physiol.* **32,** 381–388.

Boutilier, R. G., Iwama, G. K., Heming, T. A., and Randall, D. J. (1984). Acid-base nomogram and $CO_2$ solubility of rainbow trout plasma at 5–15°C. (In preparation).

Bromberg, P. A., Robin, E. D., and Forkner, C. E., Jr. (1960). The existence of ammonia in blood *in vivo* with observations on the significance of the $NH_4^+ - NH_3$ system. *J. Clin. Invest.* **39**, 332–341.

Cameron, J. N. (1971). Rapid method for determination of total carbon dioxide in small blood samples. *J. Appl. Physiol.* **31**, 632–634.

Cameron, J. N., and Heisler, N. (1983). Studies of ammonia in rainbow trout: Physico-chemical parameters, acid-base behaviour and respiratory clearance. *J. Exp. Biol.* **105**, 107–125.

Carpenter, J. H. (1966). New measurements of oxygen solubility in pure and natural water. *Limnol. Oceanogr.* **11**, 264–277.

Christoforides, C., and Hedley-Whyte, J. (1969). Effect of temperature and hemoglobin concentration on solubility of $O_2$ in blood. *J. Appl. Physiol.* **27**, 592–596.

Christoforides, C., and Hedley-Whyte, J. (1976). Solubility of $O_2$ in blood with different hemoglobin concentrations at low temperatures. *J. Appl. Physiol.* **40**, 815–818.

Christoforides, C., Laasberg, L. H., and Hedley-Whyte, J. (1969). Effect of temperature on solubility of $O_2$ in human plasma. *J. Appl. Physiol.* **26**, 56–60.

DeLaney, R. G., Lahiri, S., Hamilton, R., and Fishman, A. P. (1977). Acid-base balance and plasma composition in the aestivating lungfish (*Protopterus*). *Am. J. Physiol.* **232**, R10–R17.

Emerson, K., Russo, R. C., Lund, R. E., and Thurston, R. V. (1975). Aqueous ammonia equilibrium calculations: Effect of pH and temperature. *J. Fish. Res. Board Can.* **32**, 2379–2383.

Handbook of Chemistry and Physics (1979–1980). 60th edition. Chem. Rubber Publ. Co., Cleveland, Ohio.

Hansson, I. (1972). An analytical approach to the carbonate system in seawater. Ph.D. Dissertation, Göteborg University.

Harned, H. S., and Bonner, F. T. (1945). The first ionization of carbonic acid in aqueous solutions of sodium chloride. *J. Am. Chem. Soc.* **67**, 1026–1031.

Harned, H. S., and Davis, R., Jr. (1943). The ionization constant of carbonic acid in water and the solubility of carbon dioxide in water and aqueous salt solutions from 0 to 50°C. *J. Am. Chem. Soc.* **65**, 2030–2037.

Harned, H. S., and Robinson, R. A. (1940). A note on the temperature variation of the ionisation constants, of weak electrolytes. *Trans. Faraday Soc.* **36**, 973–978.

Heisler, N., Weitz, H., and Weitz, A. M. (1976). Extracellular and intracellular pH with changes of temperature in the dogfish, *Scyliorhinus stellaris. Respir. Physiol.* **26**, 249–263.

Kormanik, G. A., and Cameron, J. N. (1981). Ammonia excretion in animals that breathe water: A review. *Mar. Biol. Lett.* **2**, 11–23.

Lyman, J. (1956). Buffer mechanism of seawater. Ph.D. Dissertation, University of California, Los Angeles.

Maas, A. H. J., van Heijst, A. N. P., and Visser, B. F. (1971). The determination of the true equilibrium constant ($p\bar{K}_1$) and the practical equilibrium coefficient ($pK_1$) for the first ionization of carbonic acid in solutions of sodium bicarbonate, cerebrospinal fluid, plasma and serum at 25 and 38°C. *Clin. Chim. Acta* **33**, 325–343.

Malan, A., Wilson, T. L., and Reeves, R. B. (1976). Intracellular pH in cold-blooded vertebrates as a function of body temperature. *Respir. Physiol.* **28**, 29–47.

Markham, A. E., and Kube, K. A. (1941). The solubility of carbon dioxide and nitrous oxide in aqueous solution. *J. Am. Chem. Soc.* **63**, 449.

Mitchell, R. A., Herbert, D. A., and Carman, C. T. (1965). Acid-base constants and temperature coefficients for cerebrospinal fluid. *J. Appl. Physiol.* **20**, 27–30.

Murray, C. N., and Riley, J. P. (1969). The solubility of gases in distilled water and sea water. II. Oxygen. *Deep-Sea Res.* **16**, 311–320.

Murray, C. N., and Riley, J. P. (1971). The solubility of gases in distilled water and sea water. IV. Carbon dioxide. *Deep-Sea Res.* **18**, 533–541.

Peters, J. P., and Van Slyke, D. D. (1932). "Quantitative Clinical Chemistry. Methods," Vol. II. Williams & Wilkins, Baltimore, Maryland.

Pleschka, K., and Wittenbrock, I. (1971). The solubility of carbon dioxide in elasmobranch plasma between 10°C and 22°C. *Pfluegers Arch.* **329,** 186–190.

Radford, E. P., Jr. (1964). The physics of gases. *In* "Handbook of Physiology" (W. O. Fenn and H. Rahn, eds.), Sect. 3, Vol. I, p. 125. Am. Physiol. Soc., Washington, D.C.

Randall, D. J., Heisler, N., and Drees, F. (1976). Ventilatory response to hypercapnia in the larger spotted dogfish, *Scyliorhinus stellaris. Am. J. Physiol.* **230,** 590–594.

Reeves, R. B. (1976a). Temperature-induced changes in blood acid-base status: pH and $P_{CO_2}$ in a binary buffer. *J. Appl. Physiol.* **40,** 752–761.

Reeves, R. B. (1976b). Temperature-induced changes in acid-base status: Donnan $r_{Cl^-}$ and red cell volume. *J. Appl. Physiol.* **40,** 762–767.

Riley, J. P., and Skirrow, G., eds. (1975). "Chemical Oceanography," 2nd ed., Vol. 2. Academic Press, New York.

Rispens, P., Dellebarre, C. W., Eleveld, D., Helder, W., and Zijlstra, W. G. (1968). The apparent first dissociation constant of carbonic acid in plasma between 16 and 42.5°. *Clin. Chim. Acta* **22,** 627–637.

Roughton, F. J. W., and Scholander, P. F. (1943). Micro gasometric estimation of the blood gases. I. oxygen. *J. Biol. Chem.* **148,** 541–563.

Roughton, F. J. W., and Severinghaus, J. W. (1973). Accurate determination of $O_2$ dissociation curve of human blood above 98.7% saturation with data on $O_2$ solubility in unmodified human blood from 0° to 37°C. *J. Appl. Physiol.* **35,** 861–869.

Ruud, J. T. (1954). Vertebrates without erythrocytes and blood pigment. *Nature (London)* **173,** 848–850.

Scholander, P. F. (1947). Analyzer for accurate estimation of respiratory gases in one-half cubic centimeter samples. *J. Biol. Chem.* **167,** 235–250.

Severinghaus, J. W. (1965). Blood gas concentrations. *In* "Handbook of Physiology" (W. O. Fenn and H. Rahn, eds.), Sect. 3, Vol. II, p. 1475. Am. Physiol. Soc., Washington, D.C.

Severinghaus, J. W., Stupfel, M., and Bradley, A. F. (1956). Variations of serum carbonic acid p$K'$ with pH and temperature. *J. Appl. Physiol.* **9,** 197–200.

Siggaard-Andersen, O. (1962). The first dissociation exponent of carbonic acid as a function of pH. *Scand. J. Clin. Lab. Invest.* **14,** 587–597.

Siggaard-Andersen, O. (1976). "The Acid–Base Status of the Blood" (4th ed.). Munksgaard, Copenhagen.

Skirrow, G. (1975). The dissolved gases-carbon dioxide. *In* "Chemical Oceanography" (J. P. Riley and G. Skirrow, eds.), 2nd ed., Vol. 2, p. 1. Academic Press, New York.

Stumm, W., and Morgan, J. J. (1970). "Aquatic Chemistry." Wiley (Interscience), New York.

Thurston, R. V., Russo, R. C., and Emerson, K. (1974). "Aqueous Ammonia Equilibrium Calculations," Tech. Rep. No. 74-1. Fish. Bioassay Laboratory, Montana State University, Bozeman.

Thurston, R. V., Russo, R. C., and Emerson, K. (1979). Aqueous ammonia equilibrium-tabulation of percent un-ionized ammonia. *U.S. Environ. Protect. Agency [Rep.] EPA* **EPA–600/3–79–091.**

Van Slyke, D. D., and Neill, J. M. (1924). The determination of gases in blood and other solutions by vacuum extraction and manometric measurement. *J. Biol. Chem.* **61,** 523–573.

Van Slyke, D. D., and Plazin, J. (1961). "Manometric Analyses." Williams & Wilkins, Baltimore, Maryland.

Van Slyke, D. D., Sendroy, J., Jr., Hastings, A. B., and Neill, J. M. (1928). Studies of gas and electrolyte equilibria in blood. X. The solubility of carbon dioxide at 38° in water, salt solution, serum, and blood cells. *J. Biol. Chem.* **78,** 765–799.

Waddell, W. J., and Butler, T. C. (1959). Calculation of intracellular pH from the distribution of 5,5-dimethyl-2,4-oxazolidinedione (DMO): Application to skeletal muscle of the dog. *J. Clin. Invest.* **38,** 720–729.

Washburn, E. W. (1928). "International Critical Tables of Numerical Data, Physics, Chemistry and Technology, 1st ed., Vol. 3. McGraw-Hill, New York.

Weiss, R. F. (1970). The solubility of nitrogen, oxygen and argon in water and seawater. *Deep-Sea Res.* **17,** 721–735.

Weiss, R. F. (1974). Carbon dioxide in water and seawater: The solubility of a non-ideal gas. *Mar. Chem.* **2**(3), 203–215.

Wood, C. M., McMahon, B. R., and McDonald, D. G. (1979a). Respiratory gas exchange in the resting starry flounder, *Platichthys stellatus:* A comparison with other teleosts. *J. Exp. Biol.* **78,** 167–179.

Wood, C. M., McMahon, B. R., and McDonald, D. G. (1979b). Respiratory, ventilatory, and cardiovascular responses to experimental anaemia in the starry flounder, *Platichthys stellatus. J. Exp. Biol.* **82,** 138–162.

# AUTHOR INDEX

Numbers in italics refer to the pages on which the complete references are listed.

# SYSTEMATIC INDEX

*Note:* Names listed are those used by the authors of the various chapters. No attempt has been made to provide the current nomenclature where taxonomic changes have occurred. Boldface letters refer to Parts A and B of Volume X.

## A

*Acanthocybium solandri,* **A,** 20
*Acerina,* **A,** 100
*Achirus,* **B,** 119
*Acipenser,* **A,** 10, 78, 87, 136, 171
  *A. baeri,* **A,** 82, 102, 114, 115, 117, 124, 136, 156, 159; **B,** 310, 311
  *A. transmontanus,* **A,** 302
Acipenseridae, **B,** 241
Albacore, false, **A,** 23; **B,** 178
*Alosa,* **A,** 91
*Ameirus,* **A,** 89
*Amia,* **A,** 10, 15, 29, 78, 81, 87, 89, 102, 111, 113, 118, 121, 136, 159, 171; **B,** 308, 311, 319
  *A. calva,* **A,** 20, 82, 89, 203, 204, 367; **B,** 205, 214, 310, 314
Amphipnoidae, **B,** 241
*Amphipnous,* **A,** 10
*Amphiuma means,* **A,** 366
Anabantidae, **B,** 241
*Anabas,* **A,** 10, 35, 40, 63, 101
  *A. testudineus,* **A,** 35, 61
*Anguilla,* **A,** 124, 339, 341, 342
  *A. anguilla,* **A,** 82, 121, 157, 194, 203, 206, 214; **B,** 4, 5, 15, 17, 49, 71, 72, 75, 78, 79, 80, 82, 83, 85, 88, 91, 210, 244, 249, 250, 264, 267, 268, 270, 271, 274, 336, 338, 339
  *A. australis,* **B,** 339, 346, 361, 362, 372
  *A. dieffenbachi,* **B,** 17, 337
  *A. japonica,* **A,** 144, 194, 203; **B,** 72
  *A. rostrata,* **A,** 150; **B,** 17, 71, 72, 82, 83, 87, 91, 92, 179, 186, 219, 220, 251, 270, 271, 272, 337
  *A. vulgaris,* **B,** 336
Anguilliformes, **B,** 240
*Anoptichthys jordani,* **A,** 144
Apolchitonidae, **B,** 241

*Arapaima,* **A,** 101
  *A. gigas,* **A,** 101; **B,** 218
*Archirus lineatus,* **B,** 113
*Archosargus probatocephalus,* **A,** 30
Ariidae, **B,** 241
*Artemia salina,* **B,** 118
*Aruana,* **A,** 101
Aspredinidae, **B,** 241
Atlantic sea raven, *see Hemitripterus americanus*

## B

*Barbus*
  *B. conchonius,* **A,** 151
  *B. filamentosus,* **A,** 144, 150
  *B. sophor,* **A,** 169
Barracuda, **A,** 19
Bass
  black, **A,** 157; **B,** 302, 304
  small-mouthed, **A,** 43, 62, *see also Micropterus dolmieui*
  striped, **A,** 23
Batrachoididae, **B,** 241
*Bdellostoma,* **A,** 78, 86
*Belone,* **A,** 100
Belonidae, **B,** 241
Blackfish, Australian, **A,** 292
Blenniidae, **B,** 241
*Blennius,* **B,** 107, 113, 119
  *B. pholis,* **A,** 121, **B,** 118
Bluegill, **A,** 302, *see also Lepomis machrochirus*
*Bonito,* **A,** 23
Bothidae, **B,** 241
Bovichthyidae, **B,** 241
Bowfin, **A,** 82, 204; **B,** 205, 214, 218, 222, 310, *see also Amia*
Bream
  black sea, **A,** 5, 6, 302

# SUBJECT INDEX

Note: Boldface **A** refers to entries in Volume XA; **B** refers to entries in Volume XB.

## A

Acid precipitation (rain), **A,** 321, 378–382
Acid–base regulation, **A,** 315–400
  in acidic waters, **A,** 378–382
  analytical techniques for, **A,** 317–336
  counterion fluxes in, **A,** 59, 301, 382–392
  exercise and, 371–378
  extrabranchial sites of, **A,** 386–387
  in hyperoxia, **A,** 368–371
  imidazole alphastat hypothesis, **A,**
    342–343, 349
  relevant ions in, **A,** 317–318
  steady state excretion, **A,** 343–346
  in stress, **A,** 336–338, 346–382
  temperature and, **A,** 338–342, 347–358
Actinomycin D, ionic fluxes and, **B,**
  267–268
Adrenaline, *see* Catecholamines
Air bladder, *see* Swim bladder
Air breathers
  acid–base regulation in, **A,** 365–367
  gill circulation in, **A,** 101, 306
  gills of, **A,** 33–36
  labyrinthine organs, **A,** 32–33
Ammonia
  in acid–base regulation, **A,** 343–346
  and blood pH, **A,** 246
  determination of, **A,** 332–333
  equilibrium p$K$, **A,** 423–426
  excretion of, **A,** 265, 272, 386; **B,** 49–51,
    59, 223–224
  nonionic elimination of, **A,** 388–392
  solubility values, **A,** 422–425
Anastomoses, *see* Arteriovenous vascularity
Anion-dependent ATPase, **B,** 89–97, *see*
  *also* ATPases
  activity in gill, **B,** 92–93
  enzymatic properties, **B,** 93–96
  physiological role, **B,** 96–97
Arterio-arterial vascularity, **A,** 91–117, *see*
  *also* Gill vascularity

Arteriovenous vascularity, **A,** 117–138, *see*
  *also* Gill vascularity
  anastomoses, **A,** 118, 123, 124, 133–137
  central venous sinus (CVS), **A,** 118–129,
    135
  control of, **A,** 216–219
  in lower fish groups, **A,** 132–138
  of teleosts, **A,** 117–132
  venolymphatic vessels, **A,** 124–132
ATPases (gill), *see also* specific ATPases
  anion-dependent, **B,** 89–97
  biochemistry of, **B,** 66–69, 71–75
  $Ca^{2+}$-dependent, **B,** 87–89
  $(Na^+, K^+)$-dependent, **B,** 70–87

## B

Baroreceptors (branchial), **A,** 193–195
Bimodal breathing, *see* Air breathers
Blood, *see also* Gills, Plasma
  $CO_2$ solubility, **A,** 245–249
  extralamellar shunting of, **A,** 254–255
  flow and pressure relations, **A,** 272–290,
    284–285, 304–306
  hemoglobin, **A,** 290–293
  $O_2$ solubility, **A,** 244–249
  plasma skimming of, **A,** 289, 291
Blood-perfused fish, **A,** 294–295
Blood vessels, *see* Gill vasculature
Bohr shift, **A,** 290
Branchial arches (bars), **A,** 2, 78–89, *see*
  *also* Gills
  blood vessels of, **A,** 77–89
  development of, **A,** 6–9, 14
  in lower fish groups, **A,** 86–89
  muscles of, **A,** 81
  nerves of, **A,** 85, 99
  septum relations, **A,** 14
  in teleosts, **A,** 81–86
Buffering, *see* Acid–base regulation
Bunsen solubility coefficient, **A,** 415